D1499375

UNDERSTANDING MICROWAVES

UNDERSTANDING MICROWAVES

Allan W. Scott

A WILEY-INTERSCIENCE PUBLICATION

JOHN WILEY & SONS, INC.

New York / Chichester / Brisbane / Toronto / Singapore

This text is printed on acid-free paper.

Library of Congress Cataloging in Publication Data:

Scott, Allan W.
    Understanding microwaves / Allan W. Scott.
       p.   cm.
    Includes bibliographical references.
    ISBN 0-471-57567-4
    1. Microwave devices. I. Title.
  TK7876.S36   1993
  621.381'3—dc20   92-16863
                                                        CIP

Printed in the United States of America

10 9 8 7 6 5 4 3

PREFACE

Microwaves are a significant part of our lives. Television programs are transmitted by satellites using microwaves. Telephone and data signals are transmitted by microwave relay. Photos taken by space probes are sent to earth by microwaves. Cellular phone communication is by microwave. Our missiles, planes, ships, and tanks are guided and controlled by microwave radar. Commercial aircraft are guided from takeoff to landing by microwave radar and navigation systems. Microwaves are used to cook food and treat cancer patients.

Microwaves are big business, too. The microwave equipment market in the free world is $60 billion a year. The American market is $25 billion a year, which accounts for about one tenth of all electronic equipment manufactured in the United States. The microwave market will continue to grow as new equipment for communications, navigation, and radar is continually developed.

Unfortunately, conventional electronic components, such as transistors, resistors, capacitors, and integrated circuits, don't work at high microwave frequencies. Even conventional wiring and PC boards can't be used to conduct microwaves. Consequently, hundreds of special microwave components have been developed to make microwave equipment work, and, of course, a specialized "language of microwaves" has arisen. The purpose of this book is to help anyone working in the microwave industry to understand this language and how the components are used to devise communication, navigation, and radar systems.

Recent electrical engineering graduates needing training in microwave electronics, engineers who are experts in some microwave areas but need an overview of the entire field, technicians who build and test microwave equipment, mechanical, industrial, and other engineers who work with microwave equipment, sales and marketing personnel who need to understand the technical details of microwave products to better supply their customers' needs, and program managers who must understand technical details of microwave programs can all benefit from *Understanding Microwaves*. Explanations are given in terms of devices without complicated mathematics.

The book has three parts. Part I (Chapters 1–5), "Microwave Fundamentals," covers the language of microwaves. Microwave systems and devices are surveyed, and microwave fields, power (dB and dBm), insertion loss, gain, return loss, and matching with the Smith chart are discussed.

Part II (Chapters 6–13), "Microwave Devices," describes the function, operation, and important characteristics of all the unique devices used in microwave equipment, including transmission lines, signal control components, semiconductor amplifiers, oscillators, low-noise receivers, microwave integrated circuits, microwave tubes, and antennas.

Part III (Chapters 14–19), "Microwave Systems," covers all the important microwave systems: microwave relay, satellite communication, radar, electronic warfare, and navigation and other systems. Systems fundamentals, including the nature of electronic signals, modulation, multiplexing, and digitizing, and each system's purpose and operational principles are discussed. System operation is explained with block diagrams with each block representing a device already discussed in Part II.

Exercises are included at the end of each chapter, and answers to all exercises are provided at the end of the book. Each chapter also has an annotated bibliography for further study of particular topics.

Understanding Microwaves covers all important areas of the microwave industry. Careful reading of the book will enable you to fully understand microwaves and their uses. I wish you great success.

ALLAN W. SCOTT

Los Altos, California
April 1993

CONTENTS

PART II MICROWAVE DEVICES

PART III   MICROWAVE SYSTEMS

UNDERSTANDING MICROWAVES

MICROWAVE
FUNDAMENTALS

A SURVEY OF MICROWAVE SYSTEMS AND DEVICES

This chapter surveys microwave systems and devices. The relationship of microwave equipment to other electronic equipment is shown, and microwave systems are described. The reasons why low frequency conventional transistors, ICs, and circuit boards cannot be used at microwave frequencies are explained, and the seven types of microwave devices—transmission lines, signal control components, semiconductor amplifiers, semiconductor oscillators, low-noise receivers, microwave tubes, and antennas—used instead are described.

1.1 THE RELATIONSHIP OF MICROWAVES TO OTHER ELECTRONIC EQUIPMENT

Figure 1.1 shows the electromagnetic spectrum in the frequency range from 1 megahertz (MHz) to 10^{15} hertz (Hz), a span of nine orders of magnitude. Electromagnetic radiation exists at frequencies below 1 MHz and above 1 million gigahertz (GHz). Most electronic systems, however, and especially microwave systems, operate in the frequency range shown in Figure 1.1. The low end of the spectrum, from 300 kHz to 300 MHz, is called the radio-frequency (RF) range. It contains the medium frequency (MF), the high frequency (HF), and the very high frequency (VHF) ranges. These frequency ranges are all below the microwave range, and conventional transistors, tubes, and circuit wiring can be used in this frequency range. In these frequency ranges AM broadcast radio, shortwave radio, FM broadcast radio, mobile radio, and VHF television channels are located.

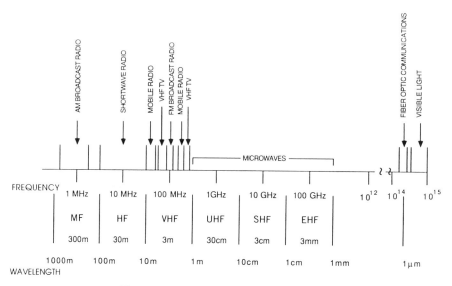

Figure 1.1   The electromagnetic spectrum.

Wavelength and frequency are shown in Figure 1.1. Wavelength is related to frequency by

$$\text{Wavelength} \times \text{Frequency} = \text{Velocity of light}$$

At 1 MHz, the wavelength is 300 m, and at 100 MHz, the wavelength is 3 m.

The microwave band extends from 300 MHz to 300 GHz. The frequency range from 300 MHz to 3 GHz is called the ultra high frequency (UHF) band, from 3 to 30 GHz is the super high frequency (SHF) band, and from 30 to 300 GHz is the extremely high frequency (EHF) band. (This upper band is also called the millimeter (mm) wave band, since the wavelength of electromagnetic radiation varies from 10 mm at 30 GHz to 1 mm at 300 GHz.)

The complete 3-decade frequency range from 300 MHz at the low end to 300 GHz at the high end is considered as the microwave band. Over this frequency range special microwave transmission lines, antennas, semiconductor devices, and tubes are required. It is to be noted that the wavelength of the electromagnetic radiation across the microwave band varies from 1 m at 300 MHz to 1 mm at 300 GHz.

Figure 1.1 also shows which systems operate above the microwave band. Visible light is three orders of magnitude in frequency above the microwave band. (Note that 100 GHz is equal to 10^{11} Hz.) Just below the visible range, at 3×10^{14} Hz, is the frequency of fiberoptic communication systems.

1.2 MICROWAVE SYSTEMS

Microwaves are desirable for communications and radar applications because of their high frequency and short wavelength. The high frequency provides wide bandwidth capability. A 10% bandwidth system at 10 GHz provides a bandwidth of 1 GHz. Into this bandwidth can be incorporated all the information in all communication systems below the microwave range, including AM and FM radio, shortwave radio, broadcast television, and mobile radio. Microwave communication systems have the capacity to handle several thousand telephone channels, several TV channels, and millions of bits of digital data.

Because of the short wavelength of microwaves, high-gain antennas with narrow beamwidths, used in radar applications, can be constructed. For example, a 1-m-diameter microwave antenna operating at 10 GHz has a 2° beamwidth. The short wavelengths also allow microwave energy to be concentrated in a small area, which makes microwave ovens practical.

These two special features, the high frequency and short wavelength of microwaves makes possible the following microwave systems:

- Communication
    UHF TV
    Microwave relay
    Satellite communication
    Troposcatter communication
    Mobile radio
    Telemetry
- Radar
    Search
    Airport traffic control
    Navigation
    Tracking
    Fire control
    Radar altimeter
    Velocity measuring
- Electronic warfare jammers
- Microwave heating
    Industrial heating
    Home microwave ovens
- Industrial, scientific, medical
    Linear accelerators

Plasma containment

Radio astronomy

• Test equipment

1.3 THE MICROWAVE SPECTRUM

Figure 1.2 is an expanded view of the microwave spectrum, showing the frequency locations of some microwave systems. Long-range military search radar operates at 450 MHz, UHF broadcast TV at 470–870 MHz, and the cellular telephone band at 900 MHz. Just above 1 GHz is the air traffic control transponder, which allows aircraft to repeat an identification code to the air traffic control radar. Just below 2 GHz are space telemetry systems, which transmit data from deep-space probes to earth. Just above 2 GHz are tropo-scatter communication systems, where the microwave signal is scattered off of the troposphere to achieve long-distance communication.

Microwave heating equipment, including home microwave ovens, is located at 2.45 GHz. Just above 3 GHz is airport search radar, and just below 4 GHz is point-to-point microwave relay, carrying thousands of telephone channels and television programs across the country. Communication satellite downlinks are at 4 GHz, and the uplinks are at 6 GHz.

Just above 7 GHz is STL, the studio transmitter link, which transmits radio and television programs from the downtown studio to the transmitter site.

Airborne fire control radar is at 10 GHz, and just above that is another microwave relay band for telephone transmission. Just below 20 GHz is another satellite communication downlink band, and just above 30 GHz is

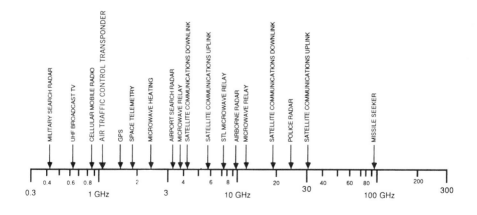

FREQUENCY

Figure 1.2 The microwave spectrum.

the corresponding uplink band. These bands are allocated for future satellite communication. Police radar operates at 24 GHz, and at 10.25 GHz.

Most microwave systems are located in the 300 MHz to 30 GHz range. High atmospheric absorption exists at some frequencies in the 30–300 GHz range, so long-range comunication and radar systems are not practical. However, between these absorption frequency ranges, there are transmission windows. Missile-seeker radars use such a window at 94 GHz. With continuing advances in microwave devices, more and more microwave systems are being developed in the millimeter portion of the microwave band.

1.4 WHY MICROWAVE DEVICES ARE NEEDED

At the very high frequency of microwaves, conventional transistors, ICs, and wiring won't work, due to *lead reactance* and *transit time,* so special microwave devices are required. Figure 1.3 illustrates the problem of lead reactance. A 10-V ac oscillator is connected to a 50-ohm (Ω) resistor by a copper wire 2.5 cm (approximately 1 in.) long and 1 mm in diameter (about the diameter of a paper-clip wire). It has a dc resistance of only 0.4 mΩ and an inductance of 0.027 microhenries (μH), which for low-frequency electronics is negligible compared with the 50-Ω load resistor.

However, the inductive reactance (X_L) of the wire increases with frequency ($X_L = 2\pi fL$). As shown in the figure at 60 Hz $X_L = 10^{-5}$ Ω, which is negligible, and the full 10-V signal from the oscillator appears across the load resistor. At 6 MHz $X_L = 1$ Ω, still small compared with the 50-Ω load resistor. But at 6 GHz, a microwave frequency, $X_L = 1000$ Ω, so almost all of the oscillator voltage is dropped across the connecting wire and never gets to the load resistor.

The lead reactance effect gets even worse at higher frequencies. Consequently, wires or printed circuit boards cannot be used to connect microwave devices. Special microwave transmission lines are required to conduct the microwave signal from one part of the equipment to another.

The second problem, transit time, is illustrated in Figure 1.4 with a field-effect transistor (FET). (Bipolar transistors and triode tubes suffer from similar problems.) In the FET, a change in gate voltage produces a change in the electron flow from the source to the drain. As the current flows through the resistor, the drain voltage becomes an amplified (but inverted) replica of the gate voltage. This idealized description of the FET (which only occurs with low frequency signals) is illustrated by the dashed curves in the left-hand sketch of Figure 1.4, where gate voltage, current, and drain voltage are shown as functions of time during a half-cycle of the input signal.

Actually, a finite time is required for the electrons to move from the source to the drain. This finite time, at microwave frequencies, becomes a large fraction of the cycle. For example, if the source-to-drain spacing is 2.5

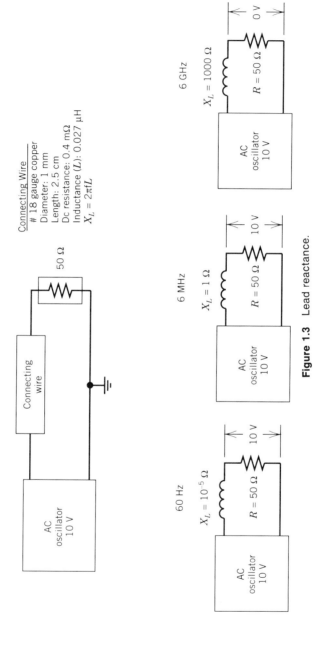

Figure 1.3 Lead reactance.

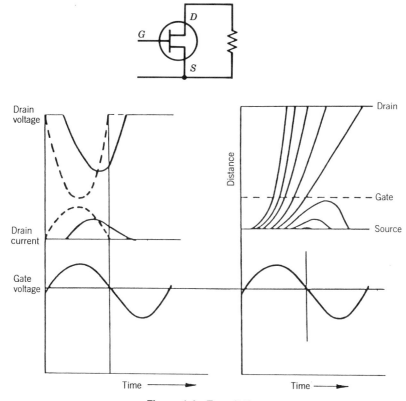

Figure 1.4 Transit time.

microns, it takes about 2.5 nanoseconds (ns) for the electrons to move from the source to the drain, which is one quarter of a cycle at 10 GHz.

The electron trajectories during the cycle, when the transit time of the electrons is one quarter of a cycle, are shown in the right-hand sketch. The resultant current and drain voltage under these conditions are shown by the solid curves in the left-hand sketch. The drain voltage and current are reduced in amplitude compared with the low frequency case.

Because of the lead reactance and transit time problems, special microwave devices must be used in place of the wiring, transistors, and ICs of low-frequency electronics. These special microwave devices are illustrated in a microwave system design in the next section, and then discussed individually in the rest of this chapter.

1.5 BASIC MICROWAVE SYSTEM DESIGN

A block diagram of a microwave communication system is shown in Figure 1.5. With the arrows reconnected, the block diagram could describe a radar

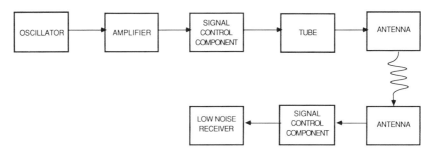

Figure 1.5 Microwave communication system.

or jamming system. The special microwave devices must be used to overcome problems of lead reactance and transit time. Each is discussed in this chapter.

1.6 MICROWAVE TRANSMISSION LINES

The three basic types of microwave transmission lines are illustrated in Figure 1.6: coaxial cable, waveguide, and microstrip. The shapes of the transmission lines are shown in the left-hand sketches, and cross sections with the electric

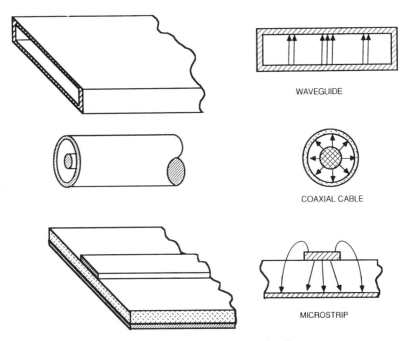

Figure 1.6 Microwave transmission lines.

fields sketched in are shown on the right. The shape and materials of each type of transmission line vary according to application. For example, the center conductor of the coaxial line may be supported by many types of dielectric materials. A waveguide with a rectangular cross section is shown in Figure 1.6 but circular, elliptical, or ridged cross sections are also used. The microstrip may have dielectric material on one side of the conductor, as shown, or both sides. The material can be plastic or ceramic.

Coaxial cable offers the advantages of large bandwidth and small size but has high attenuation and limited power-handling capability. Waveguide, on the other hand, has high power and low loss but large size and narrow bandwidth. Microstrip allows complex circuits to be fabricated easily, but it has very high loss.

1.7 SIGNAL CONTROL COMPONENTS

A variety of special signal control devices must be used at microwave frequencies, such as attenuators, phase shifters, cavities, couplers, filters, loads, circulators, isolators, and switches.

Attenuators are used to control the amplitude of the microwave signal. They can be fixed, or mechanically or electronically adjusted. Special semiconductor diodes called *pin* diodes are used for electronically adjustable attenuation.

Microwave signals are characterized by amplitude and phase. Phase shifters shift the signal's phase, and, like attenuators, they can be fixed, or mechanically or electronically adjusted.

Resonant cavities serve the same purpose as resonant circuits in low frequency electronics. Microwave semiconductor devices and/or tubes are actually built into these cavities for tuning, which can be done electronically or mechanically. Microwave filters and multiplexers are formed of microwave cavities. Like their low frequency counterparts, they separate a band of frequencies from a microwave signal.

Loads are used to terminate a transmission line during performance testing. Directional couplers sample the power flowing in one direction down a transmission line. Isolators and circulators are microwave transmission line components that, through microwave ferrites, allow microwaves to travel in only one direction down the transmission line. Microwave switches can be mechanically or electronically actuated.

1.8 SEMICONDUCTOR AMPLIFIERS AND OSCILLATORS

Microwave oscillators and low power amplifiers are made from six types of "transistor-like" devices: bipolar transistors, field-effect transistors (FETs), high electron mobility transistors (HEMTs), varactor multipliers, IMPATTS,

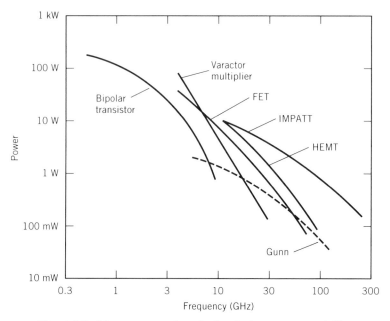

Figure 1.7 Microwave semiconductor device power capability.

and transferred electron devices (Gunn devices). The power capabilities of these devices are compared as a function of microwave frequency in Figure 1.7. Microwave bipolar transistors, FETs, and HEMTs are identical in principle of operation to their low frequency counterparts, but are specially fabricated by spacing their internal elements only fractions of microns apart to minimize transit time effects. Even so, microwave transistors only work to about the middle of the microwave band.

The other three types of microwave semiconductor devices use unique operating principles to overcome the transit time problem. The varactor multiplier uses a variable-capacitance diode to generate harmonics of a lower frequency microwave signal. The IMPATT uses avalanche breakdown of a reverse-biased *pn* junction to generate electrons, and then uses transit time effects (rather than minimizing them) to have the electrons come out of the device exactly one half-cycle after they have been generated. Transferred electron devices (Gunn devices) use the properties of gallium arsenide to generate microwaves. IMPATTS and Gunns can generate microwave signals to the upper end of the microwave band.

1.9 MICROWAVE TUBES

As shown in Figure 1.7, the power capabilities of microwave semiconductor devices are limited. Most microwave systems need more power than semiconductor sources can provide, so microwave amplifier tubes are required. Figure 1.8 compares the power capabilities of microwave tubes and semiconductor devices. The semiconductor curve is the envelope of the curves of the semiconductor devices shown in Figure 1.7. The tube curve is the envelope of a similar set of power tube curves. More than four orders of magnitude more power is available from microwave tubes than from microwave semiconductor devices.

Figure 1.9 compares characteristics of various microwave amplifier tubes. The major types are gridded tube, klystron, helix traveling wave tube (TWT), coupled-cavity TWT, crossed-field amplifier (CFA) and gyrotron.

Gridded microwave tubes are just like their low frequency counterparts, except that cathode-to-grid and grid-to-plate spacings are a minimum to reduce transit time effects. As shown in Figure 1.9, gridded tubes are usable only at the low frequency end of the microwave range. All other microwave tube types use velocity modulation techniques in which transit time effects are actually used rather than minimized. In the frequency range up to 30 GHz, the klystron provides the most microwave power. It can provide 1 megawatt (MW) of average power at the low end of the microwave range.

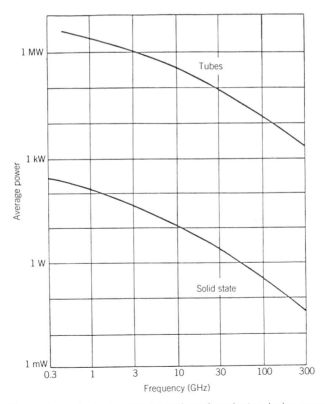

Figure 1.8 Comparison of microwave tube and semiconductor device power capability.

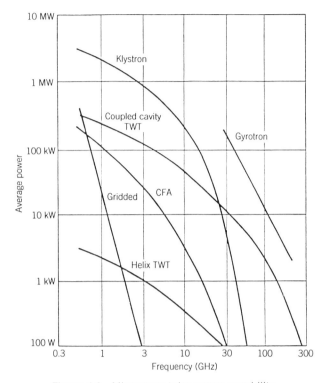

Figure 1.9 Microwave tube power capability.

Above 30 GHz the gyrotron provides the most power. Although the klystron and gyrotron provide the highest power, they are limited to only a few percent bandwidth. In contrast, the helix TWT has over 100% bandwidth. The coupled-cavity TWT, which has almost as good power capability as the klystron, has 40% bandwidth. The CFA has moderate power capability but has 80% efficiency, the highest of any microwave device.

1.10 LOW NOISE MICROWAVE RECEIVERS

The major requirement of a low noise receiver is that it amplify the very weak received microwave signal without increasing the noise level. Figure 1.10 compares the noise characteristics of various low noise microwave receivers, including mixers, bipolar transistors, FETs, HEMTs, and parametric amplifiers. A 10-fold reduction in receiver noise makes possible a 10-fold reduction in transmitter power. Hence, the proper choice of low noise receiver is a critical factor in microwave system design.

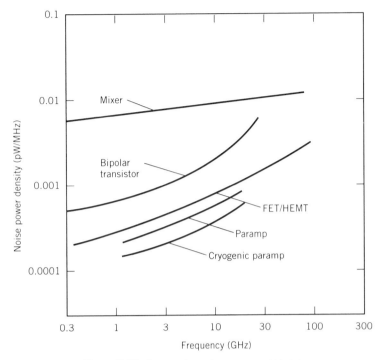

Figure 1.10 Low-noise microwave receivers.

1.11 MICROWAVE ANTENNAS

The need to transmit and receive microwaves makes antennas very important. The most common microwave antenna is the parabolic dish, several of which are illustrated in Figure 1.11. A basic parabolic dish antenna is shown in the top sketch. The antenna consists of a horn that radiates microwaves onto the parabolic surface of the dish, where they are focused into a narrow beam. By making the dish big enough or the wavelength small enough (by using higher frequency), the beamwidth of a microwave antenna can be only a fraction of a degree. Two variations of the basic parabolic dish antenna are also shown in Figure 1.11. The middle photograph shows an air traffic control search radar antenna, which is made of wire mesh to cut down on its wind resistance as it rotates to determine the angular location of the approaching aircraft. The bottom photo shows a satellite earth station antenna.

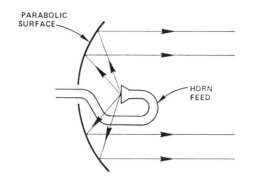

Figure 1.11 Microwave antennas.

ANNOTATED BIBLIOGRAPHY

1. W. S. Cheung and F. H. Levien, *Microwaves Made Simple: Principles and Applications,* Artech House, Dedham, MA, 1985.
2. S. Y. Liao, *Microwave Devices and Circuits,* Prentice-Hall, Englewood Cliffs, NJ, 1980.
3. T. Laverghetta, *Practical Microwaves,* Howard Sams, Indianapolis, 1984.
4. F. E. Gardiol, *Introduction to Microwaves,* Artech House, Dedham, MA, 1984.
5. T. Saad, *Microwave Engineers' Handbook,* Vol. I, II, Artech House, Dedham, MA, 1988.

References 1, 2, and 3 cover microwave principles, devices, and systems at the same level as this book. Reference 4 gives a more technical presentation of the topics of this book. Reference 5 is a handbook of all the important microwave formulas and design graphs.

EXERCISES

1.1. What is the frequency range of microwaves?

1.2. Which of the common frequency bands (HF, VHF, UHF, SHF, etc.) are in the microwave range?

1.3. What types of communication systems lie below the microwave band?

1.4. What is the wavelength at the low frequency end of the microwave band?

1.5. What is the wavelength at the middle of the microwave band?

1.6. What is the wavelength at the high end of the microwave band?

1.7. List six types of microwave communication systems, and, for each system, list a frequency at which it operates.

1.8. List four types of radar systems and the frequencies at which they operate.

1.9. What two problems prevent conventional electronic equipment from working at microwave frequencies?

1.10. What six basic types of microwave devices are used in all microwave systems?

1.11. What are the three types of microwave transmission lines?

1.12. Name eight signal control components.

1.13. What are the six major types of microwave semiconductor devices?

1.14. What is the maximum power that can be obtained from a microwave semiconductor at 10 GHz?

1.15. What is the maximum power that can be obtained from a microwave tube at 10 GHz?

1.16. What are the five types of microwave tubes?

1.17. What are the six types of low-noise microwave receivers?

2

MICROWAVE FIELDS

First, electric and magnetic fields are described. The electric field is produced by stationary electric charges, and the magnetic field is produced by moving electric charges. A time-changing magnetic field produces an electric field, and vice versa. These electric and magnetic fields are represented by Maxwell's equations. The solutions of Maxwell's equations are electromagnetic waves in which electric and magnetic fields travel together through space at the speed of light. The characteristics of electromagnetic waves are frequency, wavelength, impedance, power density, and phase. Formulas are given for calculating these characteristics.

All microwave systems, such as those described in Chapter 1, transmit microwaves from their transmitter through space to a receiver. However, inside the transmitter and the receiver, microwaves cannot be conducted by wires, so they are transmitted or broadcast from one component to another inside microwave transmission lines. The three types of transmission lines (coaxial cable, waveguide, and microstrip) are described, and the electromagnetic field configurations inside these transmission lines are shown. These electromagnetic fields have the same characteristics as electromagnetic waves traveling in space, but their values are modified when the microwaves are forced to travel through the transmission lines. Transmission lines are discussed in Chapter 7, and formulas for the various characteristics are given. However, the topic is introduced briefly here.

2.1 ELECTRIC AND MAGNETIC FIELDS

Electric fields are defined in Figure 2.1. The left-hand sketch shows two electrons. They both have negative charges, so they repel each other. The

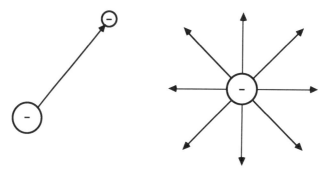

Figure 2.1 Electric fields.

force exerted on the upper electron by the lower electron can be represented by an arrow in the direction of the force. In the right-hand sketch, the effect of the lower electron on any other electron at some distance from the lower electron can be represented by arrows. The direction of the arrow again shows the direction of the force, and the strength of that force is represented by the density, or number, of arrows in a unit area. The effect of the electron as a force field is more than just a visual aid. Its presence can be detected as a form of energy, and the electric field is as real as the electron itself.

Magnetic fields are defined in Figure 2.2. The magnetic field is the force on a *moving* charge (such as an electron current flowing in a wire) due to other moving charges. The magnetic field has no effect on a stationary charge. The upper sketch shows the magnetic field around a wire, generated by the electrons flowing in the wire. The lower three sketches show the magnetic field around a loop, or several loops, of wire that are wound into a coil. Winding the wire into a coil concentrates the magnetic field along the center of the coil.

Figure 2.3 shows combined electric and magnetic fields. In the upper sketch, a 10-V battery is connected to a 100-Ω resistor through two large wires. One wire is connected to the positive terminal of the battery and thus is at $+10$ V; the other wire, connected to the negative, or ground, terminal is at 0 V. Current flows through the wire from the positive terminal of the battery, through the resistor, and returns to the battery through the other wire. The electric and magnetic fields around these wires are shown in the lower sketch. Since one wire is positively charged, and the other is negatively charged, a positive charge would be repelled from the positive wire and attracted to the negative wire, as shown by the solid arrows. The electrons are moving in opposite directions in the wires, so the magnetic fields are oppositely directed around each wire, as shown by the dotted lines.

Low frequency electronics is concerned with the voltage and currents in the wire, and is less concerned about the electric and magnetic fields. However, at microwave frequencies the voltage and current are difficult to define, and the electric and magnetic fields are dealt with directly.

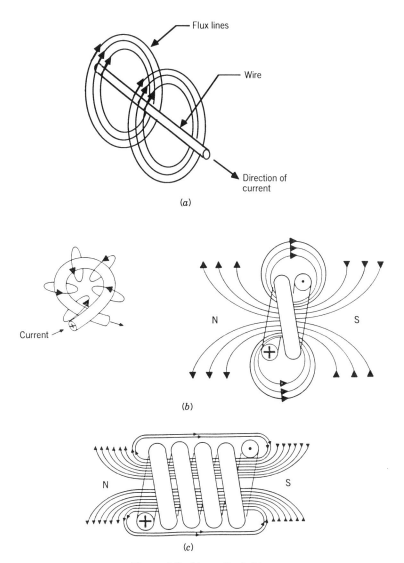

Figure 2.2 Magnetic fields.

2.2 ELECTROMAGNETIC WAVES

In Figure 2.3 even though the electrons were moving through the wires and generating a magnetic field, the field did not change but remained constant. If the magnetic field were changed, an electric field would be generated. This is the principle of operation of all motors and generators. A generator generates electricity by moving a coil of wire through a magnetic field, so

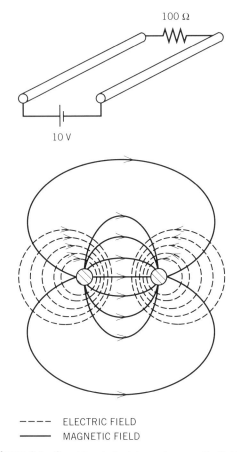

ELECTRIC FIELD
MAGNETIC FIELD

Figure 2.3 Combined electric and magnetic fields.

that the magnetic field, as seen by the coil of wire, is continually changing, and this changing magnetic field generates an electric field that will push current through an external load.

The four laws that completely describe electric and magnetic fields are called Maxwell's laws:

The electric field depends on stationary charges.

The magnetic field depends on moving charges (current).

The electric field depends on a changing magnetic field.

The magnetic field depends on a changing electric field.

The first three laws have already been discussed, and they can be demonstrated experimentally. The last law states that the magnetic field depends

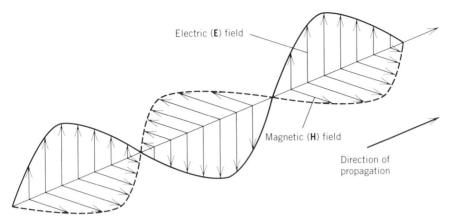

Figure 2.4   Electromagnetic waves.

on a changing electric field and was added by Maxwell to complete the symmetry.

The solutions of Maxwell's laws are electric and magnetic fields that travel together through space in a wavelike fashion at the speed of light. These waves are thus called *electromagnetic waves*. The field configurations in an electromagnetic wave traveling from left to right are shown in Figure 2.4. At the left-hand side the electric field points upward and the magnetic field points out of the plane of the paper. The electric and magnetic fields are always at right angles to each other. They decrease to zero and then increase in opposite directions, so in the middle of Figure 2.4 the electric field is pointing downward and the magnetic field is pointing into the plane of the paper. Proceeding along the wave, the electric and magnetic fields again decrease to zero and then increase in the direction that they originally had on the left-hand side, with the electric field pointing up and the magnetic field pointing out of the paper. This wave behavior of the combined electric and magnetic fields is a consequence of the simultaneous solution of Maxwell's equations, which describe electric and magnetic fields individually, based on stationary charges, moving charges, and the effect on one of the fields by changing of the other field. Maxwell's laws were conceived approximately 100 years ago, and shortly thereafter electromagnetic waves were experimentally demonstrated by the generation, transmission, and reception of radio signals.

2.3   CHARACTERISTICS OF ELECTROMAGNETIC WAVES

Characteristics of electromagnetic waves are frequency, wavelength, impedance, power density, and phase.

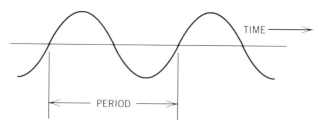

Figure 2.5   Frequency.

Frequency

Frequency and period are defined in Figure 2.5. In low frequency electronics, where electron currents flow through wires, frequency is defined as the number of oscillations that an electrical signal completes in one second. An alternative definition, which is more useful for electromagnetic waves, is that frequency is the number of electromagnetic waves that pass a given point in one second. The unit of frequency is the hertz (Hz); it is the number of oscillations per second or the number of waves per second. Typical frequencies in electronics are 1000 Hz, which is one kilohertz (kHz); 1 million Hz, which is one megahertz (MHz); and 1 billion Hz, which is a gigahertz (GHz).

The reciprocal of frequency is called the period, which is the time for an electrical signal to complete one oscillation, or the time between one electromagnetic wave and the next electromagnetic wave passing a given point. For example, if the frequency is 1 (Hz), the period is 1 second; if the frequency is 1 (kHz), which means a thousand waves come by in a second, then the period is 1 millisec (ms), which means that the duration of each wave is $\frac{1}{1000}$s.

Wavelength

Wavelength is defined in Figure 2.6 as the distance in which the fields of an electromagnetic wave repeat themselves. Frequency and wavelength are related by the formula

$$\text{Frequency} \times \text{Wavelength} = \text{Velocity}$$

In free space, electromagnetic waves travel at the velocity of light, 3×10^8 m/s, a consequence of the physical constants into Maxwell's laws describing the electric and magnetic forces on charged particles. At 1 MHz, the wavelength is 300 m; at 100 MHz, the wavelength is 3 m; at 300 MHz, the wavelength is 1 m; and at 300 GHz, the wavelength is 1 mm. Thus, across the microwave band (300 MHz–300 GHz) the wavelength varies from 1 m (about the length of a yardstick) at the low end to 1 mm (the diameter of a

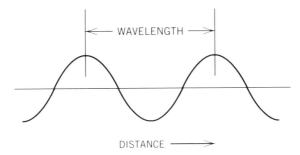

Figure 2.6 Wavelength.

paper-clip wire) at the high end. Three centimeters, which is the wavelength at the middle of the microwave band is about 1 in.

Above the microwave band at 10^{14} Hz, the wavelength is 1 micron, which is a millionth of a meter. When the wavelength of the electromagnetic wave is 0.5 micron, the electromagnetic wave can be detected by our eyes and is called *light*.

Radio broadcast originally began in the 1-MHz range where the wavelength is very large. The first microwave signal was generated at 10 GHz, and here the wavelength was extremely short, and thus these waves were called *microwaves*. However, although microwaves have a very small wavelength compared with low-frequency radio waves, their wavelength is extremely large compared with that of light.

Formulas for calculating the wavelength of microwaves in free space are given in Table 2.1.

Table 2.1 Formulas for the free-space wavelength of microwaves

Free-Space Wavelength	Frequency	
	MHz	GHz
Meters	$= \dfrac{300}{f}$	$= \dfrac{0.3}{f}$
Centimeters	$= \dfrac{30,000}{f}$	$= \dfrac{30}{f}$
Millimeters	$= \dfrac{300,000}{f}$	$= \dfrac{300}{f}$

Any one of these formula can be inverted.

For example: $f(\text{GHz}) = \dfrac{30}{\lambda(\text{cm})}$

Impedance

Impedance (Z) is defined as the ratio of the electric field to the magnetic field. The unit of the electric field is volts per meter (V/m), and the unit of the magnetic field is amps per meter (A/m). Therefore, the unit of their ratio is ohms. (Low frequency impedance is equal to voltage in volts divided by current in amps, so the unit of impedance is ohms, just as in the microwave case.)

In free space, the impedance of an electromagnetic wave is 377 Ω, as determined by Maxwell's laws. The ratio of the electric and magnetic fields is determined by the fundamental nature of electromagnetic waves, and the value of 377 Ω results from the units chosen to express the fields. In contrast, in a transmission line the impedance depends on the dimensions and material of the line. Therefore, impedance can be controlled by the transmission line designer. More will be said about this subject in later chapters.

Power Density

Power density is defined as the power carried by an electromagnetic wave and is equal to the electric field (E) multiplied by the magnetic field (H):

$$\text{Power density} = E \times H$$

The unit of power density is thus

$$(V/m)\ (A/m) = VA/m^2 = W/m^2$$

The low frequency definition of power by Ohm's law is voltage times current.

Power density can be appreciated by considering an electromagnetic field broadcast from a satellite to earth. As the microwave travels through space, it spreads out so that by the time it reaches the earth's surface it covers the entire area of the United States, which permits the signal from the satellite to be received at any location in the United States. The power density of the microwave signal defines the amount of power in a square meter at the earth's surface.

By definition, power density = $E \times H$, but $H = E/Z$, so

$$\text{Power density} = E \times H$$
$$= E \times \frac{E}{Z} = \frac{E^2}{Z}$$

or

$$\text{Power density} = E \times H$$
$$= HZ \times H = H^2 Z$$

These formulas seem very familiar. If the electric field is changed to voltage, the magnetic field to current, and the power density to power, these are just the low frequency formulas for power in terms of voltage, current, and impedance.

Example 2.1   Given a transmission line of cross section 2 cm by 1 cm (approximately the size of a waveguide operating at 10 GHz), $E = 50$ V/m, and $H = 1$ A/m. Find the impedance, power density, and total power.

Solution

$$Z = \frac{E}{H} = \frac{50 \text{ V/m}}{1 \text{ A/m}} = 50 \ \Omega$$

$$\text{Power density} = E \times H = 50 \text{ W/m} \times 1 \text{ A/m} = 50 \text{ W/m}^2$$
$$P = \text{Power density} \times \text{Cross-sectional area}$$
$$= 50 \text{ W/m}^2 \times 0.02 \text{ m} \times 0.01 \text{ m}$$
$$= 10 \text{ mW}$$

Example 2.2   The receiving antenna of a microwave relay system has an area of 2 m², $E = 377 \ \mu$V/m, and $H = 1 \ \mu$A/m. Find the impedance, power density, and total power received by the antenna.

Solution

$$Z = \frac{E}{H} = \frac{377 \ \mu\text{V/m}}{1 \ \mu\text{A/m}} = 377 \ \Omega$$

$$\text{Power density} = E \times H$$
$$= 377 \ \mu\text{V/m} \times 1 \ \mu\text{A/m} = 3.77 \times 10^{-10} \text{ W/m}^2$$
$$P = 3.77 \times 10^{-10} \text{ W/m}^2 \times 2 \text{ m}^2 = 7541 \times 10^{-12} \text{ W}$$
$$= 754 \text{ pW}$$

Phase

Phase (see Figure 2.7) is the time difference between two electrical signals at the same frequency. A single electromagnetic wave is characterized by frequency, wavelength, impedance, and power density. However, for phase to be specified, these must be two waves. The phase of one wave is relative to the other wave or to the same electromagnetic wave at another instant in time. Phase is expressed in degrees, with 360° equal to a time difference of one period. In the figure, signal A leads signal B by 90°, which means that when the two electromagnetic waves are compared on the same time scale, signal A reaches its maximum one quarter of a cycle, or 90°, before signal B reaches its maximum.

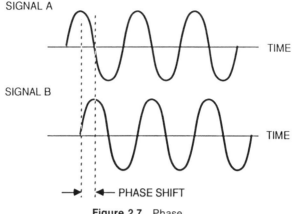

SIGNAL A

TIME

SIGNAL B

TIME

← PHASE SHIFT

Figure 2.7   Phase.

Example 2.3   If two equal microwave signals are added, what is the total signal if (a) the signals are in phase (phase difference of 0°), (b) 180° out of phase, (c) 270° out of phase, and (d) with a phase difference of 360°?

Solution   (a) The signals are in phase so their total strength is simply the sum of their individual strengths: $1 + 1 = 2$.

(b) For a 180° phase difference, one signal reaches its maximum when the other reaches its minimum. Hence, the signals cancel: $1 - 1 = 0$.

(c) For a 270° phase difference, the *vector* sum of the signals is taken: $\sqrt{1 + 1} = \sqrt{2} = 1.416$.

(d) When the phase shift is 360°, the signals are again in phase, so $1 + 1 = 2$.

Example 2.4   A 1-GHz microwave signal moves from point 1 to point 2, a distance of 3 cm. What is the phase difference between the points?

Solution   The phase difference is the time required for the signal to travel between the points divided by its period. Thus,

$$
\begin{aligned}
\text{Phase difference} &= \frac{\text{Travel time}}{\text{Period}} \times 360° \\
&= \frac{\text{Distance/Velocity}}{1/\text{frequency}} \times 360° \\
&= \frac{(3\text{ cm})/(3 \times 10^{10}\text{ cm/s})}{1/10^9\text{ s}} \times 360° \\
&= \frac{10^{-10}}{10^{-9}} \times 360° \\
&= 36°
\end{aligned}
$$

Therefore, in traveling 3 cm (slightly more than 1 in.), an electromagnetic wave wih a frequency of 1 GHz undergoes a 36° phase shift.

Example 2.5 The frequency of the ac power lines in the United States is 60 Hz. The typical distance between a household and the power plant is 10 km (6 mi). What is the phase shift between the household and the plant?

Solution

$$\text{Period at 60 Hz} = \frac{1}{60} = 1.6 \times 10^{-2}\,\text{s}$$

$$\text{Time for signal to travel between points} = \frac{10 \times 10^3\,\text{m}}{3 \times 10^8\,\text{m/sec}}$$
$$= 3.3 \times 10^{-5}\,\text{s}$$
$$\text{Phase shift} = \frac{3.3 \times 10^{-5}}{1.6 \times 10^{-2}} \times 360° = 0.7°$$

The 0.7° phase shift is insignificant and can be ignored.

2.4 MICROWAVES IN TRANSMISSION LINES

Until now, microwaves have been considered as transmitted through space from the transmitting antenna to the receiving antenna. How microwaves are transmitted from one part of the equipment to the next will now be considered.

Low frequency components like transistors and capacitors at frequencies below the microwave range are connected by wires. The electron flow in the wire carries the electrical signal from part to part. Microwaves however, cannot be conducted through wire, as was explained in Chapter 1. Inside the microwave equipment, microwaves are still considered as waves. They travel with their wavelike characteristics from part to part. Even though the parts are only a fraction of an inch apart, the microwaves travel as waves, and the microwave power must be guided from part to part. The device that does the guiding is called a microwave transmission line.

Although there are hundreds of microwave transmission lines, there are only three basic types (Figure 2.8): waveguide, coaxial cable, and microstrip. In Figure 2.8, a drawing of each type of transmission line is shown on the left, and a cross section through the transmission line is shown on the right. The arrows in the cross section are the microwave fields.

A *waveguide* is just a hollow metal pipe. It usually has a rectangular cross section but can have a circular or oval cross section. The hollow pipe guides the microwaves in the same way that a water pipe guides water. The microwaves actually travel as waves from one component to the next through the waveguide.

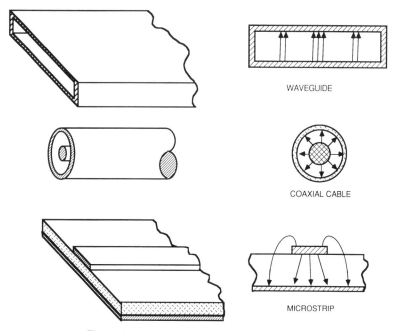

Figure 2.8   Microwave transmission lines.

Coaxial cable consists of an inner conductor surrounded by an outer conductor. The word *coaxial* means that the inner conductor is located on the axis of the outer conductor. In order to achieve this alignment some sort of insulating material is needed to support the inner conductor. The coaxial cable is filled with an insulator in the space between the inner and outer conductors, and the microwaves travel through the insulator. In contrast, the inside of waveguides, where the microwaves travel, is empty because the waveguide walls are self-supporting.

To simplify the problem of making many connections in a complex microwave circuit, microstrip transmission lines would be used. Microstrip consists of a conductor, an insulator, and a flat plate called the ground plane. Note that microstrip is like a coaxial cable that's been cut and laid out flat so that the ground plane on the bottom is like the outer conductor. The strip on the top is like the inner conductor, and the space in between is the insulator. As the arrows show, most of the microwave field is carried inside the insulator.

The advantage of microstrip is that several microwave transmission lines can be easily connected together by putting them on the top surface of the insulator and using the ground plane on the bottom as the common ground plane for all the circuits.

The electric and magnetic fields of a microwave traveling inside a waveguide transmission line at a given instant of time are shown in Figure 2.9.

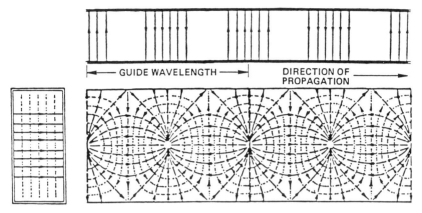

Figure 2.9 Field configurations in waveguide at a given instant of time.

Microwave power is transmitted from left to right. The electric field is shown by the solid lines, the magnetic field by the dotted lines, and the current in the waveguide walls by the dashed lines.

The upper sketch shows a cross section through the waveguide. At the left end, the electric field has a maximum value and points upward from the bottom to the top of the waveguide. The electric field varies along the waveguide. Further along the waveguide, the electric field goes to zero and then reaches a maximum with the field pointing downward. The field then reverses and reaches a maximum with the field pointing upward again.

The distance along the waveguide in which the field patterns repeat themselves is called the guide wavelength. The guide wavelength is approximately equal to the wavelength of the electromagnetic wave in free space, but is slightly modified by the presence of transmission line walls and by the dielectric material that fills the transmission line. The section of waveguide shown is two guide wavelengths long.

The lower sketch shows the magnetic fields (dotted lines) encircling the electric field lines. The dashed lines show the electron current flow in the walls of the waveguide.

The electric field, magnetic field, and the current distributions are obtained by solving Maxwell's equations in the region inside the hollow waveguide.

Figure 2.10 shows the electric field in the waveguide at several times during a microwave cycle.

The upper sketch, at the beginning of the microwave cycle, is the same as that in Figure 2.9. The electric field lines are at maximum and are pointing upward. One quarter of a cycle later, at the same reference position, the electric field configuration has moved down the waveguide, and the electric field is now zero. One-half cycle later, as the wave moves further down the guide, the maximum electric field that had previously occurred at the left side of the sketch, at the beginning of the cycle, has now moved to the center

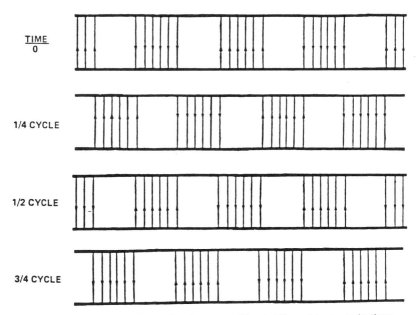

TIME 0

1/4 CYCLE

1/2 CYCLE

3/4 CYCLE

Figure 2.10 Field configuration in waveguide at different instants in time.

of the sketch, and the electric field is at maximum and pointing downward at the left edge of the guide.

Three quarters of a cycle later, as the field configuration moves further down the guide, the electric field is again zero at the reference position. Finally, one full cycle later, the electric field configuration is the same as it was at the beginning of the cycle.

The complete electric field, magnetic field, and the current distribution patterns move down the waveguide, but for simplicity, only the electric field is shown in Figure 2.10.

Figure 2.9 showed a single mode of propagation of electromagnetic waves through the waveguide. Actually, several modes of propagation exist in each type of transmission line, and each mode has a unique field configuration. Each mode begins to propagate at a particular frequency, determined by the transmission line dimensions and the dielectric support materials. The electric and magnetic field configurations of several of these modes, in rectangular and circular waveguides and in coaxial cable, are shown in Figure 2.11 and 2.12.

Figure 2.11 shows the modes in rectangular waveguides. Microwaves cannot be propagated in any mode below a critical frequency, called the mode cutoff frequency. The cutoff frequency for each mode depends on the width and height of the waveguide's cross section. The cutoff frequency for each mode is shown below each sketch for a waveguide 0.90 in. wide by 0.40 in. high (internal dimensions). Below 6.6 GHz, no modes will propagate

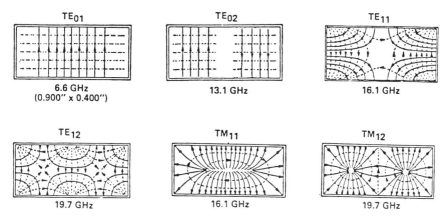

Figure 2.11 Field configurations in rectangular waveguide.

in this size waveguide, so it can't be used for transmitting microwaves below these frequencies. If microwaves must be transmitted in a waveguide at a lower frequency, a larger size of waveguide must be used.

From 6.6 to 13.1 GHz, only the single TE_{10} mode can be propagated. Microwaves start to propagate at the frequency when the width of the waveguide is one half of the free-space wavelength, which, since the width of this waveguide is 0.90 in., is 6.6 GHz.

At 13.1 GHz the TE_{02} mode begins to propagate, as well as the original TE_{01} mode. As shown, the TE_{02} mode has a different electric and magnetic field configuration. At 16.1 and 19.7 GHz, still other modes can be propagated. At 19.7 GHz, all six modes shown in Figure 2.11 can exist in the waveguide. If power is put into the waveguide in one mode, it will be coupled into the other modes and cannot be removed from the waveguide. Consequently, waveguide is normally only used in a frequency range where just one mode exists, which for this example would be 6.6 to 13.1 GHz. (Because the characteristics of the waveguide change rapidly near the ends of this range, this example waveguide is normally used only from 8.2 to 12.4 GHz.)

The labeling of the modes is as follows: The TE, or transverse electric, modes have electric fields in the transverse plane, that is, only across the waveguide. The magnetic field in the TE modes runs both transversely and axially along the waveguide. That electric fields are transverse and that magnetic fields extend across the cross section and along the waveguide itself, can be seen clearly in Figure 2.11, which showed the complete field pattern in all directions. The TM, or transverse magnetic modes, have magnetic fields that exist only in the transverse cross section of the guide, whereas the electric fields extend in the transverse and axial directions along the guide.

The subscript after the TE or TM refers to the number of electric or magnetic field variations across the height and width of the waveguide. For

CIRCULAR WAVEGUIDE

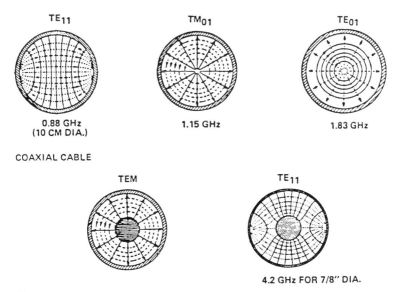

Figure 2.12 Field configurations in circular waveguide and coaxial cable.

example, in the TE_{01} mode, the electric field is uniform with no variations across the height dimension, but has one variation across the width. The field goes from zero at edge of the guide to a maximum at the center and back to zero at the opposite end of the guide.

For the TE_{02} mode, the electric field still has no variations across the height of the guide, but has two variations across its width. The field goes from zero at the one edge of the guide, to a maximum, to zero, to a maximum with the opposite field direction, and then back to zero at the other guide wall.

All standard rectangular waveguides use a width-to-height ratio of approximately 2. This is required in order that the TE_{11} and TM_{11} modes have a higher cutoff point than the TE_{02}. This gives a maximum useful bandwidth to the waveguide. To cover the frequency range from 300 MHz to 300 GHz, 34 different rectangular waveguides, each with a different width but the same width-to-height ratio, are used. The characteristics of these standard rectangular waveguides are described in Chapter 7.

Figure 2.12 shows the two lowest modes in coaxial cable. The TEM mode has electric and magnetic fields in the transverse plane only, and it propagates from zero frequency, because the coaxial cable has two conductors. Coaxial cable therefore has an extremely wide bandwidth.

Coaxial cable also has other modes of propagation, the next being TE_{11}. Like waveguide modes the TE_{11} mode in coaxial cable cannot propagate until a certain critical frequency is reached. This frequency, for a $\frac{7}{8}$-in. diameter coaxial cable is 4.2 GHz. Above this frequency, two modes exist, and this coaxial cable is normally not used.

Electric and magnetic field patterns for some of the modes in circular waveguides are also shown in Figure 2.12.

2.5 SKIN DEPTH

Microwaves travel inside the transmission line as electric and magnetic fields. They are not carried in the metal walls of the transmission lines but in the space between the walls. If the walls were made of perfect conductors, the microwaves would not penetrate at all into the walls of the guide. However, the walls are not perfect conductors, so the microwave fields penetrate slightly into the walls. The depth to which they extend is called the *skin depth*. The skin depth depends on the frequency of the microwaves and the material of the transmission line walls, and is defined as the distance in the walls at which the fields have decreased to 30% of their value at the surface. Skin depth is shown as a function of frequency for various metals in Figure 2.13. At 10 GHz the skin depth in copper is only 0.5 micron.

The walls of transmission lines need to be only about 10 skin depths thick to adequately carry microwaves, which at 10 GHz is less than 10 microns or less than 0.0004 in. This allows the use of very thin conductors on microstrip transmission lines. It also allows waveguide and coaxial cable to be made of

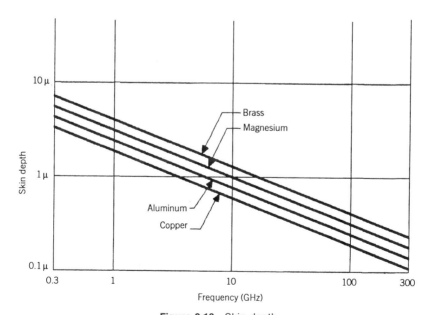

Figure 2.13 Skin depth.

any convenient material and then plated on their inner surfaces, and the microwave fields extend only into the plating.

ANNOTATED BIBLIOGRAPHY

1. T. Laverghetta, *Practical Microwaves*, Howard Sams, Indianapolis, 1984.
2. F. E. Gardiol, *Introduction to Microwaves*, Artech House, Dedham, MA, 1984.

Reference 1 provides a simple mathematical explanation of Maxwell's equations and microwave fields. Reference 2 provides a thorough mathematical analysis of microwave fields.

EXERCISES

2.1. Complete the following table.

Frequency	Period
60 Hz	_____
2 kHz	_____
5 MHz	_____
500 MHz	_____
2 GHz	_____
5 GHz	_____
10 GHz	_____
100 GHz	_____
_____	10 ms
_____	20 μs
_____	5 μs
_____	100 ns
_____	50 ns
_____	500 ps
250 kHz	_____
20 MHz	_____
_____	2.5 μs

2.2. Complete the following table.

Frequency	Wavelength
600 MHz	_____
5 GHz	_____
2 GHz	_____
15 GHz	_____
75 GHz	_____

Frequency	Wavelength
150 GHz	_____
_____	3 cm
_____	10 mm
_____	3 mm
_____	50 cm
_____	300 mm
6.8 GHz	_____
_____	28 mm
15.5 GHz	_____
_____	25 cm
3.5 GHz	_____
1.5 GHz	_____
_____	1.5 mm

2.3. Draw arrows whose length is equal to one wavelength at the following frequencies:

Frequency (GHz)	Wavelength (mm)	Arrow Representing Wavelength
4		
8		
15		
60		

2.4. Complete the table.

Electric Field (V/m)	Magnetic Field (A/m)	Impedance (Ω)	Power Density (W/m²)
1	0.02		
754×10^{-3}	2×10^{-3}		
0.025		50	
	0.04	377	
		50	10^{-3}
		377	10^{-3}

2.5. How much microwave power is received by a 3-m² antenna if the electric field at the antenna is 10^{-5} V/m and the magnetic field is 2.6×10^{-8} A/m?

2.6. What is the phase relationship of the following signals?

B _____ A by _____ degrees
    leads or lags

C _____ A by _____ degrees
    leads or lags

D _____ A by _____ degrees
    leads or lags

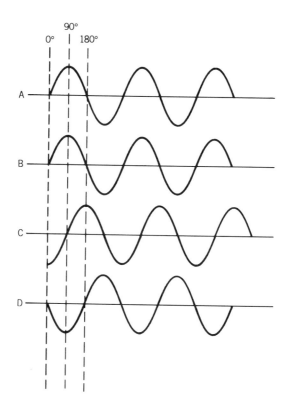

2.7. Complete the following table.

Frequency (GHz)	Material	Skin Depth (microns)
2	Copper	
6	Gold	
10	Aluminum	
10	Copper	
30	Copper	

3

MICROWAVE POWER—dB AND dBm

At low frequencies, the strength of the electrical signal is specified by voltage or current. These two quantities are easily measured. Voltage and current are related by the impedance of the circuit (voltage = impedance × current), and power is the product of voltage and current.

As shown in Chapter 2, at microwave frequencies the equivalents to voltage and current are electric field and magnetic field, respectively. Unlike their low frequency equivalents, electric and magnetic fields cannot be directly measured. At microwave frequencies, only power can be measured, and the electric and magnetic fields are calculated from the measured power.

The range of power in microwave equipment, the use of the dB number system to specify microwave power, and the equipment to measure microwave power are discussed in this chapter.

3.1 MICROWAVE POWER

Figure 3.1 shows the range of power encountered in microwave equipment. Power is the strength or amplitude of the electromagnetic wave and is measured as electrical power in units of watts. At microwave frequencies, however, the reference level of power is not 1 W, but, as shown in Figure 3.1, 1 mW, which is $\frac{1}{1000}$ W. The reason is that a milliwatt of power is enough to operate a telephone, or video display, or computer, so the standard power reference in the telecommunication and radar industries (the industries that use microwave equipment) is a milliwatt. Most microwave test equipment is designed to operate at the milliwatt level and is calibrated in milliwatts.

However, microwave systems operate at powers much greater than or

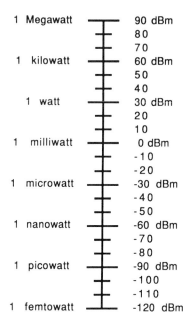

Figure 3.1 Microwave power.

much less than 1 mW. Each line on the scale of Figure 3.1 represents a change in power by 10 times. Starting at the reference level of 1 mW and going up in steps of 10, the scale lines represent 10 mW, 100 mW, 1000 mW (which is 1 W), and then 10 W, 100 W, 1000 W (which is 1 kW), and finally 10 kW, 100 kW, and 1000 kW (which is 1 MW). The kilowatt and megawatt powers are the levels encountered in radar and communications transmitters.

Going down below a milliwatt, the scale lines represent a tenth of a milliwatt, a hundredth of a milliwatt, and a thousandth of a milliwatt, which is a microwatt. Notice that a microwatt is one millionth of a watt or one thousandth of a milliwatt. Further down from a microwatt is a nanowatt, the lowest microwave signal that can be measured. Still lower is a picowatt, which is approximately the level of the microwave signal in radar and communication receivers. Finally, below a picowatt is a femtowatt, which is approximately the level of noise power generated in all microwave receivers, and it sets the limit on the lowest possible power that a signal can have and still be detected. If a signal is less than a femtowatt, it will be lost in the noise, and when the signal is amplified, the noise is amplified with it, and the amplified signal will always stay lost in the amplifed noise.

Table 3.1 defines the prefixes used to specify all these very high and very low powers.

Table 3.1   Units

Giga	= 1,000,000,000 (Billion)
Mega	= 1,000,000 (Million)
Kilo	= 1,000 (Thousand)
Milli	= 1/1000 (Thousandeth)
Micro	= 1/1,000,000 (Millioneth)
Nano	= 1/1,000,000,000 (Billioneth)
Pico	= 1/1,000,000,000,000 (Trillioneth)
Femto	= 1/1,000,000,000,000,000 (Thousandeth of a millioneth of a millioneth!)

3.2   dB TERMINOLOGY

All the zeros before the decimal point for high powers or after the decimal point for low powers make calculations cumbersome. For convenience, the dB and dBm system of units is used. (All microwave equipment performance is specified in dB and dBm, and all microwave measurement equipment have their scales calibrated in dB and dBm.)

The dB notation compresses the wide range of power values that occur in microwave equipment into a practical range of numbers. It allows picowatts and megawatts to be dealt with in the same calculation. The dB notation also allows addition to be used instead of multiplication, when tracing a microwave signal through a microwave system or test setup.

Figure 3.2 is a block diagram of a microwave communications system that illustrates the advantages of dB and dBm. The system is similar to the HBO satellite-to-earth transponder. The three upper boxes represent some of the equipment in the satellite transmitter. The three lower boxes represent some of the equipment in the earth station receiver. The equipment designer must determine, as shown in the circle, the received power. Starting at the upper left, the transmitter in the satellite generates 10 W of power (10,000 mW) at 4 GHz, the downlink frequency. This signal suffers a loss of 0.4 in going through the cable from the transmitter to the transmitting antenna in the satellite. The signal, just before transmission is increased 300 times by the gain in the transmitting antenna. Then it is reduced by 4×10^{-19} in travelling through space from the satellite to the earth.

This reduction in the signal as the microwave power spreads to cover the entire surface of the United States is called the free space path loss, and it would be written as a decimal point with 18 zeros before the 4. The path loss accounts for the fact that the power from the HBO transponder is spread over the entire surface of the United States, so that it can be received anywhere. Only a tiny fraction, as expressed by the 4×10^{-19}, can be picked up at a given location, because the rest of the power is someplace else.

The lower three boxes are the parts of the system in the earth terminal. The 10-foot diameter receiving antenna increases the signal by 3000 times.

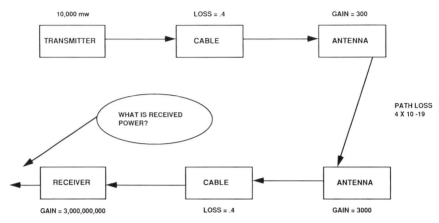

Figure 3.2 Microwave communication system.

The signal is reduced by 0.4 in travelling through the cable from the antenna to the low noise receiver. And in the low noise receiver, the signal is increased by 3 billion times.

The received power is 2 milliwatts. It would take several minutes of calculations to get this answer, provided all the "zeros" could be kept straight. This problem will be worked later to show how easily it can be solved using dB terminology.

Three number systems are compared in Table 3.2. The decimal system is shown in the left column. Scientific notation is shown in the middle column, and the dB number system is shown in the right column. There are three different ways of representing each number. The decimal number system is well understood. Its problem is that very large numbers have many zeros before the decimal point, and very small numbers have many zeros after the

Table 3.2   Comparison of decimal numbers, scientific notation, and dB

Number	Scientific Notation	dB
10	10^1	10
100	10^2	20
1,000	10^3	30
10,000	10^4	40
0.1	10^{-1}	-10
0.01	10^{-2}	-20
0.001	10^{-3}	-30
0.0001	10^{-4}	-40
1	10^0	0

Table 3.3   dB equivalents for other numbers

10	10^1	10 dB
20	$10^{1.3}$	13 dB
50	$10^{1.7}$	17 dB
100	10^2	20 dB

decimal point. It is very hard to do arithmetic with big and small numbers, and keep all the zeros straight.

The problem with keeping all the zeros straight is usually solved by scientific notation, shown in the middle of the column. Note that the dB representation is just 10 times the exponent of the scientific notation. That's all there is to dB. To convert a decimal number to its dB representation, just represent the number as 10 to some exponent and multiply the exponent by 10.

To multiply numbers, add their dB values. For example, to multiply 10 times 10, add their dB equivalents: 10 dB + 10 dB = 20 dB. Again, multiply 10 × 0.1: adding the dB equivalents gives 10 dB − 10 dB = 0 dB, the dB equivalent of 1.

To divide numbers, subtract their dB values. Note that 100 divided by 10 is 10, which is 20 dB − 10 dB = 10 dB.

The previous examples have been for powers of 10. What about other numbers? For instance, what is the dB equivalent of 20? As shown in Table 3.3, in scientific notation 20 is $10^{1.3}$ This exponent is the logarithm of 20. (The logarithm of a number is the exponent to which 10 is raised to get the number.) Thus, for any number N,

$$dB = 10 \log N$$

The dB equivalent of any number can always be obtained by finding the logarithm of the number on a calculator and multiplying the logarithm by 10.

But, as shown in Table 3.4, the dB equivalent of most numbers can be estimated without a calculator, just by remembering that the dB equivalent

Table 3.4   Number to dB conversion

Number	dB
②	③
③	⑤
10	10
4 = 2 × 2	3 + 3 = 6
5 = 10/2	10 − 3 = 7
80 = 2 × 2 × 2 × 10	3 + 3 + 3 + 10 = 19
500 = 5 × 10^2	7 + 20 = 27
= 1000/2	30 − 3 = 27
0.02 = 2/100	3 − 20 = −17

Table 3.5 dB to number conversion

dB	Number
③	②
⑤	③
6	4
10	10
15	30
33	2000
−17	−0.2
−15	0.03
−24	0.004

of 2 is 3 dB and the dB equivalent of 3 is approximately 5 dB (circled in Table 3.4). Hence, the dB equivalent of 4 is 4 = 2 × 2 = 3 dB + 3 dB = 6 dB.

What is the dB equivalent of 5? Since 5 = 10/2, and the dB equivalent of 10 is 10, and the dB equivalent of 2 is 3, 10 dB − 3 dB = 7 dB.

What about 80? 80 is 2 times 2 times 10, so the dB equivalent as shown is 3 + 3 + 3 + 10, or 19 dB.

Finally, what is the dB equivalent of 0.02, which is 2/100? It is 3 dB − 20 dB, which is −17 dB.

The conversion of a dB equivalent to a number can easily be done on a calculator. Enter the dB and divide by 10, which converts the dB value to a logarithm. Then use the inverse logarithm function to get the number.

The conversion of dB equivalents to numbers can also be done just by remembering that 3 dB is 2. For example, as shown in Table 3.5, to convert 6 dB to a number, note that 6 dB is 3 dB + 3 dB. Since 3 dB is 2, 2 × 2 = 4, so 6 dB is 4.

3.3 dBm TERMINOLOGY

The dB terminology is used for the ratio between two power levels in microwave equipment, for example, to compare microwave power coming out of a component to that going into it. The dB number system can also be used to express absolute values of microwave power as dBm. As discussed previously, the reference level for microwave power measurements is 1 mW. Power measured in dBm is just the power value that is being measured referenced to 1 mW. That's what the m stands for. To see how dBm is used, look at Figure 3.1. On the left-hand scale microwave power is shown in watts,

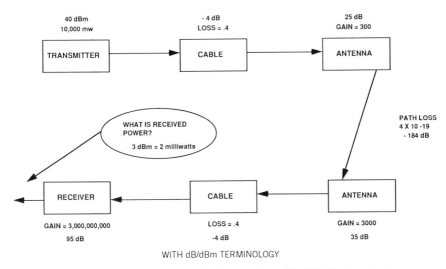

Figure 3.3 Microwave communication system with dB/dBm terminology.

kilowatts, megawatts, and so on. On the right-hand scale, the same power levels are shown in dBm. The reference level of power is 1 mW, which is 0 dBm. For example, a power of 10 dBm is 10 times 1 mW or 10 mW; 20 dBm is 20 dB above 1 mW, and 20 dB is 100 (Table 3.2), so 20 dBm is 100 mW.

Now dB and dBm notation will be used to solve the communication system problem discussed at the beginning of this chapter. Figure 3.3 shows the microwave communication system again, but with dB and dBm values of power, gain, and loss: The transmitter power of 10,000 mW is 40 dBm; the loss of 0.4 in the cable is -4 dB; the gain of the transmitting antenna is 300, which is 25 dB; the path loss is 4×10^{-19}, or -184 dB; The gain of the receiving antenna is 3000, which is 35 dB; the loss of the cable in the receiving system is -4 dB; the low noise receiver gain of 3 billion is 95 dB. The received power is calculated by taking the transmitter power and adding the gains and subtracting the losses, which gives a received power of 3 dBm, which is 2 mW. It is much easier to trace the microwave signal through the system by adding and subtracting dBs than by multiplying numerical gains and dividing numerical losses as was done at the beginning of the chapter.

3.4 EQUIPMENT FOR MEASURING MICROWAVE POWER

Figure 3.4 shows a microwave power meter. It has a sensor, which converts the microwave power into a voltage that is proportional to the power; the meter itself determines what microwave power the sensor voltage represents and displays it.

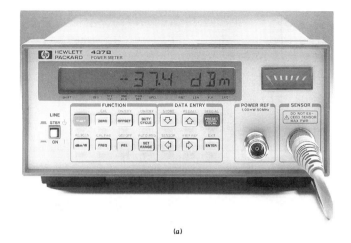

(a)

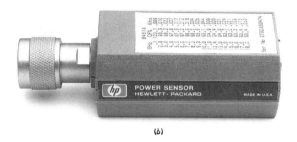

(b)

Figure 3.4   Power meter and power sensor. (Photograph courtesy of Hewlett Packard.)

Three types of power sensors are used: thermistors, thermocouples, and microwave diodes. Thermistors and thermocouples are heated by the micro-wave power, which raises the temperature of the sensor. The heated sensor develops a voltage that is read by the power meter and converted to a microwave power display. The microwave diodes convert the microwave power directly to a dc voltage, which is read and displayed as a power by the power meter. The microwave diodes give rapid response and greater sensitivity than the thermal sensors.

Figure 3.5 shows a microwave frequency counter, which is used to deter-mine and display the frequency of the microwave signal. The power meter measures the total microwave power. If several frequencies are present, the meter measures the total power of all the signals but cannot distinguish the power levels at different frequencies. The frequency counter can only count a single frequency. If multiple frequencies are present, the counter cannot make a measurement.

Figure 3.5 Frequency meter. (Photograph courtesy of Hewlett Packard.)

Many microwave signals consists of several frequencies each with a differ-ent power level. The spectrum analyzer shown in Figure 3.6 is used to distinguish the frequencies and measure and display the microwave power at each frequency. The power is displayed vertically, and the frequency is displayed horizontally. The spectrum analyzer controls allow the frequency and power ranges of the measurement to be adjusted.

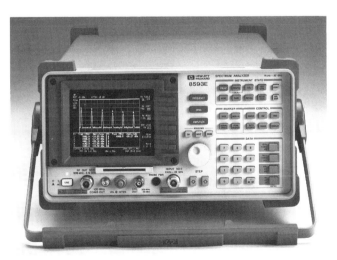

Figure 3.6 Spectrum analyzer. (Photograph courtesy of Hewlett Packard.)

ANNOTATED BIBLIOGRAPHY

1. W. S. Cheung and F. H. Levien, *Microwaves Made Simple: Principles and Applications*, Artech House, Dedham, MA, 1985, pp. 7–17.
2. T. S. Laverghetta, *Modern Microwave Measurements and Techniques*, Artech House, Dedham, MA, 1988.

Reference 1 provides a good explanation of dB and dBm. Reference 2 describes equipment and techniques for measuring microwave power.

EXERCISES

3.1. Convert the following numbers to dB (approximately) without using a calculator.

Numbers	dB
6	
8	
400	
3000	
0.05	
0.002	

3.2. Convert the following dB values to numbers (approximately) without using a calculator.

Numbers	dB
	7
	13
	25
	− 13
	− 57

3.3. Convert the following powers to dBm (approximately) without using a calculator.

Power	dBm
1 W	
2 kW	
50 W	
30 mW	
200 nW	

3.4. Convert the following dBm values to power (approximately) without using a calculator.

Power	dBm
_____	13
_____	24
_____	55
_____	-13
_____	-40

3.5. Convert the following phrases to the equivalent numerical change in dB.

a. Ten times larger   dB = _____
b. One-half as large   dB = _____
c. Decreased by a factor of 12   dB = _____
d. Twice as large as   dB = _____
e. Three times larger than   dB = _____
f. Smaller by 700 times   dB = _____
g. One hundred fifty times greater than   dB = _____
h. A 60:1 reduction   dB = _____
i. Eight orders of magnitude larger   dB = _____
j. Three orders of magnitude smaller   dB = _____

Note: One order of magnitude means a factor of 10, two orders of magnitude means a factor of 100, and so on.

3.6. Express the following numbers in dB notation without using a calculator.

20	_____
5	_____
500	_____
40	_____
3000	_____
0.25	_____
0.20	_____
0.05	_____

6   ———————
30   ———————

3.7. Express the following dB values as numbers without using a calculator

3 ———————
23 ———————
− 3 ———————
35 ———————
46 ———————
− 13 ———————
− 20 ———————
17 ———————

3.8. Use a calculator for these conversions.

Ratio	dB
14	
560	
7.2	
984	
0.013	
10^{-5}	
10^4	
	38.2
	14.7
	51
	− 17.8
	− 34
	− 25

3.9. Use a calculator for these conversions.

Power	mW	dBm
1.5 kW		
250 W		

Power	mW	dBm
18 W		
4 kW		
600 mW		
20 μW		
32 nW		
15 pW		
482 pW		
5 mW		

3.10. Use a calculator for these conversions.

mW	dBm
	20
	-14
	34
	56
	72.5
	18
	25
	-13
	-55
	-75
	21

3.11. Determine from the context of each statement whether the answer should be dB or dBm. Remember, dB indicates a change of power (or a ratio), and dBm indicates an absolute power level.

a. The lab detector is rated for a maximum power of 25 _____.

b. An attenuator has an insertion loss of 10.4 _____.

c. The coupled power level of a 20-dB directional coupler is nominally 20 _____ less than the input power.

d. When tested at input power of 33 dBm, the return loss of the *pin* attenuator was 16.5 _____.

e. The reflected power level from the circulator was − 18.5 _____.

f. The output power of a TM-812 amplifier is always 37 _____ higher than the input power.

4

INSERTION LOSS, GAIN, AND RETURN LOSS

Insertion loss describes the reduction of a microwave signal as it passes through a component. *Gain* describes the increase of a microwave signal as it passes through an amplifier. The insertion loss and gain of microwave components, and their net effect on components in cascade, are determined.

Whenever a microwave transmission line is connected to a microwave device, the electric and magnetic fields in the transmission line and in the device do not match, so some of the microwave power is reflected at the connection. The most common way of specifying this reflected power is in terms of *return loss*. Other ways include percent reflected power, SWR, and reflection coefficient. The relationship between these quantities is defined, and conversion among them is studied.

Yet another way of specifying insertion loss, gain, and return loss is S-parameters, which are defined in this chapter. The chapter concludes with a discussion of test equipment for measuring insertion loss, gain, and return loss.

4.1. INSERTION LOSS AND RETURN LOSS

The concept of insertion loss and return loss is illustrated in Figure 4.1. A microwave component, which might be a filter, an isolator, an attenuator, an amplifier, is shown. Microwave power is sent down a transmission line from the left and it reaches the component. This power is the *incident power*. When it reaches the component, a portion is reflected back down the transmission line where it came from and never enters the component. The power is reflected because the microwave field configurations in the transmission

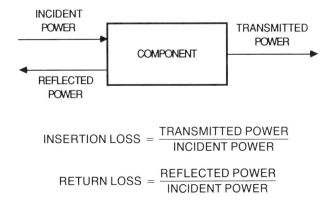

$$\text{INSERTION LOSS} = \frac{\text{TRANSMITTED POWER}}{\text{INCIDENT POWER}}$$

$$\text{RETURN LOSS} = \frac{\text{REFLECTED POWER}}{\text{INCIDENT POWER}}$$

Figure 4.1   Insertion loss and return loss.

line and the component are different. Some of the microwave power doesn't want to fit into the new field configuration in the component, so it is reflected back into the transmission line. Every microwave component has this problem. Unless two identical sections of transmission lines, or two identical components are being connected, not all of the power will get from the transmission line into the component. There is a mismatch between the transmission line and the component. Actually there is a mismatch between the microwave field configuration in the transmission line and in the component.

The power not reflected gets into the component. There some of it gets absorbed and the remainder passes through the component into the transmission line on the other side. The power that actually comes out of the component is called the *transmitted power,* and it is less than the incident power for two reasons: (1) some of the power got reflected and never got into the component in the first place; (2) some of the power that entered the component was absorbed inside.

The ratio of the transmitted power to the incident power, in dB terminology, is the insertion loss. The ratio of the reflected power to the incident power, in dB terminology, is the return loss.

$$\text{Insertion loss} = \frac{\text{Transmitted power}}{\text{Incident power}}$$

$$\text{Return loss} = \frac{\text{Reflected power}}{\text{Incident power}}$$

Table 4.1 **Comparison of incident power and insertion loss**

Incident Power Transmitted (%)	Insertion Loss (dB)
1	20
2	17
5	13
10	10
50	3

These two quantities completely describe the performance of the component. Insertion loss is just the dB equivalent of the percent power that gets through the component relative to what was trying to get in. Insertion loss, when used in calculations, will always be a negative quantity because the dB expression of any percent less than 100 is a negative number. But because the word *loss* is used, insertion loss always specified as a positive number.

Table 4.1 lists some percentages of incident power transmitted along with the corresponding insertion loss.

The term *attenuation* is often used incorrectly to mean insertion loss. Attenuation is the power loss inside the component, but the power that gets out is reduced not only by what is lost inside the component due to attenuation but also by what is reflected and never gets in in the first place.

Only insertion loss can be measured. To determine the attenuation, the component must be matched perfectly, so that all the incident power gets inside. In that case, there will be no reflected power, so the insertion loss, which can be measured, will equal the attenuation.

Table 4.2 shows some insertion loss calculations. In the first set of calcula-

Table 4.2 **Insertion loss calculations**

Incident Power	Insertion Loss	Transmitted Power
	Calculation of Insertion Loss	
10 W	10 dB	1 W
10 mW (10 dBm)	7 dB	2 mW
23 dBm	10 dB	13 dBm
5 μW (− 23 dBm)	7 dB	− 30 dBm
	Calculation of Transmitted Power	
1 W (30 dBm)	3 dB	0.5 W
		27 dBm
5 mW (7 dBm)	6 dB	1 dBm
20 W (43 dBm)	10 dB	33 dBm
17 dBm	20 dB	− 3 dBm

tions the incident power and the transmitted power are given and the insertion loss is to be determined. In the second set, the incident power and the insertion loss are given, and the transmitted power is to be determined.

The calculation of the first line of the first set is made as follows. The incident power into the component is 10 W, but only 1W comes out, which is the transmitted power. The insertion loss is the ratio of 1 W to 10 W, or $\frac{1}{10}$. In dB terminology, this is -10 dB, but the minus sign is ignored because the term *loss* is used.

The first line of the second set of calculations is done as follows. The power into the component is 1 W, and the component has an insertion loss of 3 dB. What is the transmitted power? Since 3 dB is 2, a 3-dB loss means that the power is reduced to one half. An easier way to work the problem is to convert the powers to dBm. The 1 W of input power is equal to 30 dBm. The insertion loss is 3 dB, so the transmitted power is 30 dBm $-$ 3 dB = 27 dBm, which is 500 mW or 0.5 W. Note that when insertion loss values are used in calculations, the minus sign must be included with the insertion loss value.

4.2 INSERTION LOSS OF COMPONENTS IN CASCADE

Figure 4.2 shows a filter, isolator, and attenuator connected together. The microwave power must pass through all three components. The insertion loss of each component is shown in the block. If 1 W of power is incident on this chain of components, how much power comes out of the chain? The calculation will first be made with the power in watts. This is the harder way to make the calculation. The powers in watts are shown above the arrows. One watt of power is put into the filter, and the filter has an insertion loss of 3 dB, so half of the power, 0.5 W, comes out of the filter. The power then goes into the isolator, which has an insertion loss of 3 dB, so half of the power that goes into the isolator comes out, 0.25 W. The power then goes to the attenuator, which has a 10-dB insertion loss, which means that only one tenth of the power that goes into the attenuator comes out. Thus, only 0.025 W, or 25 mW, comes out. Along the bottom of the figure the calculation is shown in dBm notation. The input power of 1 W is 30 dBm, and the insertion loss of the filter is 3 dB, so the power coming out of the filter is 30 dBm $-$ 3 dB, or 27 dBm. This power goes into the isolator, whose insertion loss is 3 dB, so the power coming out of the isolator is 27 dBm $-$ 3 dB = 24 dBm. This power then goes into the attenuator, which has a 10-dB insertion loss, so the power coming out of the attenuator is 24 dBm $-$ 10 dB or 14 dBm, and 14 dBm is 25 mW. It is easy to determine the total insertion loss of the three components in cascade and represent the combination as a single insertion loss simply by adding the insertion losses of the individual components. The total insertion loss of the three components is 16 dB, and if the input power is 30 dBm and the total insertion loss is 16 dB the transmitted power is 30 dBm $-$ 16 dB or 14 dBm.

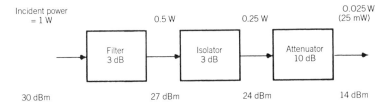

30 dBm   27 dBm   24 dBm   14 dBm

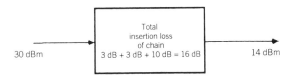

Figure 4.2 Components in cascade.

4.3 GAIN

If the component is an amplifier, then the transmitted power from the component is greater than the incident power because the incident power is amplified. Therefore, the ratio of the transmitted power to the incident power is
defined as the *gain* of the amplifier. Figure 4.3 shows input power entering
an amplifier, being amplified, and output power leaving the amplifier. The
terms *input power* or *incident power* can be used interchangeably, and the
terms *output power* or *transmitted power* can be interchanged. The gain of
the amplifier is the output power divided by the input power. In dB notation,

$$\text{Gain (dB)} = 10 \log \frac{\text{Output power}}{\text{Input power}}$$

or, in dBm,

$$\text{Gain (dB)} = \text{Output power (dBm)} - \text{Input power (dBm)}$$

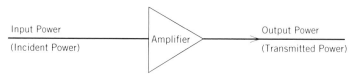

Figure 4.3 Gain

Example 4.1 Calculate the gain of an amplifier with an input power of 10 mW and an output power of 200 mW.

Solution 1

$$
\begin{aligned}
\text{Gain (dB)} &= 10 \log \frac{\text{Output power}}{\text{Input power}} \\
&= 10 \log \frac{200 \,\text{mW}}{10 \,\text{mW}} \\
&= 10 \log 20 \\
&= 10\,(1.3) = 13 \,\text{dB}
\end{aligned}
$$

Solution 2

$$
\begin{aligned}
10 \,\text{mW} &= 10 \,\text{dBm} \\
200 \,\text{mW} &= 23 \,\text{dBm} \\
\text{Gain (dB)} &= \text{Output power} - \text{Input power} \\
&= 23 \,\text{dBm} - 10 \,\text{dBm} \\
&= 13 \,\text{dB}
\end{aligned}
$$

Example 4.2 Calculate the output power of an amplifier with a 15-dB gain for an input of 3 dBm.

Solution

$$
\begin{aligned}
\text{Output power} &= \text{Gain} + \text{Input power} \\
&= 15 \,\text{dB} + 3 \,\text{dBm} = 18 \,\text{dBm}
\end{aligned}
$$

4.4 CASCADED INSERTION LOSS AND GAIN

In most microwave systems insertion loss and gain must be considered together. An example is shown in Figure 4.4 where the microwave input power passes through an attenuator that has an insertion loss of 3 dB, an amplifier with a gain of 10 dB, a connecting cable with an insertion loss of 1 dB, a second amplifier with a gain of 6 dB, and, finally, a filter with an insertion loss of 2 dB. The input power to this chain is 6 dBm, which is 4 mW. What is the output power from the chain? The chart in the figure shows the gains and losses of the signal as it passes through the system, which can be calculated by adding the losses and gains of the components, to give a total gain of 10 dB. The output power is thus 16 dBm.

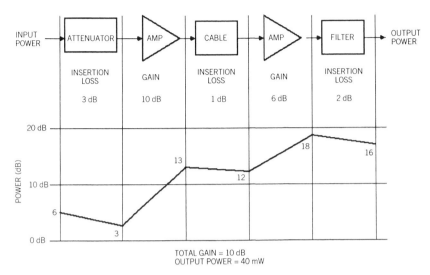

Figure 4.4 Cascaded insertion loss and gain.

4.5 MISMATCHES AND RETURN LOSS

Mismatches are defined in Figure 4.5. Microwave power travels down a transmission line from the left to enter a microwave component. However, as shown, some power enters, but some is reflected because the fields between the transmission lines and the component don't match. Mismatch is a measure of the microwave power that gets reflected when it reaches a microwave component.

Mismatch occurs *every* time a transmission line is connected to a microwave part. In the simplest microwave subassemblies there may be 10 connections where mismatches occur. In a microwave subsystem there are probably 100 connections, and a mismatch occurs at every one. In a radar system, there are probably 10,000 connections, and at every connection a mismatch occurs! Mismatches must be measured at every connection.

To understand better how mismatches occur, refer back to Figure 2.8, which shows a cross section of a coaxial cable and a microstrip with the microwave field configuration sketched in. When the microwave power traveling down the coaxial cable reaches the microstrip, all the power is supposed to go from the coaxial cable into the microstrip, but the fields in these two transmission lines don't match. In the coaxial cable, the microwave fields extend radially in all directions from the center conductor. In the microstrip, most of the fields extend from the conductor into the insulator. Hence, at the junction between the two transmission lines, the fields in the bottom part of the coaxial cable approximately match the fields in the microstrip, so that part of the power will go right from the coaxial cable into the microstrip. But

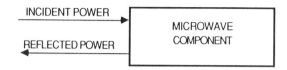

Incident Power: The microwave power trying to enter the component.
Reflected Power: The microwave power reflected from the component.
Mismatch: The ratio of reflected power to incident power.
Return Loss: The dB expression of the ratio of reflected power to incident power (1% reflected power = 20-dB return loss).

Figure 4.5 Return loss.

the fields that point upward in the coaxial cable find no matching fields for them in the microstrip, so they do not leave the transmission line because there is no matching field configuration in the next line. Consequently, they are reflected back into the coaxial transmission line.

The incident power, as defined in Figure 4.6, is the microwave power trying to enter the component. The reflected power is the microwave power reflected from the component. The mismatch is the ratio of reflected power to incident power. For example, assume the incident power is 1 mW and half gets reflected, so half a milliwatt enters the component. The mismatch is then

$$\frac{0.5 \text{ mW}}{1 \text{ mW}} = 0.5 = 50\%$$

Thus, 50% of the power is reflected.

Normally the mismatch is not specified as a percentage of incident power. Instead, dB terminology is used. The mismatch is specified as return loss,

Table 4.3 Return loss calculations

Incident Power	Return Loss	Reflected Power
Calculation of Return Loss		
1 mW	10 dB	0.1 mW
20 W (43 dBm)	13 dB	1 W (30 dBm)
50 mW (17 dBm)	7 dB	10 dBm
−16 dBm	14 dB	−30 dBm
Calculation of Reflected Power		
0 dBm	10 dB	−10 dBm
10 mW (10 dBm)	13 dB	−3 dBm
100 μW (−10 dBm)	17 dB	−27 dBm
25 W (44 dBm)	20 dB	24 dBm

and return loss is the dB expression of the ratio of the reflected power to the incident power. For example, if 1% of the power is reflected, the return loss is 20 dB.

The actual return loss is a negative quantity because the reflected power is always less than the incident power. However, on a specification or test data sheet, it's given as a positive number because the word *loss* is used.

Table 4.3 shows some return loss calculations. In the first set of calculations, the incident power and the reflected power are given and the return loss is to be determined. In the second set, the incident power and return loss are given, and the reflected power is calculated.

4.6 ALTERNATIVE WAYS OF SPECIFYING REFLECTED POWER

Figure 4.6 shows several other ways of specifying the power reflected at a mismatch. In the upper sketch a transmission line is connected to a component. At the connection, part of the incident power is reflected, and the rest is transmitted *into* the component. (All of the power transmitted into the component is not necessarily transmitted out of the component, due to attenuation.)

Various ways of specifying reflected power are given in Table 4.4. The simplest way to express the reflected power is as a percent of the incident power, as shown in column 1.

The most common way of specifying mismatch is as return loss, which is simply 10 times the logarithm of the percent reflected power, given in column 3.

The fourth column gives the standing wave ratio (SWR). Standing wave ratio can be understood from the Figure 4.6, which shows the electric field of the microwaves along the input transmission line. The total electric field

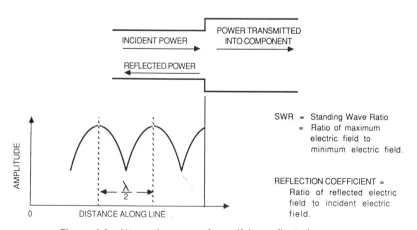

Figure 4.6 Alternative ways of specifying reflected power.

Table 4.4 Ways to specify reflected power

Power Reflected (%)	Power Transmitted into Component (%)	Return Loss (dB)	SWR	Reflection Coefficient
1	99	20	1.25	0.10
5	95	13	1.58	0.22
10	90	10	1.95	0.32
50	50	3	5.80	0.71

is the sum of the incident electric field, which is traveling down the transmission line toward the mismatch point, and the reflected electric field, but these two fields add up in phase and out of phase. If the reflected electric field is in phase with the incident electric field, then the two fields add, resulting in a maximum electric field in the transmission line. However, if the two fields are out of phase, they subtract, resulting in a minimum total electric field. Between these points where the fields exactly add or exactly subtract are intermediate values, depending on the phase relationship.

The ratio of the maximum electric field to the minimum electric field is the standing wave ratio, and its value is related to the percent of the power reflected at the mismatch. As column 4 shows, if 1% of the power is reflected, the standing wave ratio is 1.25, which means that the maximum of the total electric field is 1.25 times greater than the minimum value.

The SWR can actually be measured. A small slot is cut in the input transmission line and the field is measured by inserting a small sensing probe in the slot. Note that the pattern of field maxima and minima repeats itself every half wavelength. The pattern might have been expected to repeat every full wavelength, but note that, in moving a half wavelength down the transmission line, the incident electric field has traveled a half wavelength and the reflected electric field has traveled a half wavelength in the other direction. Therefore, the total phase difference between the two fields changes by a full wavelength.

The last way to specify reflected microwave power is by the reflection coefficient. The reflected field or the incident field can't actually be measured separately, but if they could, their ratio would be the reflection coefficient. Reflection coefficient can only be calculated from other measured quantities. As shown in column 5 of Figure 4.4, if 1% of the power is reflected, then the reflection coefficient is 0.10, which means that the reflected electric field is 10% of the incident electric field.

The relationships between reflection coefficient and the other ways of specifying the reflected microwave signals are shown in Figure 4.7. The incident electric field is labeled E_i, and the reflected electric field is labeled E_r. The ratio E_r/E_i is the reflection coefficient ρ, which can take any value between 0 (i.e., none of the incident electric field is reflected) to 1 (i.e.,

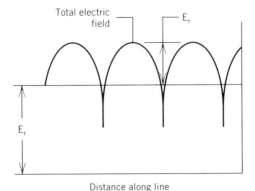

$$\text{Reflection coefficient} = \rho = \frac{E_r}{E_i}$$

$$\frac{\text{Reflected power}}{\text{Incident power}} = \left(\frac{E_r}{E_i}\right)^2 = \rho^2$$

$$\text{Return loss} = \left[10 \log \frac{\text{Reflected power}}{\text{Incident Power}}\right]$$

$$= \mid 10 \log \rho^2 \mid \; = \; \mid 20 \log \rho \mid$$

$$\text{SWR} = \frac{1 + \rho}{1 - \rho} \quad \text{And} \quad \rho = \frac{\text{SWR} - 1}{\text{SWR} + 1}$$

Figure 4.7 Conversion between ways of specifying reflected power.

all of the incident electric field is reflected). Since microwave power is proportional to the electric field squared (recall that in low frequency ac circuits power is proportional to voltage squared), the expression for reflected power over incident power is proportional to the reflected electric field squared divided by the incident electric field squared, or ρ^2. So percent reflected power is just equal to ρ^2. The return loss is 10 times the log of the percent reflected power; the absolute value is taken because the reflected power is less than the incident power, and taking the log gives a minus sign, which has already been taken into account by using the word *loss*. Since the reflected power divided by incident power equals ρ^2, the return loss is just 10 log ρ^2 or 20 log ρ.

At any point in the transmission line, the total electric field is equal to the sum of the incident field and the reflected field, but how they add up depends on their phase relationship. As shown in Figure 4.7, along the transmission line the total electric field has maxima and minima as the incident and reflected fields go in and out of phase. This field variation is called a standing wave pattern. The ratio of the maximum field to the minimum field is called the standing wave ratio (SWR), and is equal to $1 + \rho$ (which is the maximum electric field when the fields are in phase) divided by $1 - \rho$ (which is the minimum electric field when the fields are out of phase).

Example 4.3   If 10% of the microwave power is reflected at the mismatch, find the return loss, reflection coefficient, and SWR.

Solution

$$\text{Return loss} = \left| 10 \log \frac{\text{Reflected power}}{\text{Incident power}} \right|$$
$$= | 10 \log 0.10 | = 10 \, dB$$

$$\text{Reflection coefficient} = \sqrt{\frac{\text{Reflected power}}{\text{Incident power}}}$$
$$= \sqrt{0.10} = 0.32$$

$$\text{SWR} = \frac{1 + 0.33}{1 - 0.33} = 1.92$$

Example 4.4   If the return loss is 20 dB, find the percent reflected power, reflection coefficient, and SWR.

Solution

$$\% \text{ Reflected power} = \text{Inverse log} \left| - \frac{\text{Return loss}}{10} \right|$$
$$= \text{Inverse log } 2$$
$$= 0.01 = 1\%$$

$$\text{Reflection coefficient} = \sqrt{\frac{\text{Reflected power}}{\text{Incident power}}}$$
$$= 0.01 = 0.1$$

$$\text{SWR} = \frac{1 + 0.1}{1 - 0.1} = \frac{1.1}{0.9} = 1.2$$

Return loss, SWR, and reflection coefficient are just different ways of specifying the ratio of the reflected power to the incident power at a mismatch. To characterize a mismatch completely, however, the amplitude of the mismatch (as expressed by return loss, SWR, or reflection coefficient) and the phase of the reflected microwave field relative to the incident field must be

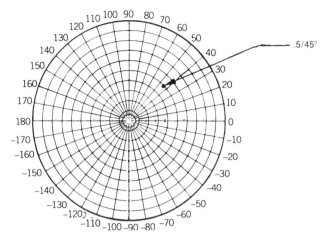

Figure 4.8 Reflection coefficient shown on polar plot.

specified. The discussion up until now has covered only the amplitude of the reflected and incident signals.

When the amplitude and phase of the mismatch are to be specified, it is customary to use reflection coefficient terminology. Consider a mismatch with a reflection coefficient of 0.5/45°. The first number (0.5) is the ratio of the amplitudes of the reflected and incident electric fields. It states that half of the electric field is reflected. As discussed previously, if half of the electric field is reflected, then half of the magnetic field is also reflected, since the electric and magnetic fields are related by the impedance of the circuit. Since power is the electric field times the magnetic field, and half of each are reflected, one quarter of the power is reflected, which is a return loss of 6 dB. The second number (45°) states that the phase of the reflected electric field leads the incident electric field by 45°.

The reflection coefficient is often displayed on a circular graph as shown in Figure 4.8. The distance out from the center of the graph is the amplitude of the reflection coefficient; the angle around the chart is the phase of the reflection coeficient.

4.7 S-PARAMETERS

S-parameters are another way of specifying return loss and insertion loss. *S*-parameters are defined in Figure 4.9, which shows microwave signals entering and exiting a microwave component in both directions. If a microwave signal is incident on the input side of the component, some of the signal is reflected and some is transmitted through the component. The ratio of the reflected electric field to the incident electric field is the reflection coefficient. The

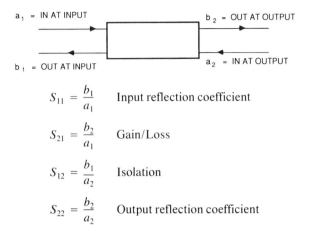

$$S_{11} = \frac{b_1}{a_1} \qquad \text{Input reflection coefficient}$$

$$S_{21} = \frac{b_2}{a_1} \qquad \text{Gain/Loss}$$

$$S_{12} = \frac{b_1}{a_2} \qquad \text{Isolation}$$

$$S_{22} = \frac{b_2}{a_2} \qquad \text{Output reflection coefficient}$$

S-Parameters have amplitude and phase.

Figure 4.9 *S*-parameters.

ratio of the transmitted electric field to the incident electric field is the transmission coefficient.

The signals entering the input and leaving the input and output are easily understood in terms of reflection and transmission coefficients. But how can there be an input signal at the output? The answer is that as the microwave signal travels down the output transmission line some of it is reflected at some point in the line and comes back into the component's output.

To characterize the component completely, the reflection and transmission coefficients must be specified in both directions. In words, such expressions as the "input going into the output," the "input going into the input," or "the output coming out of the input," must be used. *S*-parameter terminology is designed to avoid these cumbersome descriptions.

S-parameters are defined as follows. Microwave signals going into or coming out of the input port are labeled by a subscript 1. Signals going into or coming out of the output port are labeled by a subscript 2. The electric field of the microwave signal going into the component ports is designated *a;* that leaving the ports is designated *b*. Therefore,

a_1 is the electric field of the microwave signal entering the component input.

b_1 is the electric field of the microwave signal leaving the component input.

b_2 is the electric field of the microwave signal leaving the component output.

a_2 is the electric field of the microwave signal entering the component output.

By definition, then,

$$S_{11} = \frac{b_1}{a_1} \qquad a_2 = 0$$

$$S_{21} = \frac{b_2}{a_1} \qquad a_2 = 0$$

$$S_{12} = \frac{b_1}{a_2} \qquad a_2 = 0$$

$$S_{22} = \frac{b_2}{a_2} \qquad a_1 = 0$$

Therefore, S_{11} is the electric field leaving the input divided by the electric field entering the input, under the condition that no signal enters the output. Since b_1 and a_1 are electric fields, their ratio is a reflection coefficient.

Similarly, S_{21} is the electric field leaving the output divided by the electric field entering the input, when no signal enters the output. Therefore, S_{21} is a transmission coefficient and is related to the insertion loss or the gain of the component.

In like manner, S_{12} is a transmission coefficient related to the isolation of the component and specifies how much power leaks back through the component in the wrong direction. S_{22} is similar to S_{11}, but looks in the other direction into the component.

4.8 EQUIPMENT FOR MEASURING INSERTION LOSS AND RETURN LOSS

The power meters described in Chapter 3 could be used to measure insertion loss and return loss. With directional couplers, which sample the power traveling in one direction down a transmission line but not in the other direction, incident and reflected powers could be separated and measured with power meters. The ratio of the two powers could then be calculated to determine return loss, and further calculations could be made for SWR, reflection coefficient, or S-parameters.

However, because so many return loss and insertion loss measurements are made on microwave equipment, a *network analyzer* is used. The network analyzer integrates the power meters, the directional couplers, a signal source, a computer, and a CRT to display the measured and calculated quantities as a function of frequency.

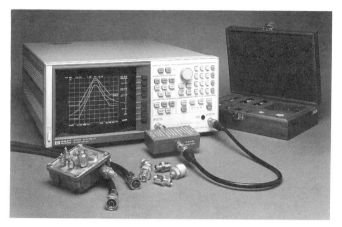

Figure 4.10   Network analyzer. (Photograph courtesy of Hewlett Packard.)

A typical network analyzer is shown in Figure 4.10. The device under test is connected between the two microwave connections on the front panel. The controls set the frequency range over which the measurements are made and the quantities to be displayed. The analyzer in the figure is set up to measure and display the amplitude and phase of the insertion loss of a filter. Such a network analyzer, which measures both amplitude and phase, is called a *vector* analyzer. Other, less expensive, *scalar* network analyzers only measure signal amplitude.

ANNOTATED BIBLIOGRAPHY

1. W. S. Cheung and F. H. Levien, *Microwaves Made Simple: Principles and Applications*, Artech House, Dedham, MA, 1985, pp. 48–63.
2. S. Y. Liao, *Microwave Devices and Circuits*, Prentice-Hall, Englewood Cliffs, NJ, 1980.
3. G. D. Vendelin, *Design of Amplifiers and Oscillators by the S Parameter Method*, Wiley, New York, 1981, pp. 1–39.
4. T. S. Laverghetta, *Modern Microwave Measurements and Techniques*, Artech House, Dedham, MA, 1988.

References 1 and 2 provide good explanation of insertion loss, gain, and return loss. Reference 3 explains *S*-parameters. Reference 4 describes equipment and techniques for measuring insertion loss, gain, and return loss.

EXERCISES

4.1. Calculate the insertion loss.

Incident Power	Insertion Loss	Transmitted Power
6 mW		2 mW
100 W		2 W
7 dBm		13 dBm
10 mW		− 30 dBm

4.2. Calculate the transmitted power.

Incident Power	Insertion Loss	Transmitted Power
0.4 mW	3 dB	
15 mW	6 dB	
20 W	16 dB	
27 dBm	20 dB	

4.3. What is the difference between insertion loss and attenuation?

4.4. Complete the table for three components in cascade.

P_{IN} — Insertion Loss 1 — P_A — Insertion Loss 2 — P_B — Insertion Loss 3 — P_{OUT}

P_{IN} (dBm)	Insertion Loss 1 (dB)	P_A (dBm)	Insertion Loss 2 (dB)	P_B (dBm)	Insertion Loss 3 (dB)	Total Insertion Loss of Chain (dB)	P_{OUT} (dBm)
0	3		10		7		
25	15		5		8		
− 17	14		25		3		
3	10		8		5		
− 35	3		3		8		

4.5. Calculate the gain.

Input Power	Gain	Output Power
20 mW		200 mW
−13 dBm		1 mW
500 W		2000 W
−33 dBm		10 dBm

4.6. Calculate the output power.

Input Power	Gain	Output Power
3 μW	10 dB	
5 kW	6 dB	
10 dBm	3 dB	
1 nW	20 dB	

4.7. Complete the table.

Input Power	Insertion Loss of Filter (dB)	Gain of Amplifier (dB)	Insertion Loss of Cable (dB)	Total Insertion Loss	Output Power
1 mW	3	10	1		
20 W	6	10	1		
1 μW	10	12	1		
23 dBm	10	20	1		

4.8. Convert the following percent reflected powers to return loss.

Reflected Power (%)	Return Loss (dB)
1	
5	
10	
20	
50	
100	

4.9. Convert the following return loss values to percent reflected power.

Reflected Power (%)	Return Loss (dB)
	30
	20
	17
	13
	10

4.10. Convert the following return loss values to percent reflected power, reflection coefficient, and SWR.

Return Loss (dB)	Reflected Power (%)	Reflection Coefficient	SWR
0			
10			
13			
17			
20			
30			

4.11. Rank the following mismatches from least reflected power (1) to most reflected power (6).

Return loss	= 0 dB	_____
Return loss	= 20 dB	_____
Reflected power	= 10%	_____
Reflected power	= 5%	_____
SWR	= 1.4	_____
SWR	= 1.1	_____

5

MATCHING WITH THE SMITH CHART

Matching at the transmission line–component interface to reduce or eliminate reflected microwave power involves measuring the reflected power and calculating where to add a "matching element" that compensates for the field mismatch and forces the power from the transmission line into the component. The *Smith chart* allows this calculation to be made.

A Smith chart is shown in Figure 5.1*b*. It is derived from the reflection coefficient chart of Figure 5.1*a*. The reflection coefficient chart shows all the information about the mismatch. The distance out from the center of the chart is the amplitude of the reflection coefficient p, and the angle around the chart is the phase. In the figure $p = 0.5/45°$. Recall that this means the reflected electric field is half of the incident electric field, and the reflected field is shifted in phase by 45° from the incident field at the mismatch.

One way to design the element that would force all the microwaves into the component would be to solve Maxwell's equations for the electric and magnetic fields in the transmission line, in the component, and in the transition region, and then get consistent solutions where the field configurations match from one region to the other, quite a difficult task.

An easier way is to use the Smith chart. The measured microwave mismatch is plotted in the same location, but the Smith chart scales allow this mismatch to be represented as a low frequency equivalent of the microwave problem. In other words, the Smith chart represents the mismatch as an "equivalent circuit" of a resistor and inductor or a resistor and capacitor.

Figure 5.2*a* shows a mismatch represented by an equivalent series circuit of a 50-Ω resistor and an inductor with a reactance of 50 Ω. The voltage across the resistor is in phase with the current, but the voltage across the inductor is 90° out of phase. The well-known vector triangle is used to solve for the total impedance of 71 Ω.

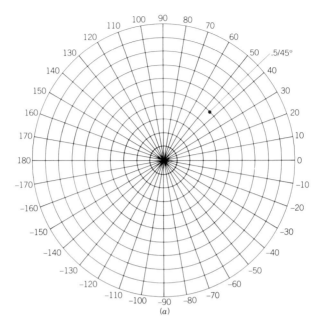

(a)

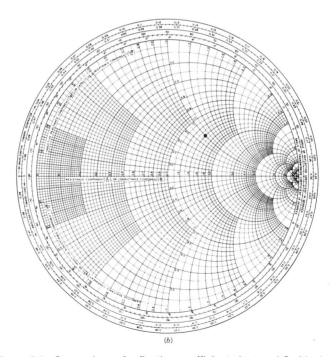

(b)

Figure 5.1 Comparison of reflection coefficient chart and Smith chart.

There would be no mismatch (i.e., no reflected power) if the equivalent circuit of the transition between the transmission line and the component was just a 50-Ω resistor, because the component would match the 50-Ω characteristic impedance of the transmission line. The equivalent circuit of the mismatch could be changed to a 50-Ω resistor, as shown in Figure 5.2b, by adding a capacitor to cancel out the inductance. The capacitor must have a reactance of 50 Ω at the frequency of operation, and since its reactance is 180° out of phase with the inductor reactance, the reactances cancel.

Thus, the Smith chart transforms a difficult microwave problem into a simple low frequency equivalent circuit. This chapter shows how the Smith chart is derived, how to plot mismatches on the chart, and how to use it to design various matching elements, including lumped capacitors and inductors in series and in parallel with the transmission line, matching stubs, and quarter-wave transformers.

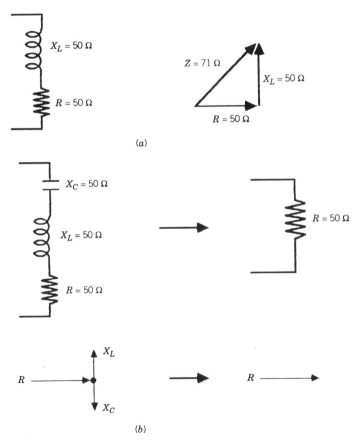

Figure 5.2 Low frequency equivalent circuit.

5.1 DERIVATION OF THE SMITH CHART

The measured mismatch, as shown in the reflection coefficient graph of Figure 5.1a, is the amplitude and phase of the reflected electric field relative to the incident electric field. The Smith chart provides an equivalent impedance for the mismatch. This impedance can be derived mathematically from the reflection coefficient (see the bibliography). In this section, this imped-ance–reflection coefficient relationship is described in physical terms.

The impedance of a transmission line is defined in Table 5.1. Impedance is the ratio of the electric and magnetic fields in the transmission line. The units of electric field are volts per meter, and the units of magnetic field are amps per meter, so the ratio is volts per amps, or ohms. For an infinitely long transmission line, or one with a perfectly matched termination, there is no reflected signal, so the electric field and the magnetic field are both constant all along the line. Therefore, their ratio, and hence impedance, is constant all along the line. This constant impedance is called the *characteristic imped-ance* of the transmission line, and is denoted Z_0. The characteristic impedance depends on the dimensions of the transmission line and on the dielectric material that fills it, so it's completely under the designer's control. In free space, the impedance of an electromagnetic wave (the ratio of the electric field to the magnetic field) is 377 Ω.

Coaxial cable and microstrip are usually designed to have an impedance of 50 Ω. In waveguide the impedance varies with frequency, but in the center of the band of standard waveguide the impedance is about 150 Ω.

Table 5.1   Impedance of a transmission line

The impedance of a microwave transmission line is the ratio of the electric field to the magnetic field.

$$Z = \frac{E}{H} = \frac{\text{volts/meter}}{\text{amps/meter}} = \text{ohms}$$

With no reflections, Z is constant along line and is called the characteristic impedance = Z_0.

The Characteristic impedance is determined by the dimensions of the transmission line and the dielectric material in the transmission line.

Transmission Line	Z_0 (ohms)
Free space	377
Coax cable	50
Waveguide	150
Microstrip	50

If the transmission line is not matched and part of the microwave signal is reflected, then the electric field in the transmission line is the sum of the incident and reflected electric fields. At different locations along the transmission line, these add up differently, depending on their phase relationship, and a standing wave pattern is obtained.

The incident and reflected magnetic fields also add, but their phase relationship is different from that between the electric fields, so the maximum and minimum of the magnetic fields occur at different positions in the transmission line. This is illustrated in Figure 5.3. Since the total electric fields and the total magnetic fields vary differently along the transmission line, their ratio, which is impedance, also varies.

Figure 5.4b shows a polar plot of the reflection coefficient, which is the amplitude and the phase of the reflected electric field relative to the incident electric field.

The reflected magnetic field can also be shown on a reflection coefficent chart, and is the dotted line in Figure 5.4. Its amplitude is equal to the electric field reflection coefficient (because the same percentage of magnetic field as electric field is reflected), but the phase of the magnetic fields is 180° different.

Figure 5.5a shows the vector addition of the electric fields and magnetic

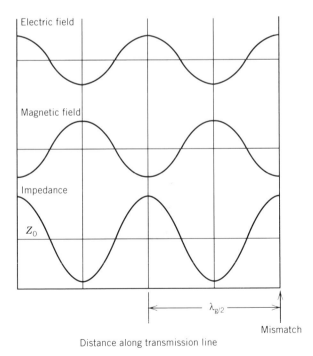

Figure 5.3 Impedance as a function of distance along a mismatched transmission line.

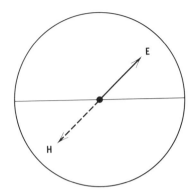

Figure 5.4 Polar plot of reflection coefficient.

fields at the mismatch. Remember that the total electric field is the vector sum of the incident and reflected fields, and the amplitude and phase of the reflected field relative to the incident field is given by the reflection coefficient. The total magnetic field is calculated in exactly the same way. In the example, the reflected electric field is half the incident electric field and leads it by 45°.

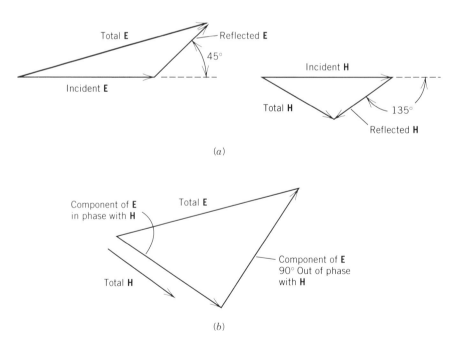

Figure 5.5 Vector combination of incident and reflected fields: (a) total electric and magnetic fields; (b) electric field components relative to total magnetic field.

The reflected magnetic field is also equal to half the incident magnetic field, but its phase angle lags the incident field by 135°.

Remember that impedance is the ratio of the total electric field to the total magnetic field. The total electric and magnetic fields are vector quantities, having amplitude and phase. In Figure 5.5b the total electric field is divided into two components, one in phase with the total magnetic field and the other 90° out of phase with it. The impedance then consists of two components: one component in which the fields are in phase, and one in which the fields are 90° out of phase. In Figure 5.5b the in-phase component has a value of about 1.5 (i.e., the strength of the electric field is about 1.5 times the strength of the magnetic field), and the out-of-phase component also has a value of about 1.5. Compare these impedances to the impedances of an ac circuit.

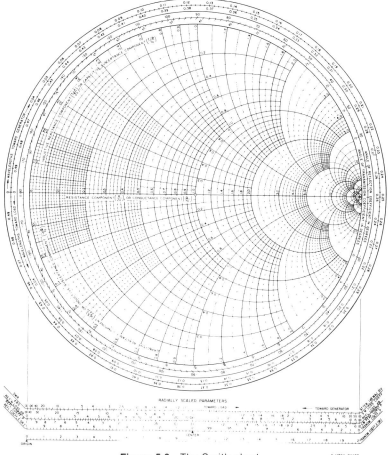

Figure 5.6 The Smith chart.

The impedance of an ac circuit consists of a resistive part (where the voltage and current are in phase) and a reactive part (where the voltage and current are 90° out of phase). It's helpful to remember, when contrasting microwave problems to low frequency ac circuit problems, that the electric field of a microwave signal and the voltage of an ac circuit are related, and the magnetic field of a microwave signal and the current of an ac signal are related.

An electric field can be generated by applying a voltage across a pair of capacitor plates, and a magnetic field can be generated by passing a current through an inductor coil. Often the terms *voltage* and *current* are used to describe the microwave fields along the transmission line, but actually *electric field* and *magnetic field* are correct.

For every value of reflection coefficient (that is, every different amplitude and phase combination), there is a specific value of impedance. From calculations like those of Figure 5.5, the resistive and reactive parts of the impedance can be calculated. This is exactly how the Smith chart is made. For hundreds of reflection coefficient points the resistive and reactive parts of the impedance were calculated. Then all points of equal resistance and all points of equal reactance were connected, and the result is the Smith chart shown in Figure 5.6.

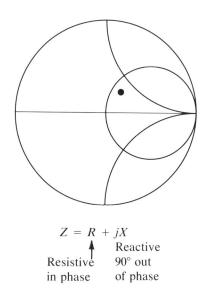

$$Z = R + jX$$

Resistive ↑ Reactive
in phase   90° out
            of phase

All Smith chart points are normalized to the characteristic impedance of transmission line.

$$\text{Normalized impedance} = \frac{Z}{Z_0} = z = \frac{R}{Z_0} + j\frac{X}{Z_0}$$

Figure 5.7 Normalization.

The distance out from the center of the Smith chart is the amplitude of the reflection coefficient (that is, the amplitude of the reflected electric field relative to the incident electric field), and the angular location around the chart is the phase of the reflected electric field relative to the incident electric field. However, instead of amplitude and phase coordinates, the Smith chart uses coordinates of resistance and reactance. Then if the reflection coefficient of a mismatch is known, it can be plotted on the Smith chart to determine the resistive and reactive parts of its impedance.

All lines of equal resistance and equal reactance are "normalized," and the meaning of this normalization is shown in Figure 5.7. The actual resistances and reactances are not plotted but are divided by the characteristic impedance of the transmission line. For example, if $R = 75 \ \Omega$ and $X = 25 \ \Omega$ in a transmission line where $Z_0 = 50 \ \Omega$, the normalized impedance z is

$$z = \frac{75 \ \Omega}{50 \ \Omega} + j \frac{25 \ \Omega}{50 \ \Omega}$$
$$= 1.5 + j0.5$$

The reactive part of the expression is preceded by the letter j to indicate that the voltage and current are 90° out of phase.

5.2 PLOTTING MISMATCHES ON THE SMITH CHART

Figure 5.8 explains the Smith chart scales. Some resistance and reactance lines have been darkened and numbered. Circle 1 is called the unity-resistance circle; every point lying on that circle has the resistive part of its impedance equal to 1. Since the Smith chart is normalized, this means 1 times the characteristic impedance of the transmission line. With coax or microstrip, which have characteristic impedances of 50 Ω, every point on circle 1 has 1 times 50 Ω or 50 Ω resistance. The points all have different values of inductive or capacitive reactance, but they all have 50 Ω resistance. Correspondingly all the points along circle 2 have 2 times the characteristic impedance of the transmission line for the resistive component of their impedance. Thus, for a 50-Ω line, all points along circle 2 have 100 Ω resistive impedance. Similarly, on circle 3, which has a normalized resistance of 0.5, all points have a 25-Ω resistance in a 50-Ω line. All the points along curve 4, where the electric and magnetic fields are 90° out of phase, have a normalized reactive component of 1, which in a 50-Ω system means that they have a reactance of 50-Ω. All points in the upper half of the chart have a positive reactance, which means they look like inductors (electric field leads the magnetic field). In curve 5 in the lower part of the chart the reactance is also 1, but is capacitive because the electric field lags the magnetic field. Curve 6 is a normalized inductive reactance of 0.5, and curve 7 is a normalized capacitive reactance of 0.5.

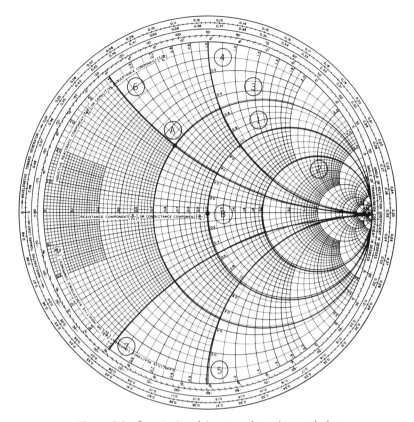

Figure 5.8 Constant resistance and reactance circles.

Point A, which is a mismatch point lying at the intersections of curve 3 and 6 has a resistive component of 0.5 and an inductive reactive component of 0.5, so in a 50-Ω system a mismatch represented by this point has 25 Ω resistance and 25 Ω inductive reactance. The resistance and reactance of any other mismatch point can be determined in the same way.

Consider point B at the center of the chart. Since the distance out from the center represents the magnitude of the reflected wave, which is zero at the center, this point represents a perfect match—the reactance is zero, and the resistive component is Z_0.

Figure 5.9 shows some additional points of interest on the Smith chart. Part (a) shows the pure resistance line. Every point on this line has resistance only, the reactance is zero, and the electric and magnetic fields are in phase. However, $R = Z_0$ only at the center, so although there are no inductive or capacitive reactive components, points along this line are still mismatched. Mismatches along the pure resistance line are at SWR maximums to the right of center, since the incident and reflected electric fields are in phase, and

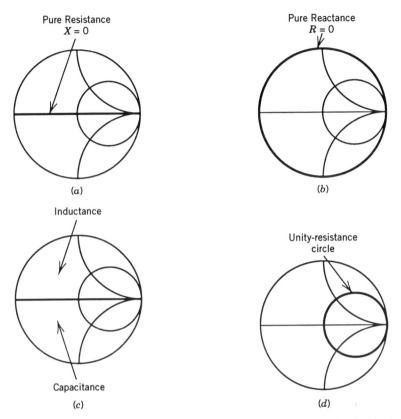

Figure 5.9 Resistance, inductance, and capacitance represented on the Smith chart.

mismatches to the left of the center are SWR minimums, since the reflected field is 180° out of phase with the incident electric field.

Figure 5.9b shows that the outer boundary of the Smith chart is a pure reactance; that is, the resistance is zero. This outer boundary of the chart represents a reflection coefficient of 1, which means that all the power is reflected. The reflected wave is equal to the incident wave, and the electric field is 180° out of phase with the magnetic field. Figure 5.9c again shows that the upper half of the Smith chart has inductive reactances values and the lower part has capacitive reactance values. Part (d) again shows the unity-resistance circle, where although the impedance has an inductive or capacitive component, the resistance componant is equal to the characteristic impedance of the transmission line. The unity-resistance circle is an important part of the chart for matching.

Figures 5.10 and 5.11 show how a mismatch is plotted on the Smith chart. A transistor is mounted in a microstrip test fixture. The mismatch at the junction of the microstrip line and the transistor, measured as a reflection

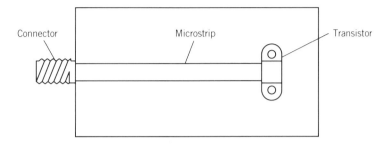

At microstrip-to-transistor junction, the reflection coefficient is 0.44/117°. The impedance, determined from Smith chart is

$$0.5 + j0.5$$
$$R = 25 \ \Omega$$
$$X = 25 \ \Omega$$

Figure 5.10   Example of a microwave mismatch.

coefficient, is 0.44/117°. This mismatch is plotted on the Smith chart of Figure
5.11. Note that the distance from the chart center to a mismatch point is the
amplitude of the reflection coefficient, and the direction or angle around the
Smith chart is the phase angle. The phase angle is shown around the outer
circumference, labeled "angle of reflection coefficient in degrees." The angle
is positive around the upper half of the chart, indicating that the reflected
wave leads the incident wave, and negative around the lower part of the
chart, indicating that the reflected wave lags the incident wave. This angle
scale is the same as that on the polar plot of the reflection coefficient in Figure
5.1a. In this example the phase angle is 117°. The exact amplitude of the
reflection coefficient is measured by using the reflection coefficient scale
along the bottom of the Smith chart; it is the third scale down on the left-
hand side and is marked "reflection coefficient" or p. Note that it is a linear
scale from the center to the outside of the chart. With a compass or divider,
the length of the reflection coefficient arrow from the center out to the
reflection coefficient point can be measured along the reflection coefficient
scale. The mismatch point therefore lies along the 117° phase angle line, at
a distance of 0.44 from the center, as shown in Figure 5.11.

The equivalent values of resistance and reactance of the mismatch can
then be read directly from the Smith chart's scale. The mismatch point lies
on the 0.5 normalized resistance circle and the 0.5 normalized inductive
reactance circle. As tabulated in Figure 5.10, the resistance in a 50-Ω transmission line is therefore 25 Ω and the inductive reactance is 25 Ω. The
mismatch can be represented as a 50-Ω resistor in series with an inductance
that, at the frequency of operation, has a reactance of 50 Ω.

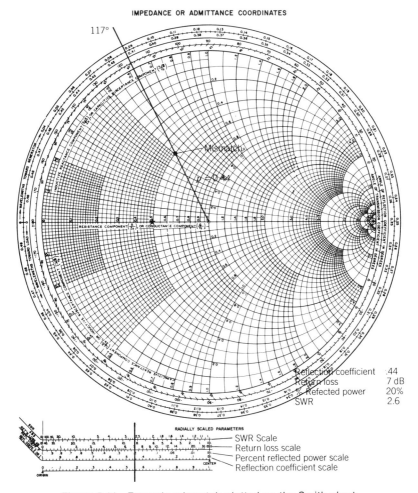

Figure 5.11 Example mismatch plotted on the Smith chart.

Various ways of matching to correct this mismatch by using the Smith chart will be shown later.

The scales along the bottom left of the Smith chart can be used to convert reflection coefficient to percent reflected power, return loss, and SWR. These conversions could be made with the formulas of Chapter 4, but they are provided for convenience along the bottom of the Smith chart. The values of SWR, return loss, and reflected power for the example mismatch point, taken from these scales, are tabulated in the right-hand corner of Figure 5.11.

The scales at the lower right bottom of the Smith chart, which have been covered over in Figure 5.11 by the labels describing the left-hand scales, involve transmitted power and other less commonly used quantities. These other scales are explained in the books given in the bibliography and will not be discussed further.

5.3 MATCHING CALCULATIONS WITH THE SMITH CHART

Matching means reducing or eliminating the reflected power, which means moving the mismatch point to the center of the Smith chart where the reflection coefficient is zero. To illustrate the use of the Smith chart for matching design calculations, the mismatch plotted in Figure 5.11 will be matched out by adding a capacitor in series with the transmission line. (Other matching designs for this same mismatch are described later.)

The reflection coefficient amplitude of the mismatch of Figure 5.11 is 0.44, which is equivalent to an SWR of 2.6 or a return loss of 7 dB, which means that 20% of the power is reflected at the mismatch. This is really a terrible match. This mismatch could also be expressed as an impedance consisting of a 25-Ω resistance and a 25-Ω inductive reactance. It's still a terrible match; converting a mismatch from a reflection coefficient to an impedance doesn't make it better.

A capacitor could be added to cancel the effect of the mismatch inductance. In this example the capacitor must have a reactance of 25 Ω at the operating microwave frequency. The mismatch, now only the 25-Ω resistance, would then appear at the point marked by the star in Figure 5.11. The new mismatch is slightly better, but still bad because the resistive component is not 50 Ω.

A perfect match could be obtained by moving back down the transmission line, away from the junction where the transmission line joins the component, to a location at which the resistance is 50 Ω; and adding a capacitor to cancel the inductance. To find this point, consider what happens to the impedance along the transmission line. Along the transmission line, the amplitude of the reflection coefficient remains the same, because the amount of reflected power remains the same, but the phase relationship between the reflected and incident fields changes constantly, so the total electric field and the total magnetic-field change, so the resistive and reactive components of the impedance change. This was illustrated in Figure 5.3. The mismatch doesn't improve along the transmission line, but a point exists where the resistance is exactly equal to Z_0. At that point a capacitance can be added to cancel the inductance. Refer to Figure 5.12. Moving along the transmission line, the amplitude of the reflection coefficient remains constant, since the reflected power stays constant, so the move is represented as traveling around the constant ρ circle, that is, the constant reflection coefficient circle. This traveling is equivalent to rotating around that circle, and the phase of the reflection coefficient, and therefore the impedance, changes. Matching is achieved by moving along the transmission line to the intersection of the ρ circle and the unity-resistance circle. At this intersection, the resistive part of the mismatch

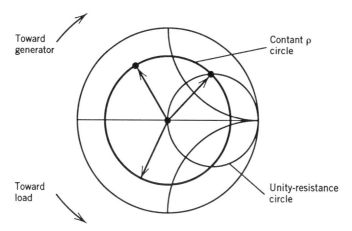

Figure 5.12 Matching with the Smith chart.

is the same as Z_0 and the inductance can be canceled with a matching capacitance.

Figure 5.13 relates movement along the transmission line with rotating around the Smith chart. Rotating once around the Smith chart is equivalent to moving half a wavelength down the transmission line. In moving half a wavelength down the line, the incident wave travels half a wavelength and the reflected wave travels half a wavelength in the other direction, so the reflected wave moves a full wavelength *relative* to the incident wave.

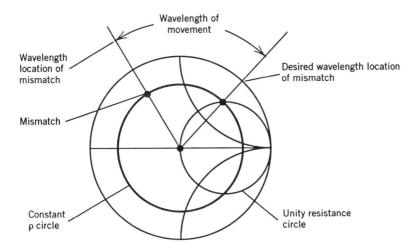

Figure 5.13 Rotation around the Smith chart.

Rotating halfway around the chart is equivalent to moving a quarter of a wavelength down the transmission line. Moving away from the mismatch toward the generator corresponds to a clockwise rotation around the Smith chart, and moving from the mismatch toward the load corresponds to a counterclockwise rotation. Note the outermost set of numbers around the circumference of the Smith chart labeled at the left-hand side of the chart as "wavelengths towards generator." Starting at the left side, these numbers are 0.04, 0.05, 0.06, and so forth. One quarter of the way around the chart the wavelength is 0.125; halfway around the chart, the wavelength is 0.25, which is a quarter of a wavelength. Three quarters of the way around the chart in a clockwise direction the wavelength is 0.375, and finally back to the left-hand side again the wavelength is half a wavelength.

Here is a summary of the matching procedure:

1. Locate mismatch on the Smith chart.
2. Determine wavelength position of mismatch
3. Move down transmission line toward generator until ρ circle intersects the unity resistance circle.
4. Determine the number of wavelengths moved down the transmission line.
5. Add normalized capacitance to cancel inductance.
6. Determine capacitance.

Example 5.1 A transistor mounted in a 50-Ω microstrip line, as shown in Figure 5.14a, presents a mismatch at its input of $0.5 + j0.5$. This means that the resistive component is 25 Ω and the reactive component is 25 Ω. This is the same mismatch shown in Figure 5.11. The return loss of this mismatch is 7 dB. The first step in matching this transistor is to plot its mismatch on the Smith chart, which is shown as point 1 of Figure 5.14 b. The next step is to move down the transmission line, away from the transistor, to the unity-resistance circle. At this point, the resistive component is 50 Ω. The inductive reactance can be cancelled with a capacitor, which is shown added into the transmission line in Figure 5.14a.

To determine how far to move down the transmission line to reach the unity circle, the location of the mismatch in wavelengths is determined, and as shown by point 2 on the Smith chart in Figure 5.14b, it is 0.088 wavelengths toward the generator. This is the starting point. Moving down the transmission line means moving around the constant ρ circle to point 3 on the Smith chart, where the p circle and the unity-resistance circle intersect. At this point, the resistive component of the impedance is 1, and is exactly equal to Z_0. The distance moved down the transmission line from 0.088 to 0.162, as shown by point 4, is 0.074 wavelength. The Smith chart calculation shows how far to move in wavelengths. This wavelength value must be converted

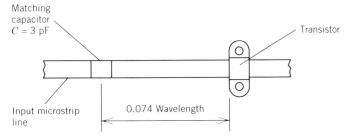

Figure 5.14a   Matching of example mismatch.

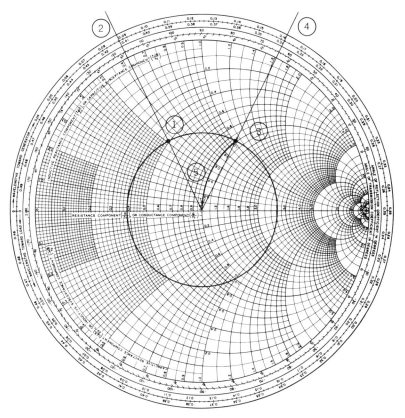

Figure 5.14b   Smith chart of example matching.

to actual physical dimensions, so the guide wavelength of the transmission line must be known. At point 3, where the constant-p circle and the unity circle intersect, the correct value of capacitance must be added to cancel the inductance. At point 3 in Figure 5.14, the normalized inductive reactance is 1, so a capacitance must be added whose normalized capacitive reactance is

1. The actual capacitive reactance is 1 times 50 Ω so the value of the capacitance is

$$C = \frac{1}{2\pi f X_C}$$
$$= \frac{1}{2\pi (10^9)(50)}$$
$$= 3.2\ pF$$

at an operating frequency of 1 GHz. Adding this capacitance is equivalent to moving, as shown by arrow 5 in Figure 5.14, along the unity-resistance circle from the point where the reactive component is 1 down to the point where the reactive component is 0, which is the center of the chart, a perfect match.

The matching capacitor must be located at exactly the right distance along the transmission line, and the right amount of capacitance must be added to obtain a perfect match. Figure 5.15 shows what happens if a wrong location or a wrong value of capacitance is used. Point *A* shows the result if the

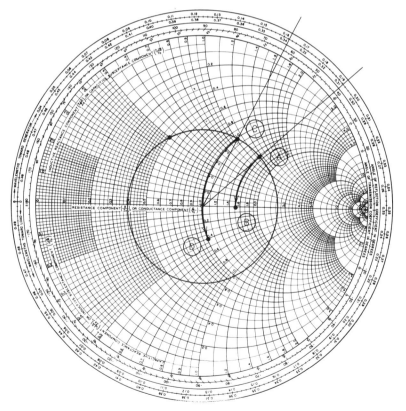

Figure 5.15 Effect of incorrect location and incorrect value of matching capacitor.

location is 0.03 wavelength too far down the transmission line. The location is beyond the unity-resistance circle, at the 1.4-resistance circle, so that the resistive component is 1.4 times 50 Ω or 70 Ω. Even if a normalized capacitive reactance of 1 is added to cancel the inductive reactance, the resultant match is at point *B*, which is not at the center of the chart. The match has been improved, but it's not perfect. Or suppose, as shown by point *C* the location is the right distance down the transmission line, but a capacitive reactance of 1.4 is added instead of 1.0. The resultant mismatch is down the unity-resistance circle to point *D* and again the match is not perfect.

5.4 MOVING TOWARD THE LOAD

A transistor is shown in Figure 5.16*a* mounted in a microstrip fixture. The mismatch has previously been specified at the transistor. Probably the mis-

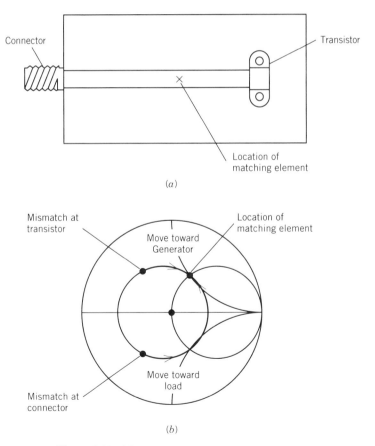

Figure 5.16 Moving toward the load.

match was measured at the fixture connector, and the mismatch at the transistor calculated by plotting the measured mismatch on the Smith chart and rotating from the connector to the transistor. These two mismatch points at the connector and transistor are shown in the Smith chart in Figure 5.16*b*.

Once the mismatch at the transistor is known, a movement along the constant ρ circle *toward the generator* intersects the unity-resistance circle. This point is shown by the $\times$ on the microstrip line in Figure 5.16*a*. At that point a matching capacitance must be added. Alternatively, a movement from the mismatch at the connector *toward the load* also intersects the unity-resistance circle. Note that whether movement was from the transistor toward the generator or from the connector toward the load the same point along the microstrip line is reached.

5.5 LUMPED INDUCTANCE IN SERIES

Previously, using a series capacitance, matching was achieved by moving down the transmission line to the intersection of the constant ρ circle and the unity-resistance circle in the upper half of the Smith chart. At this point the resistance of the mismatch equals Z_0, and the remaining inductance is canceled with a capacitance.

As shown in Figure 5.17, an alternative matching technique is to move from the mismatch toward the generator to the unity-resistance circle in the lower half of the Smith chart. At this point the resistance is equal to Z_0, but there is an extra capacitance, which is canceled by adding a lumped series inductance.

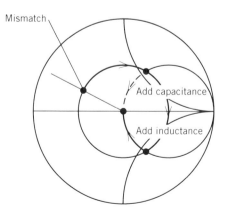

Figure 5.17 Matching with a lumped series inductance.

Example 5.2 Matching with a series inductance. A transistor is mounted in a 50-Ω microstrip line as shown in Figure 5.18. The mismatch at the transistor is $0.5 + j0.5$. The transistor is to be matched at 1 GHz with a series inductor. Note that this is the same mismatch used as the example of matching with a series capacitor. The following procedure is used.

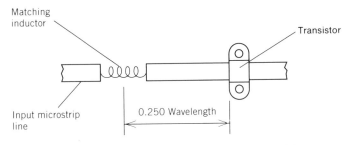

Figure 5.18*a* Example of series inductance matching.

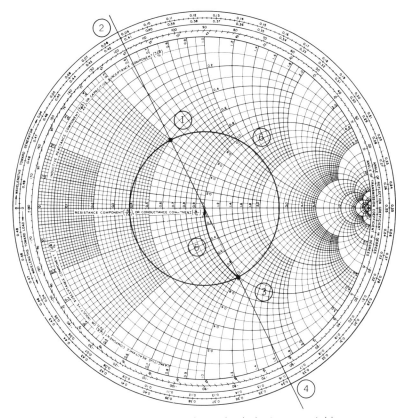

Figure 5.18*b* Smith chart for series inductance matching.

1. Locate mismatch on Smith chart. This is point 1 in Figure 5.18b.
2. Determine wavelength position of mismatch, which is 0.088 wavelengths toward generator.
3. Move down transmission line toward generator until ρ circle intersects $R = 1$ circle in lower half of Smith chart. This is point 3 in Figure 5.18b.
4. Determine number of wavelengths moved down transmission line: $0.338 - 0.088 = 0.250$. This is point 4 in the figure.
5. Add normalized inductance of 1.0 to cancel capacitance. This is shown by arrow 5 as a movement along the $R = 1$ circle.
6. Determine inductance.

$$X_L = 2\pi f L = 1.0 \times 50\Omega = 50\ \Omega$$

$$L = \frac{X_L}{2\pi f} = \frac{50}{2\pi \times 10^9} = 8\ \text{nH}$$

5.6 MATCHING ELEMENTS IN PARALLEL

In the matching techniques considered so far, matching elements have been added in series with the transmission line. This is not always easy to do, because it requires breaking the transmission line. The microwave signal travels down the transmission line through the matching capacitor or inductor and reenters the transmission line.

In the next four matching methods, the matching element is added in parallel (shunt), and the original transmission line is undisturbed. However, designing must now be done in the admittance plane of the Smith chart, as shown in Figure 5.19. Admittance is the reciprocal of impedance, and therefore is equal to the magnetic field divided by the electric field.

$$Y = \frac{1}{Z} = \frac{\text{Magnetic field}}{\text{Electric field}}$$

Admittance is used when microwave circuit elements are combined in parallel. To see the significance of using admittance when elements are combined in parallel, recall the formula for combining resistors in parallel:

$$\frac{1}{R_T} = \frac{1}{R_1} + \frac{1}{R_2} = G_T = G_1 + G_2$$

Notice that the reciprocals of the resistances are combined to get the reciprocal of the total resistance. Alternatively, conductances could have simply been added to get the total conductance of the circuit.

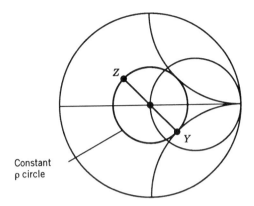

Figure 5.19 Admittance.

As might be expected, it is more difficult to determine the admittance of a mismatch from the impedance of the circuit, because the impedance is a vector quantity with in-phase and out-of-phase parts. Fortunately, conversion of impedance to admittance is already done on the Smith chart. As shown in Figure 5.19 to convert impedance to admittance a rotation from the imped- ance point 180° (i.e., halfway around the Smith chart) on the constant ρ circle is made. The validity of this conversion could be illustrated by using the electric- and magnetic-field phase vectors as was done previously. The trans- formation intuitively appears reasonable considering that the phase of the reflected magnetic field (relative to the incident magnetic field) is 180° differ- ent from the phase of the reflected electric field (relative to the incident electric field).

Impedance has resistive and inductive components, so admittance also has resistive and inductive components. Note that for admittances, the lower half of the Smith chart is inductive and the upper half is capacitive, the reverse of when the Smith chart is used to show impedances.

Admittance is designated by the letter Y. Its two components are conduc- tance G (where the magnetic field is in phase with the electric field) and susceptance B (where the magnetic field is 90° out of phase with the electric field). The unit of admittance is Siemens (S), which is equal to 1/ohms.

The characteristic admittance Y_0 of a transmission line is equal to $1/Z_0$. Thus, if Z_0-50 Ω, then $Y_0 = 1/50 \ \Omega = 0.02$ S.

An example of the conversion of impedance to admittance is shown in Figure 5.20. The mismatch in a 50-Ω transmission line has a resistance of 15 Ω and an inductive reactance of 25 Ω. The normalized impedance (z) is $0.3 + j0.5$, and this point is plotted as usual on the Smith chart. To convert to admittance the constant ρ circle is drawn and the admittance point is halfway around the chart. The normalized admittance (y) is then $0.9 - j1.5$, which is inductive. The characteristic admittance of the 50-Ω transmission line is 0.02 S, so the conductance (which is the magnetic field in phase with

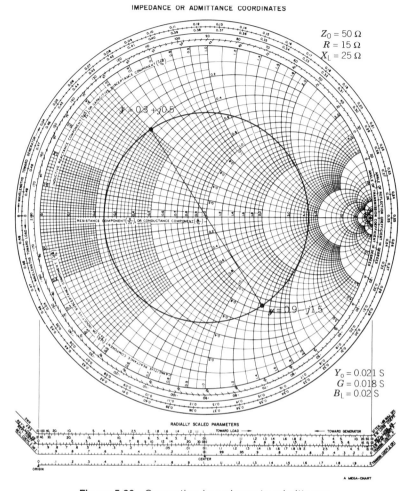

Figure 5.20 Converting impedance to admittance.

the electric field) is 0.018 S and the susceptance (which is the component of the magnetic field that is 90° out of phase with the electric field) is 0.03 S.

Converting from admittance to impedance is done in this same way. The admittance is plotted on the Smith chart, and impedance is obtained by rotating 180° around the chart on the constant ρ circle.

Another way of converting impedance to admittance is to use two Smith chart scales, with one rotated 180° from the other. Then for any mismatch point on the Smith chart, impedance could be read on one set of scales and admittance on the other set. This combined impedance-admittance Smith chart will be shown later.

Figure 5.21 shows matching techniques using a capacitance or inductance *in parallel* with the transmission line. The following procedure is used:

1. Locate the impedance on the Smith chart.
2. Determine the admittance by a 180° rotation around the constant ρ circle.
3. Move toward the generator to the intersection of the $G = 1$ circle and the ρ circle in the lower (inductive) or upper (capacitive) half of the chart.
4. Add capacitance (inductance) to cancel inductance (capacitance).

In Figure 5.21*a*, matching is done with lumped capacitance; in Figure 5.21*b*, with lumped inductance.

Shunt capacitance

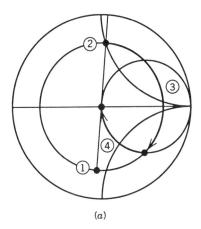

(a)

Shunt inductance

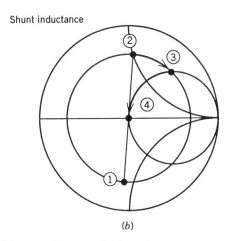

(b)

Figure 5.21 Matching with a parallel (shunt) capacitance or inductance.

As discussed previously in this chapter, the points shown in Figure 5.21 could have been reached by moving around the Smith chart toward the load, if the physical constraints of the matching problem allowed this.

Example 5.3   For the 50-Ω transmission line of Figure 5.22a, $p = 0.80/-82°$ at 1 GHz. Calculate the matching capacitance.

Solution   Plotting the mismatch on the Smith chart in Figure 5.22c shows that $z = 0.5 - j1.0$. Therefore, a 180° rotation around the constant ρ circle gives $y = 0.38 + j0.8$. The wavelength location of the admittance at the mismatch is 0.114 wavelengths toward the generator. All of these values are plotted in Figure 5.22c.

To match out this point with a shunt capacitor, use the following procedure.

1. Move down the transmission line from the wavelength location of the admittance to the intersection of constant ρ and unity conductance circles in the lower (inductive) half of Smith chart. The wavelength of the movement is $0.322 - 0.114 = 0.208$.
2. At this point the normalized conductance is 1.0 and the normalized inductive susceptance is 1.6.
3. Add capacitance to cancel the inductance.

$$B_c = 1.6Y_0 = 1.6 \times 0.02 \text{ S} = 2\pi f C$$

$$C = \frac{B_c}{2\pi f} = \frac{1.6 \times 0.02}{2\pi \times 10^9} = 5 \text{ pF}$$

Example 5.4   For the transmission line of Figure 5.22b, calculate the matching inductance.

Solution   From Example 5.3,

$$z = 0.5 - j1.0$$
$$y = 0.38 + j0.8$$

Wavelength location of mismatch admittance $= 0.114$

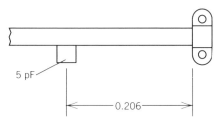

5 pF

0.206

Figure 5.22a   Example of matching with a shunt capacitor.

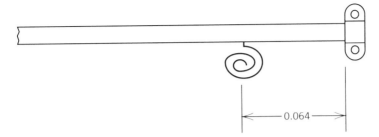

Figure 5.22*b* Example of matching with a shunt inductor.

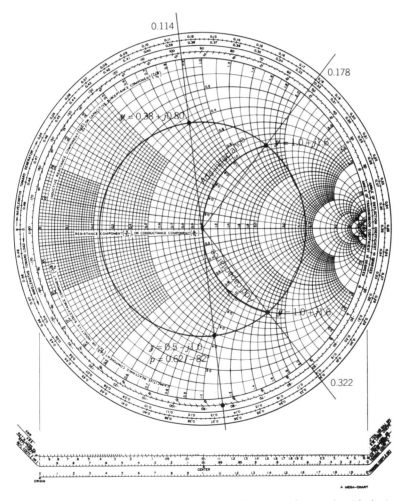

Figure 5.22*c* Smith chart for matching with a shunt capacitor or shunt inductor.

1. Move down the transmission line from the wavelength location of the admittance to the intersection of constant ρ and unity conductance circles in the upper (capacitance) half of Smith chart. The wavelength of movement is $0.178 - 0.114 = 0.064$.

2. Normalized capacitive susceptance is 1.6.

3. Add inductance to cancel capacitance.

$$B_L = 1.6Y_0 = 1.6 \times 0.02 \, \text{S} = \frac{1}{2\pi f L}$$

$$L = \frac{1}{2\pi f B_L} = \frac{1}{2\pi \times 10^9 \times 1.6 \times 0.02} = 5 \, \text{nH}$$

Figure 5.22b shows a spiral inductor, which can easily be printed on the microstrip substrate. The center of the inductor is connected through the substrate to the ground plane.

5.7 MATCHING STUBS

At microwave frequencies above a few GHz the achievement of a lumped inductor or capacitor becomes difficult. Therefore, one of the most common matching techniques is to use stubs to provide the shunt capacitance or inductance. Matching stubs are shown in Figure 5.23 and are simply sections of transmission line branching off from the main line. At the branching point the microwave signal could take one of two parallel paths: toward the mismatch or down the stub, and so the stub is a shunt element.

As shown in Figure 5.23, a shorted stub provides an inductance, whereas an open-circuit stub provides a capacitance. Examples of stubs in coax line and in microstrip are shown. Why a stub provides inductance or capacitance can be determined by looking at the microstrip configuration. If the stub is shorted, the end of the stub must be connected to the ground plane, and the stub essentially forms a one-turn coil. The open-circuit stub simply ends on the upper surface of the substrate and acts as a capacitor plate with the lower ground plane.

The relationship between the length of the stub and the amount of inductance or capacitance added in parallel with the main transmission lines is shown in Figure 5.24.

An example of matching with inductive or capacitive stubs is shown in Figure 5.25. The stubs are used to match out the same mismatch of Figure 5.22, where lumped capacitors or inductors were used. From the calculations of Figure 5.22, it was determined that a matching capacitance whose normalized susceptance was 1.6 was required, and it had to be located at a distance of 0.208 wavelengths from the mismatch. The capacitor is now replaced with an open-circuit stub.

Short-Circuit Stub (Inductive)

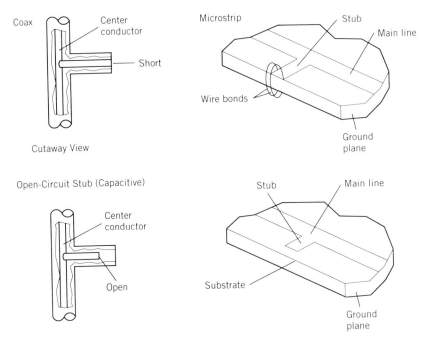

Open-Circuit Stub (Capacitive)

Figure 5.23   Matching stubs.

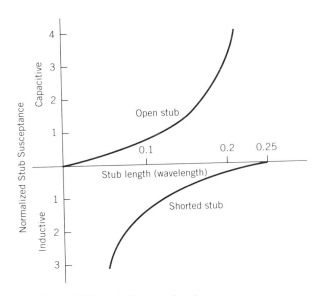

Figure 5.24   Admittance of stubs.

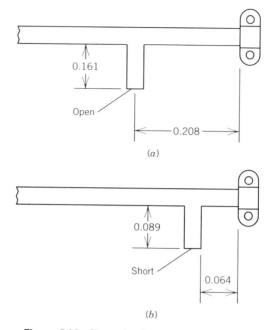

Figure 5.25 Example of matching with stubs.

The length of open stub that will provide the required normalized capacitive susceptance of 1.6 is determined from Figure 5.24 to be 0.161 wavelengths. Therefore, as shown in Figure 5.25a, a stub whose length is 0.161 wavelengths is placed at a distance of 0.208 wavelengths from the mismatch.

Matching with the inductive stub is done in the same way. As shown in Figure 5.25b the shunt inductance must be located 0.064 wavelengths from the mismatch. The normalized inductive susceptance that must be added is 1.6. A shorted stub 0.089 wavelengths long provides this susceptance.

5.8 QUARTER-WAVE TRANSFORMER

The next matching technique to be considered uses a quarter-wave transformer. With transformer matching, the resistance of the mismatch is transformed into the characteristic impedance of the transmission line. Matching is done in the impedance plane, and no inductance or capacitance is involved. A sketch of the matching transformer and a Smith chart are shown in Figure 5.26.

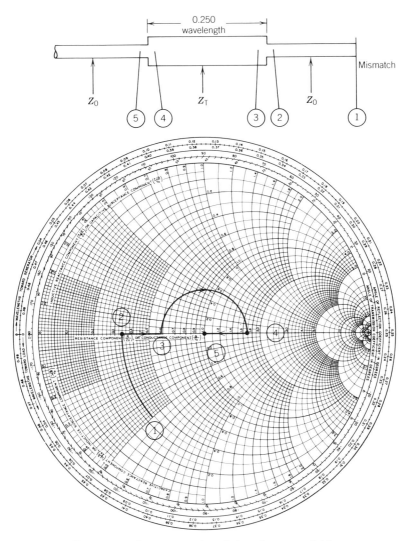

Figure 5.26 Quarter wavelength transformer matching.

The impedance of the mismatch is shown at point 1. The first step is to rotate to point 2 along the resistive axis, so that the mismatch has no inductive or capacitive component. Even though the mismatch is a pure resistance, the match is still bad, because Z_2, the impedance at point 2, is not equal to Z_0.

As shown in Figure 5.26, at point 2 a different transmission line—the transformer—is used, which has a characteristic impedance Z_T, which is selected to be equal to $\sqrt{Z_2 Z_0}$. The normalized impedance at the beginning of the transformer section is Z_2/Z_T. This impedance is point 3 on the Smith chart.

A quarter-wavelength movement is made down the transformer section toward the generator to point 4. The normalized impedance in the transformer at point 4 is equal to Z_T/Z_2. This is the key to the quarter-wave transformer matching technique. Movement in the transformer by a quarter-wavelength transforms the normalized impedance to its reciprocal. At point 4 the impedance in the transformer is $Z_T/Z_2 \times Z_T = Z_0$ since $Z_T = \sqrt{Z_2 Z_0}$. At the end of the transformer section, a change back to the original transmission line is made, and a perfect match is obtained.

Example 5.5   Refer to Figure 5.27. The mismatch is shown at point 1 with a normalized impedance of $0.9 - j0.8$.

1. Move down the transmission line 0.148 wavelengths to the resistive axis at point 2 where $z = 0.44$ (a pure resistance). At this point the resistance of the mismatch is $0.44 \times 50\ \Omega = 22\ \Omega$.

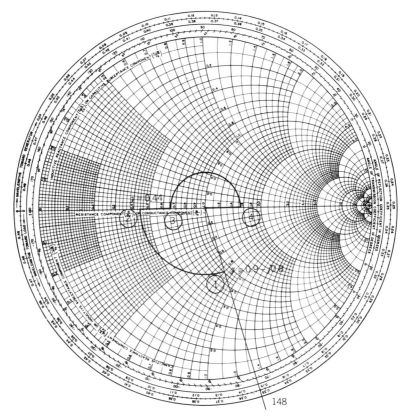

Figure 5.27   Example of matching with a quarter wave transformer.

2. Change to a transformer where $Z_T = \sqrt{22 \times 50} = 33\ \Omega$. The characteristic impedance of the transformer was selected based on the impedance of the mismatch itself. Moving into this transformer changes the impedance from point 2 to point 3, and the normalized impedance at point 3 will be 22/33 = 0.67.

3. Move a quarter-wavelength down the transformer to point 4, where the normalized impedance is 1/0.67 = 1.5.

4. At the end of the quarter-wave transformer the impedance in the transformer is 1.5 × 33 = 50 Ω and this is a perfect match when the change back to the original transmission line is made.

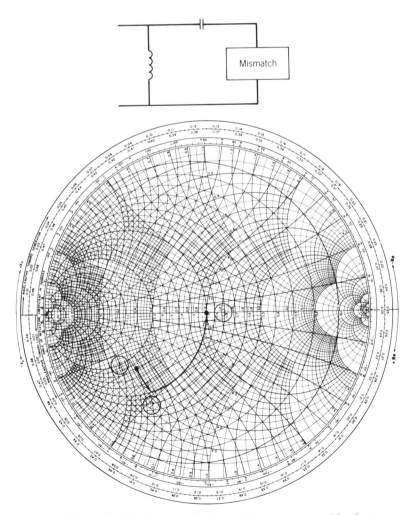

Figure 5.28 Matching with lumped elements in combination.

5.9 LUMPED ELEMENTS IN COMBINATION

In all previous matching techniques using inductors or capacitors, there was movement down the transmission line until the location of the unity resistance or conductance circle was reached, where the resistive part of the mismatch exactly equaled the characteristic impedance or admittance of the transmission line. Then an inductor or capacitor was added to cancel the unwanted capacitance or inductance. Another matching technique uses an inductor or capacitor to move the mismatch point to the unity conductance circle rather than moving down the transmission line. As shown in Figure 5.28. a lumped capacitance is used to move the mismatch point to the unity conductance circle, where a parallel inductance is used to cancel the remaining capacitive component to achieve a perfect match. Remember that if a matching element is being used in series, the impedance chart is used for the matching calculation, and if a matching element is being added in parallel the admittance chart is used. Here both series and parallel matching elements are being used, so both charts must be used. In Figure 5.28 a special Smith chart, which has two sets of scales, is utilized. For any point on this Smith chart, reading on one set of scales gives the impedance, and reading the same point on the other set of scales give the admittance. Impedance and admittance can be determined from a single point, and rotating 180° across the chart is not required, because the second scale of the chart has already been rotated 180°. In the match shown, the series element is used to transform the mismatch to the unity-conductance circle where the parallel element can cancel the remaining susceptance. The series element takes the place of rotating down the transmission line.

5.10 SELECTING THE BEST MATCHING TECHNIQUE

Any of the matching techniques discussed will accomplish a perfect match at any specific microwave frequency. The choice of which technique to use for a matching problem is determined by the physical limitations of the microwave parts and the need for broadband matching (matching over a range of frequencies). Consider the physical limitations. Using lumped capacitors or inductors in series with the transmission line requires breaking the transmission line. This can be accomplished in coax or microstrip with some difficulty, but cannot be accomplished in waveguide. Using capacitors or inductors in parallel with the transmission line is readily achieved in waveguide with capacitive screws, inductive posts, or capacitive or inductive irises, and is the main technique used for matching waveguide. Lumped inductors and capacitors work well at the low end of the microwave band, but are difficult to achieve at frequencies above a few gigahertz. Capacitive stubs and quarter-wave transformers are easy to implement in microstrip line, being simply a modification of the printed microstrip pattern.

Although some matching techniques are more easily accomplished physically than others, the major factor determining the choice of matching technique is broadband matching. The problem of broadband matching is illustrated in Figure 5.29, where the mismatch point has a normalized impedance
of 0.5 + j0.5. This was the original mismatch point considered for matching
with a series capacitor. Matching this point with a series capacitor requires
moving from the mismatch point toward the generator to the intersection of
the constant ρ and unity resistance circles. This is accomplished, as shown
in Figure 5.29 at a frequency of 1 GHz by moving 0.074 wavelength. A
physical movement to accomplish a movement of this many wavelengths
at 1 GHz will result in a movement of 0.059 wavelengths at 0.8 GHz and
0.084 wavelengths at 1.2 GHz, as shown in Figure 5.29. Only the mismatch
at 1.0 GHz lies on the unity-resistance circle. Adding a series capacitance
whose normalized capacitive reactance is 1.0 to cancel the inductance pro-

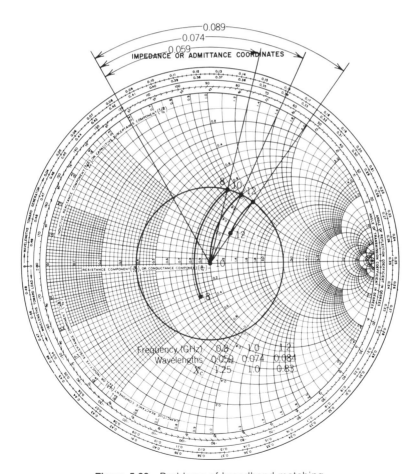

Figure 5.29 Problems of broadband matching.

vides a perfect match at 1.0 GHz, but is not enough at 1.2 GHz and is too much at 0.8 GHz. The final match at 0.8, 1.0, and 1.2 GHz is shown in Figure 5.29, and at 0.8 and 1.2 GHz a residual mismatch with a return loss of about 13 dB remains.

Several recommendations for broadband matching are shown in Figure 5.30, depending on the Smith chart location of the original mismatch. The best way to achieve a broadband match is to use a technique that requires the least rotation around the Smith chart, because movement of a given physical distance results in different wavelengths of rotation and is a major problem with broadband matching. If the mismatch points are grouped as shown in Figure 5.30a, the best matching scheme is a shunt capacitive stub. Changing the mismatch in the impedance plane to the admittance plane would group all the points around the unity-conductance circle in the admittance plane, and very little movement would be required.

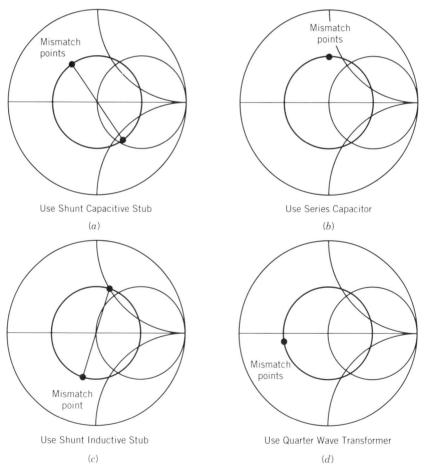

Use Shunt Capacitive Stub

(a)

Use Series Capacitor

(b)

Use Shunt Inductive Stub

(c)

Use Quarter Wave Transformer

(d)

Figure 5.30   Broadband matching.

If the mismatch points are grouped as shown in Figure 5.30b, around the unity-resistance circle in the impedance plane, then a series matching capacitor should be used.

If the mismatch points are distributed as shown in Figure 5.30c, a shunt inductive stub should be used, because rotating 180° across the Smith chart will locate the points near the unity-conductance circle and a minimum movement along the transmission line is required. If the mismatch points are located as shown in Figure 5.30d, a quarter-wave transformer would be the best matching technique.

To obtain a broadband match, many Smith chart calculations must be made at different frequencies with various matching techniques. These calculations are made with a variety of available computer-aided design (CAD) programs.

ANNOTATED BIBLIOGRAPHY

1. W. S. Cheung and F. H. Levien, *Microwaves Made Simple: Principles and Applications*, Artech House, Dedham, MA, 1985, pp. 78–112.
2. S. Y. Liao, *Microwave Devices and Circuits*, Prentice-Hall, Englewood Cliffs, NJ, 1980.
3. P. H. Smith, *Electronic Applications of the Smith Chart*, McGraw Hill, New York 1969 (available from Analog Instruments, New Providence, NJ).

References 1 and 2 present a simple explanation of matching with the Smith Chart and many examples. Reference 3 is Smith's original book on his chart and presents basic and advanced uses.

EXERCISES

5.1 Plot the following mismatch points on a Smith chart

	Amplitude	Phase
a	0	Not defined
b	0.58	144°
c	0.60	3°
d	0.85	−52°
e	1	−64°
f	0.56	−45°
g	0.60	−134°
h	0.42	55°

5.2 The amplitude and phase of the reflection coefficient of a mismatch are 0.57, $-144°$ respectively.
Match out the mismatch using a series capacitor.

a. How many wavelengths toward the generator from the mismatch must the capacitor be placed?

b. How much capacitive reactance must be added?

c. If the frequency is 3 GHz, what is the value of the capacitor?

5.3 If the matching capacitor of Exercise 5.2 is located at the correct position, but has a capacitive reactance of 1.0 (instead of the correct value) what is the resulting return loss?

Match the following mismatch in Exercises 5.4 through 5.6.
$$p = 0.34/120°$$

5.4 Match out the mismatch using a capacitive stub.

a. How many wavelengths toward the generator from the mismatch must the stub be placed?

b. What must the normalized susceptance of the stub be?

c. How long (in wavelengths) must an open stub be to provide the required capacitive susceptance?

5.5 Match out the mismatch using a quarter-wave transformer.

a. How many wavelengths toward the generator from the mismatch must the starting edge of the transformer be placed?

b. What must the normalized impedance of the transformer be?

c. Sketch the matching transformer.

5.6 Match out the mismatch using a lumped series capacitor and a lumped parallel inductor.

a. What is the normalized reactance of the capacitor?

b. What is the normalized susceptance of the inductor?

II

MICROWAVE DEVICES

MICROWAVE TRANSMISSION LINES

The three basic types of microwave transmission lines, as shown in Figure 6.1, are coaxial cable, waveguide, and microstrip. Examples of each type are given, and their performance characteristics are compared with each other and with other types of transmission lines. The guide wavelength and the characteristic impedance of transmission lines will be calculated. Transmission line connectors and adaptors are also discussed.

6.1 COMPARISON OF TRANSMISSION LINES

The factors to be considered in comparing transmission lines are:

Frequency range
Bandwidth
Power-handling capability
Attenuation
Size
Ease of fabrication

Unfortunately, no transmission line has optimal performance in all these areas, so for any application the transmission line that is the best trade-off must be chosen.

The advantages and disadvantages of the three basic types of transmission lines are compared in Table 6.1. Coaxial cables have large bandwidth and small size, but they have high attenuation and low power-handling capability.

Waveguide, in contrast, has extremely low loss and very high power-

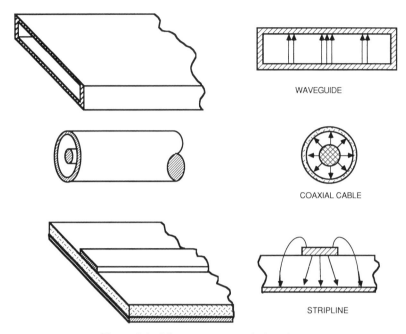

Figure 6.1 Microwave transmission lines.

handling capability, but is limited in bandwidth. Actually, 34 waveguides are required to cover the complete microwave band. At the low end of the microwave band the waveguide is very large. Size is determined by the frequency of operation and is independent of the required power-handling capability or attenuation.

The major advantage of microstrip is that it can be fabricated by photoetching techniques, so that complicated microwave circuits can be readily made. Like coaxial cable, microstrip has a large bandwidth and is small, but it has very high attenuation and low power-handling capability.

Thirteen transmission lines will now be compared, including 4 types of coaxial cable, 8 types of waveguide, and 1 microstrip, to show how different

Table 6.1 Comparison of transmission lines

Type	Advantage	Disadvantage
Waveguide	Low attenuation High power	Limited bandwidth Large size
Coax	Large bandwidth Small size	High attenuation Low power
Microstrip	Easy to connect multiple lines together	Very high attenuation Low power

transmission lines of the same type compare and how the different types compare. The transmission lines are shown in Figures 6.2 through 6.4. Each transmission line is numbered, and the numbers are used in the comparison chart of Figure 6.5.

Coaxial cables are shown in Figure 6.2. Cable 1 is a small semirigid coaxial cable. It has a solid center conductor made of copper, a solid Teflon dielectric support, and a solid outer conductor made of copper. Its outer diameter is 0.141 in. Even though the cable is small, it is still only semiflexible because of its solid outer conductor. Cables of this type are made with outer diameters of 0.008 to 0.390 in.

Cable 2 is similar to cable 1 as far as the inner conductor and dielectric support are concerned, but it uses a braided outer conductor with a plastic protective jacket. Braiding the outer conductor makes the cable flexible. Such cables are made with outer diameters of 0.080 to 0.870 in., so their sizes

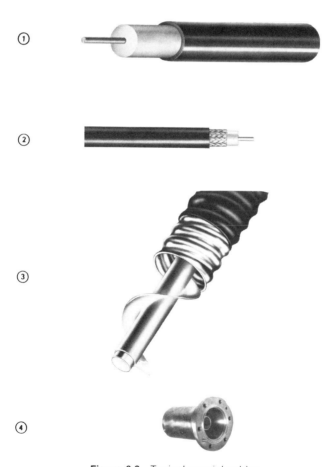

Figure 6.2 Typical coaxial cables.

vary over a 10:1 range. The larger the cable diameter, the greater is its power-handling capability, but the lower is its frequency range of operation. For this example, the outer diameter of the cable is 0.42 in.

Cable 3 is a semirigid cable with a large outer diameter of $\frac{7}{8}$ in. To make the cable semiflexible, a helical dielectric support of Teflon and a corregated outer conductor of copper are used. This construction makes the cable type semiflexible even up to $1\frac{5}{8}$ in. in diameter.

Cable 4 is a rigid coaxial transmission line with an outer diameter of $3\frac{1}{8}$ in. The outer conductor is a hollow tubing, and the inner conductor is a hollow

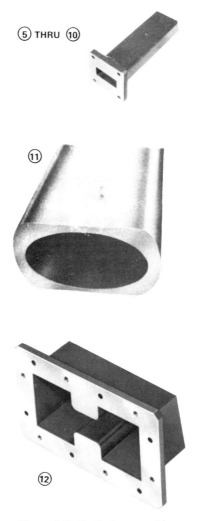

Figure 6.3 Typical waveguides.

tubing supported on small periodically spaced Teflon posts. The thickness of the Teflon posts is so small that the cable is considered as having an air dielectric, and so is called *air line*.

Figure 6.3 shows several types of waveguide. Six different rectangular waveguides (5 through 10) are compared. The only difference between them is their size. The useful bandwidth of waveguide is only about 1.5:1. For example, waveguide 7, whose inner dimensions are 0.9 in. wide by 0.4 in. high, will not begin to propagate microwaves until a frequency of 6.55 GHz is reached. It then propagates microwaves in a single mode until 13.1 GHz is reached. To avoid high attenuation at the low end of this range and to ensure that none of the power gets into a higher-order mode, the waveguide is normally used only from 8.2 to 12.4 GHz. If transmission at other frequencies is desired, another standard waveguide would be chosen. Standard rectangular waveguide is made of aluminum or brass. The brass is often silver-plated to reduce attenuation.

Waveguide 11 is an elliptical waveguide made of extruded aluminum. In spite of its large size, it is semiflexible and can be rolled onto cable reels.

Waveguide 12 is a ridged waveguide. Recall that the frequency range of rectangular waveguide is determined by the waveguide's width. The ridge effectively increases the width for the fundamental mode of propagation so that it can begin at a lower frequency. The ridge has only a small effect on the cutoff frequency of the next higher mode, since no electric field exists in the higher mode in the region of the ridge. Therefore, the ridged waveguide has approximately twice the bandwidth of standard rectangular waveguide.

Figure 6.4 shows microstrip transmission lines. When the dielectric material is placed on both sides of the strip conductor, as shown in A, the transmission line is called *stripline*. When the dielectric support material is on only one side of the conductor and the dielectric on the other side is air, the transmission line is called *microstrip*, as shown in B. Figure C is a microstrip circuit that illustrates the major advantage of stripline and microstrip—allowing complicated circuits consisting of several microwave transmission lines connected together to be fabricated easily by photoetching. The microwave signal is actually carried, as shown in Figure 6.1 in the dielectric material between the conductor and the ground plane. Stripline and microstrip are the microwave equivalents of printed circuit wiring used at low frequencies. For comparison with other transmission lines, a microstrip circuit with a 0.025-in.-wide gold conductor on an alumina ceramic substrate is considered. This is a standard microstrip configuration used in microwave integrated circuits.

The characteristics of these 13 transmission lines are compared in Figure 6.5. The graph shows the attenuation of the transmission lines in dB per foot as a function of frequency, and thus shows the frequency range or bandwidth of the transmission lines as well as their attenuation. The overall size of the transmission lines (their outer dimensions) and their power-handling capability are also shown. Transmission lines 1–4 are coaxial cables, shown

A. STRIPLINE

B. MICROSTRIP

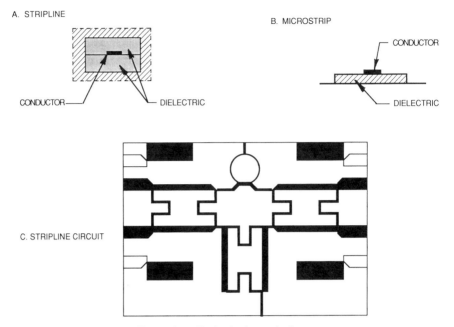

C. STRIPLINE CIRCUIT

Figure 6.4 Typical microstrip lines.

in Figure 6.2. Cable 1 is only 0.141 in. in outer diameter, has solid inner and outer conductors, and is semiflexible. It operates across the whole microwave frequency range up to 32 GHz, where a higher-order mode of propagation exists and the cable is no longer useful. Notice that the attenuation of each coaxial cable increases with frequency, and at 32 GHz cable 1 has an attenuation of almost 1 dB/ft. The power-handling capability of coaxial cables decreases with frequency, and the values shown in the table are the lowest values (that is, at the highest frequency of operation). At 32 GHz, cable 1 can handle 50 W of power. Cable 1 is perhaps the most commonly used coaxial cable in microwave equipment, because of its small size, wide frequency range, and fair power-handling capability.

Cable 2 is a flexible cable 0.42 in. in outer diameter. Because this cable is larger, it has lower loss at the low end of the microwave band, but as it reaches its maximum frequency of operation the microwave power begins to leak through the braided outer conductor and the attenuation rises rapidly. Normally, the larger the cable, the greater its power-handling capability, but the plastic jacket around the outside of the cable which protects the braid, impairs its heat-transfer ability, so its power-handling capability is less than cable 1. Cable 3 is the large, $\frac{7}{8}$-in.-diameter cable with corregated outer conductor and helical Teflon support. It has low loss because of its large size and very high power-handling capability, 700 W at its highest frequency. However, because of its large size, it only operates up to 5 GHz.

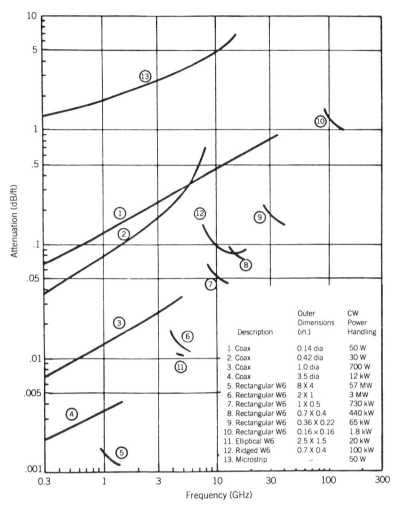

Figure 6.5 Comparison of transmission lines.

Coaxial cable 4, the large, $3\frac{1}{8}$ in. diameter air line, has a power-handling capability of 12 kW and very low attenuation. Its attenuation is less than .005 dB/ft, but the cable operates only to 1.3 GHz because of its size.

The six rectangular waveguides compared in Figure 6.5 operate from 1 to 100 GHz. Their size varies from 8 in. by 4 in. for waveguide 5, which operates around 1 GHz and can handle 57 MW of power, to 0.16 in. by 0.16 in. for waveguide 10, which operates at 100 GHz. The inner cross section of waveguide 10 is 1 mm by 2 mm (a millimeter is about the size of a paper-clip wire). Even with its small size, waveguide 10 can handle 1.8 kW of power, but its attenuation is extremely high, over 1 dB/ft. Waveguide 6 operates between

4 and 6 GHz, waveguide 7 between 8 and 12 GHz, waveguide 8 from 12 to 18 GHz, and waveguide 9 from 26 to 40 GHz. The power-handling capability of these waveguides varies from 3 MW in the 5 GHz range to 96 kW in the 30 GHz range. Note that the power-handling capability of these waveguides is much larger than that of the coaxial cable, primarily because the microwave power absorbed in a coaxial cable is absorbed mostly in the inner conductor and there is no way for this heat to leave the cable. In contrast, the power absorbed in the waveguide is absorbed in the outer conductor and can be readily removed. Note also the order-of-magnitude less loss in waveguide compared with coaxial cable. The attenuation of coaxial cable increases with frequency, but the attenuation of waveguide decreases with frequency, because the attenuation is extremely high near the cutoff frequency and then decreases as the frequency of operation moves away from the cutoff frequency.

The loss of the elliptical waveguide 11 is lower than the loss of the rectangular waveguide that operates in the same frequency range. The loss of the ridged waveguide 12 is higher than rectangular waveguides 7 and 8, but the ridged waveguide covers as much bandwidth as the two rectangular waveguides together.

Refer to coaxial cable 1 and waveguide 9. Both of these transmission lines operate at 30 GHz. The waveguide has one fourth the loss of the coaxial cable, and 2000 times the power-handling capability.

The loss of the microstrip transmission line 13 is extremely high, 5 dB/ft at 10 GHz. This is not really a problem, because microstrip is used for making complex interconnections between different transmission lines and is not intended for conducting microwaves over long distances.

The outer dimensions of the 13 transmission lines are given in Figure 6.5. Cross sections of some of these transmission lines are shown in Figure 6.6.

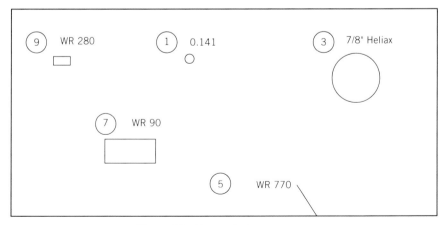

Figure 6.6 Transmission line size.

Waveguide 5 is 4 in. by 8 in. in outer dimensions, and the large size is required to permit the waveguide to operate at 1 GHz. Waveguide 7, which operates around 10 GHz, has approximately $\frac{1}{10}$ the width, and wave-guide 9, which operates around 30 GHz has about $\frac{1}{30}$ the width of the 1 GHz waveguide. Note that coaxial cable 1 operates at 1 GHz as does waveguide 5, but has an outer diameter of only 0.141 in.

The factors determining the dimensions of waveguide and coax are tabulated in Table 6.2. For waveguide, the width determines the frequency range. In the operating band of the waveguide, the width is about three fourths of a free-space wavelength. The ratio of the height to the width of the waveguide is approximately 1:2. Keeping the height equal to about half of the width raises the frequency of higher-order modes and maximizes the waveguide bandwidth. The height also determines the power-handling capability and impedance of the waveguide. The greater the height, the greater the power that the waveguide can handle and the greater is its characteristic impedance.

The outer diameter of coaxial cable determines its power-handling capability, its loss, and its maximum frequency of operation: the greater the outer diameter, the greater the power-handling capability, the smaller the loss, and the lower the maximum frequency of operation. The ratio of the inner to outer diameter determines the characteristic impedance of the coaxial transmission line.

Small semirigid coaxial cables, like cable 1, are used for most microwave connections at power levels up to 10 W.

Higher-power coaxial cables and waveguides are primarily used to handle large microwave powers from the transmitters of radar and communication systems, and to conduct the small received microwave power from the antenna to the first low noise amplifying stage.

In the millimeter-wave frequency range (30 to 300 GHz), the attenuation of coaxial cable is so high that waveguide is the best choice of transmission line, even for power levels of only a few watts.

Table 6.2   Factors determining transmission line dimensions

Waveguide
   Width determines frequency range
   Height determines
      Power handling
      Characteristic impedance
Coaxial cable
   Outer diameter determines
      Power handling
      Maximum frequency
Ratio of outer to inner diameter determines
   characteristic impedance

For microwave integrated circuits, stripline and microstrip are the best choices of transmission line, so that complex circuits can be fabricated by photoetching.

6.2 GUIDE WAVELENGTH AND CHARACTERISTIC IMPEDANCE

The guide wavelength and characteristic impedance of microwave transmission lines are controlled by the dimensions and materials of the line.

Guide wavelength is the distance that a microwave signal travels in one cycle. It is related to the free-space wavelength, but is modified, depending on the dimensions and materials of the transmission line.

The characteristic impedance (in ohms) of a transmission line is the ratio of the electrical field of the microwave signal to the magnetic field of the microwave signal. Formulas for calculating guide wavelength and characteristic impedance of coaxial cable, waveguide, and microstrip are summarized in Figure 6.7. The formulas relate guide wavelength to free-space wavelength, and characteristic impedance to free-space characteristic impedance, which is 377 Ω.

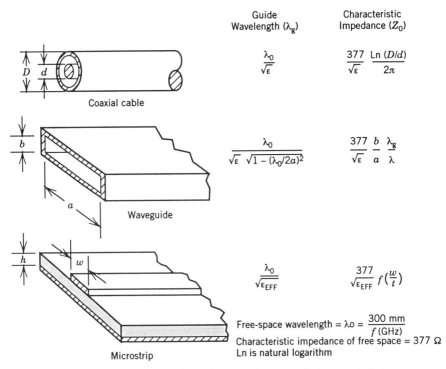

Figure 6.7 Formulas for calculating guide wavelength and characteristic impedance.

For coaxial cable, the guide wavelength is equal to the free-space wavelength divided by the square root of the dielectric constant of the support material. If the coaxial cable is an air line, then the guide wavelength is equal to the free-space wavelength. The guide wavelength of waveguide is more complicated, involving the ratio of the free-space wavelength to the width of the guide. It is also affected by the dielectric constant of the material inside the waveguide, but normally no dielectric is used inside the waveguide, so the dielectric constant can be ignored. The formula for the guide wavelength in microstrip is complicated because there is a different dielectric material on each side of the conductor—plastic or ceramic on one side and air on the other side. Exact formulas for calculating the guide wavelength and characteristic impedance of stripline and microstrip are discussed later.

The formulas for the characteristic impedance of the three different types of transmission line are all related to 377 Ω. When the microwave signal is forced to travel inside a transmission line, the characteristic impedance (the ratio of the fields) depend on the dimensions and dielectric material of the transmission line. Coaxial cable is normally made with the diameter of its outer conductor selected, relative to the diameter of the inner conductor, so that the characteristic impedance is 50 Ω. Microstrip is often made in this fashion also, so that the microstrip circuit can directly connect to a 50-Ω coaxial line. The characteristic impedance of waveguide depends not only on the height-to-width ratio but also on the guide wavelength, and thus varies with frequency across the band of the waveguide. In contrast, the characteristic impedance of coaxial cable and microstrip is not frequency-dependent. Note that all dimensions used in the formulas of Figure 6.7 are the inside dimensions of the transmission line.

Before a sample calculation using the formulas of Figure 6.7, is made, the types of dielectric support materials used in coaxial cable, stripline, and microstrip should be considered. The most commonly used dielectric materials, along with their properties, are tabulated in Table 6.3. The dielectric materials used for coaxial cables are most commonly Teflon and polyethylene. Both materials have a dielectric constant around 2 and good flexibility, which is important for making flexible or semirigid cable. Teflon has a maximum operating temperature of 200°C, polyethylene slightly lower, which means that Teflon has a higher power-handling capability. Polyethylene has slightly lower microwave loss.

Materials used for stripline and microstrip are Teflon fiberglass, boron nitride, beryllia, alumina, and epsilam 10 (a Teflon fiberglass material). The Teflon fiberglass is flexible and easy to machine, but its maximum operating temperature is 200°C, which limits the attachment of transistors and diodes by soldering. Boron nitride, beryllia, and alumina are ceramic materials. They can attain high temperatures, but their flexibility ranges from poor to very poor, and being hard ceramics they are difficult to machine. Their dielectric constant is higher than Teflon fiberglass, ranging from 4 to 10. The

Table 6.3 Microwave dielectric materials

Material	Dielectric Constant	Maximum Temperature (°C)	Flexibility
Teflon	2.04	200	Good
Polyethylene	2.25	150	Good
Teflon fiberglass	2.55	200	Good
Boron nitride	4.4	500	Poor
Beryllia	6.6	500	Very poor
Alumina	9.6	500	Very poor
Epsilam 10	10	150	Good

greater the dielectric constant, the smaller the parts become, which for many applications is an advantage. Consequently, a ceramic-loaded Teflon fiberglass called epsilam 10 has been especially developed with the same microwave characteristics as the alumina ceramic, but it is flexible and easily machined.

Examples of guide wavelength and impedance of the three different transmission lines, calculated at 10 GHz by using the formulas of Figure 6.7, are shown in Table 6.4. The free-space wavelength at 10 GHz is 30 mm. The first example is WR 90 waveguide, which is the standard waveguide operating

Table 6.4 Examples of transmission line guide wavelength and impedance

at 10 GHz and $\lambda_0 = 30$ mm

WR 90 Waveguide

$a = 0.900$ in. $b = 0.400$ in.

$\lambda_g = 40$ mm $Z_0 = 220\ \Omega$

RG 141 Semirigid Cable

$D = 0.118$ in. $d = 0.036$ in. $\varepsilon = 2.04$

$\lambda_g = 21$ mm $Z_0 = 50\ \Omega$

Microstrip

$w = 0.025$ in. $h = 0.025$ in. $\varepsilon = 10.0$ $\varepsilon_{EFF} = 6.7$

$\lambda_g = 12$ mm $Z_0 = 50\ \Omega$

around 10 GHz. Its width is 0.900 in. and its height is 0.400 in. According to Figure 6.7, the guide wavelength is 40 mm and the characteristic impedance is 220 Ω. Note that the guide wavelength in waveguide is greater than the free-space wavelength. This was waveguide 7 in the comparison of transmission lines shown in Figure 6.5.

The next example is the 0.141-in. diameter semirigid cable, which was cable 1 in the comparison. The inner diameter of the outer conductor is 0.118 in., the diameter of the inner conductor is 0.036 in., and the support material is Teflon, which has a dielectric constant of 2.04. The guide wavelength is 21 mm. The diameters of the inner and outer conductors were deliberately selected to give an impedance, with the Teflon dielectric used, of 50 Ω. Note that the guide wavelength of the coaxial cable is less than the guide wavelength in free space, due to the effect of the dielectric support material.

The third example is microstrip, transmission line 13 in the comparison. It has a line width of 0.025 in., a ceramic thickness of 0.025 in., and an alumina ceramic substrate with a dielectric constant of 9.6. Later the calculation of the characteristic impedance and guide wavelength of microstrip is discussed. By those formulas, the guide wavelength of the microstrip is only 12 mm, compared with the free-space wavelength of 30 mm or the guide wavelength in the Teflon-filled semirigid coaxial cable of 21 mm. The width of the microstrip line and the thickness of the ceramic support were deliberately chosen to give a 50 Ω impedance for the microstrip.

6.3 COAXIAL CABLE

The formulas of Figure 6.7, along with the properties of the dielectric support materials of Table 6.3, provide a method of designing coaxial cables. Sometimes coaxial cables need to be designed. Normally, however, standard coaxial cables are simply used to connect microwave devices together, so special cables will not have to be designed. More than 100 different coaxial cables are available with different power capability, attenuation, and flexibility, and the designer's job is to select the best cable for a particular use. Some of the most commonly used coaxial cables, both flexible and semirigid, are compared in Table 6.5. Except for the requirement of carrying large amounts of power with low attenuation at the low end of the microwave band, where air line or special cable with helical dielectric support are used, most coaxial cable requirements can be met by one of the cables listed. The RG cables are all flexible. Note their higher attenuation and lower power-handling capability compared to semirigid cables of approximately the same size.

Table 6.5 Standard coaxial cables

Type	Flexibility*	Outer Diameter (in.)	Characteristic Impedance (ohms)	Attenuation (dB/100 ft)		Power Handling (W)	
				3 GHz	10 GHz	3 GHz	10 GHz
085	SR	0.085	50	34	73	115	48
RG 196	F	0.080	50	78	172	41	14
141	SR	0.141	50	21	45	310	160
RG 58A	F	0.195	50	41	—	22	—
250	SR	0.250	50	14	29	600	280
RG 214	F	0.425	50	19	47	95	37
RG 59A	F	0.242	75	25	—	40	—
RG 62A	F	0.242	93	9	30	40	15

* f = flexible; sr = semirigid

6.4 WAVEGUIDE

The formulas of Figure 6.7 allow the characteristic impedance and guide wavelength of waveguide to be calculated. They can be used to design special waveguide matching sections. Normally, however, standard waveguides are used. Table 6.6 lists standard rectangular waveguides. Note that 34 waveguides are required to cover the full microwave band.

As discussed previously, the major factor affecting the choice of waveguide is the frequency range of operation. Table 6.6 shows this range for all the different waveguides, along with their outer dimensions, attenuation, and power-handling capability. Waveguides are specified by a WR number, which is the inner width of the waveguide in hundredths of an inch. For example, WR 90 waveguide is 1 in. by $\frac{1}{2}$ in. in outer dimensions, but the inner width is 0.9 in. or $\frac{9}{100}$ in., hence the designation WR 90. The older JAN designation is also shown because waveguide is still sometimes specified in this way.

6.5 STRIPLINE AND MICROSTRIP

In contrast to coaxial cable or waveguide, where standard coaxial cables and waveguides are used to connect the parts of the equipment together, microstrip is used inside of the microwave devices themselves. Consequently, special stripline and microstrip transmission lines need to be designed. Examples of stripline and microstrip design are given in Figures 6.8 and 6.9 and in Examples 6.1 to 6.4.

Table 6.6 Standard rectangular waveguides

EIA Waveguide Designation (Standard RS-261-A)	JAN Waveguide Designation (MIL-HDBK-216, 4 January 1962)	Outer Dimensions and Wall Thickness (in.)	Frequency Range (GHz) for Dominant (TE$_{1,0}$) Mode	Theoretical Attenuation, Lowest to Highest Frequency (dB/100 ft)	Theoretical Power Rating (MW) for Lowest to Highest Frequency
WR-2300	RG-290/U	23.250 × 11.750 × 0.125	0.32–0.49	0.051–0.031	153.0–212.0
WR-2100	RG-291/U	21.250 × 10.750 × 0.125	0.35–0.53	0.054–0.034	120.0–173.0
WR-1800	RG-201/U	18.250 × 9.250 × 0.125	0.425–0.620	0.056–0.038	93.4–131.9
WR-1500	RG-202/U	15.250 × 7.750 × 0.125	0.49–0.740	0.069–0.050	67.6–93.3
WR-1150	RG-203/U	11.750 × 6.000 × 0.125	0.64–0.96	0.128–0.075	35.0–53.8
WR-975	RG-204/U	10.000 × 5.125 × 0.125	0.75–1.12	0.137–0.095	27.0–38.5
WR-770	RG-205/U	7.950 × 4.100 × 0.125	0.96–1.45	0.201–0.136	17.2–24.1
WR-650	RG-69/U	6.660 × 3.410 × 0.080	1.12–1.70	0.317–0.212	11.9–17.2
WR-510	—	5.260 × 2.710 × 0.080	1.45–2.20	—	—
WR-430	RG-104/U	4.460 × 2.310 × 0.080	1.70–2.60	0.588–0.385	5.2–7.5
WR-340	RG-112/U	3.560 × 1.860 × 0.080	2.20–3.30	0.877–0.572	—
WR-284	RG-48/U	3.000 × 1.500 × 0.080	2.60–3.95	1.102–0.752	2.2–3.2
WR-229	—	2.418 × 1.273 × 0.064	3.30–4.90	—	—
WR-187	RG-49/U	2.000 × 1.000 × 0.064	3.95–5.85	2.08–1.44	1.4–2.0
WR-159	—	1.718 × 0.923 × 0.064	4.90–7.05	—	—
WR-137	RG-50/U	1.500 × 0.750 × 0.064	5.85–8.20	2.87–2.30	0.56–0.71
WR-112	RG-51/U	1.250 × 0.625 × 0.064	7.05–10.00	4.12–3.21	0.35–0.46

(Continued)

Table 6.6 (continued)

EIA Waveguide Designation (Standard RS-261-A)	JAN Waveguide Designation (MIL-HDBK-216, 4 January 1962)	Outer Dimensions and Wall Thickness (in.)	Frequency Range (GHz) for Dominant ($TE_{1,0}$) Mode	Theoretical Attenuation, Lowest to Highest Frequency (dB/100 ft)	Theoretical Power Rating (MW) for Lowest to Highest Frequency
WR-90	RG-52/U	1.000 × 0.500 × 0.050	8.20–12.40	6.45–4.48	0.20–0.29
WR-75	—	0.850 × 0.475 × 0.050	10.00–15.00	—	—
WR-62	RG-91/U	0.702 × 0.391 × 0.040	12.40–18.00	9.51–8.31	0.12–0.16
WR-51	—	0.590 × 0.335 × 0.040	15.00–22.00	—	—
WR-42	RG-53/U	0.500 × 0.250 × 0.040	18.00–26.50	20.7–14.8	0.043–0.058
WR-34	—	0.420 × 0.250 × 0.040	22.00–33.00	—	—
WR-28	RG-96/U	0.360 × 0.220 × 0.040	26.50–40.00	21.9–15.0	0.022–0.031
WR-22	RG-97/U	0.304 × 0.192 × 0.040	33.00–50.00	31.0–20.9	0.014–0.020
WR-19	—	0.268 × 0.174 × 0.040	40.00–60.00	—	—
WR-15	RG-98/U	0.228 × 0.154 × 0.040	50.00–75.00	52.9–39.1	0.0063–0.0090
WR-12	RG-99/U	0.202 × 0.141 × 0.040	60.00–90.00	93.3–52.2	0.0042–0.0060
WR-10	—	0.180 × 0.130 × 0.040	75.00–110.00	—	—
WR-8	RG-138/U	0.140 × 0.100 × 0.030	90.00–140.00	152–99	0.0018–0.0026
WR-7	RG-136/U	0.125 × 0.0925 × 0.030	110.00–170.00	163–137	0.0012–0.0017
WR-5	RG-135/U	0.111 × 0.0855 × 0.030	140.00–220.00	308–193	0.00071–0.00107
WR-4	RG-137/U	0.103 × 0.0815 × 0.030	170.00–260.00	384–254	0.00052–0.00075
WR-3	RG-139/U	0.094 × 0.0770 × 0.030	220.00–325.00	512–384	0.00035–0.00047

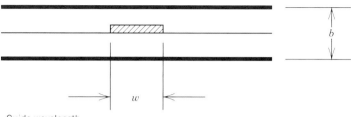

Guide wavelength

$$\lambda_g = \frac{\lambda_0}{\sqrt{\varepsilon}} \quad (\lambda_0 \text{ is free-space wavelength})$$

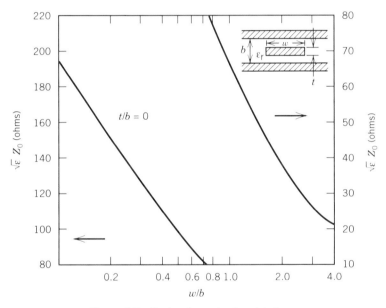

Figure 6.8 Design formulas for stripline.

Recall that the dielectric support material is on both sides of the stripline conductor. In stripline the guide wavelength λ_g is equal to the free-space wavelength λ_0 divided by the square root of the dielectric constant ε of the stripline support material.

$$\lambda_g = \frac{\lambda_0}{\sqrt{\varepsilon}}$$

The dielectric constant of the support material can be obtained from Table 6.3. Normally Teflon fiberglass is used for stripline, and its dielectric constant is 2.55.

The characteristic impedance Z_0 of the stripline depends on the thickness b of the dielectric support and on the width w of the conductor strip, and this relationship is graphed in Figure 6.8. There $Z_0\sqrt{\varepsilon}$ is plotted versus w/b. Note

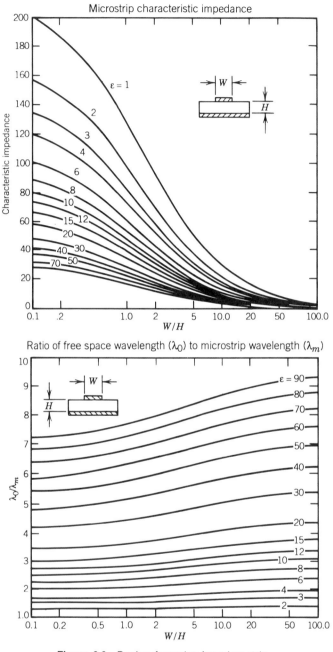

Figure 6.9 Design formulas for microstrip.

that b is the total thickness of the two support laminates—one on top of the strip and one on the bottom. The thickness of the strip itself is only 0.0007 in. and is negligible. The support material laminates are usually 0.030 or 0.062 in. thick. Therefore, if 0.030-in.-thick laminates were used, then $b = 2 \times 0.030$ in. $= 0.060$ in.

Example 6.1   Design a 50-Ω stripline using 0.03-in.-thick Teflon fiberglass (ε-2.55). Determine the guide wavelength at 3 GHz.

Solution

$$Z_0\sqrt{\varepsilon} = 50\sqrt{2.55} = 79.8 \ \Omega$$

From Figure 6.8,

$$\frac{w}{b} = 0.75$$

so

$$w = (0.75)(2 \times 0.03) = 0.045 \ \text{in.}$$

Then

$$\lambda_g = \frac{\lambda_0}{\sqrt{\varepsilon}}$$
$$= \frac{10 \ \text{cm}}{\sqrt{2.55}}$$
$$= 6.26 \ \text{cm}$$

Design graphs for microstrip transmission lines are shown in Figure 6.9. Formulas for guide wavelength and impedance of microstrip are complicated because the dielectric constant on top of the line is different from that on the bottom of the line. Consequently, the design must be done graphically. The upper graph of Figure 6.9 shows the characteristic impedance of the microstrip line as a function of the line width to dielectric support thickness ratio for various values of support material dielectric constant. The lower graph shows the reduction of the free-space wavelength in the microstrip line as a function of the width-to-thickness ratio (w/h) for various dielectric constants. Commonly used values from Figure 6.9 are

Material	Dielectric Constant	w/h for 50-Ω Impedance	Reduction of Free-Space Wavelength
Teflon fiberglass	2.55	3	1.5
Alumina	9.6	1	2.5

Example 6.2 Find the line width and guide wavelength for a 50 Ω microstrip line on 0.025 in. thick alumina ($\varepsilon = 9.6$) at 10 GHz.

Solution From the upper graph of Figure 6.9,

$$\frac{w}{h} = 1.0 \qquad (Z_0 = 50\,\Omega, \varepsilon = 9.6)$$

$$w = 1.0\,h = 1.0 \times 0.025 = 0.025\,\text{in.}$$

From the lower graph of Figure 6.9,

$$\frac{\lambda_0}{\lambda_m} = 2.5 \qquad \left(\frac{w}{h} = 1.0, \varepsilon = 9.6\right)$$

Thus,

$$\lambda_m = \frac{\lambda_0}{2.5} = \frac{30/f}{2.5} = \frac{30/10}{2.5} = 1.2\,\text{cm}$$

Example 6.3 What is the line width and guide wavelength for a 30 Ω microstrip line on 0.025-in.-thick alumina ($\varepsilon = 9.6$) at 10 GHz?

Solution From the upper graph of Figure 6.9,

$$\frac{w}{h} = 2.5 \qquad (Z_0 = 30\,\Omega, \varepsilon = 9.6)$$

$$w = 2.5 \times 0.025 = 0.0625\,\text{in.}$$

From the lower graph of Figure 6.9,

$$\frac{\lambda_0}{\lambda_m} = 2.6$$

$$\lambda_m = \frac{\lambda_0}{2.6} = \frac{3.0}{2.6} = 1.15\,\text{cm}$$

Example 6.4 What is the line width and guide wavelength for a 50-Ω microstrip line on 0.062-in.-thick Teflon fiberglass ($\varepsilon = 2.55$) at 10 GHz?

Solution From Figure 6.9,

$$\frac{w}{h} = 3.0 \qquad \text{so} \qquad w = 3.0 \times 0.062 = 0.186\,\text{in.}$$

$$\frac{\lambda_0}{\lambda_m} = 1.5 \qquad \text{so} \qquad \lambda_m = \frac{3.0}{1.5} = 2.0\,\text{cm}$$

6.6 CONNECTORS AND ADAPTORS

Typical transmission line connectors and adaptors between different transmission lines are shown in Figure 6.10.

Waveguides are easy to connect because they have no inner conductor. They are simply bolted together at the flanges attached to the waveguide ends (Figure 6.10). The simple flat flange is used for most connections. The

Figure 6.10 Transmission line connectors.

choke flange, on the upper right, reduces microwave leakage through any gaps where the flanges don't mate perfectly, and also allows for the use of gaskets between the flanges so that a gas-tight joint can be made. The waveguide can then be filled with an inert gas, such as nitrogen or sulfur hexafluoride, to increase its peak power-handling capability.

The middle four figures are coaxial connectors. Making a coaxial connection is difficult because the inner and outer conductors must be joined. The coaxial N and TNC connectors consist of a plug and a jack, and the inner conductors slide one inside the other to ensure a perfect fit, as the outer conductors are connected. The actual connectors are the small inside regions in the photographs, and the outer threaded parts are the clamping mechanism. The precision 7-mm connector has a spring-loaded inner conductor, and the inner conductors simply touch and do not slide one inside of the other. The precision connector then achieves a perfect alignment of the transmission lines and minimum reflected power at the connection. The N, TNC, and 7-mm precision coaxial connectors all have a 7-mm inner diameter of the outer conductor. The N and the precision connectors operate mode-free to 18 GHz. The TNC connector is filled with Teflon and is suitable for high-altitude operation, but the coaxial material inside the connector limits its mode-free frequency to 16 GHz. The SMA connector is half the size of the TNC connector (3.5-mm inner diameter), and it operates mode-free to 32 GHz. It is compatible with the 0.141-in. semirigid cable, which was used as cable 1 of the comparison cables and operates to 32 GHz. The coaxial EIA connector shown at the bottom left is used for joining large air line coaxial cables.

A ridged-waveguide-to-coaxial adaptor is shown at the bottom right of Figure 6.10, and illustrates one of the many combinations for connecting one transmission line type to another.

ANNOTATED BIBLIOGRAPHY

1. T. Saad, *Microwave Engineers' Handbook,* Vols. I, II, Artech House, Dedham, MA, 1988.
2. T. Moreno, *Microwave Transmission Design Data,* Artech House, Dedham, MA, 1988.
3. H. Howe, Jr., *Stripline Circuit Design,* Artech House, Dedham, MA, 1974.
4. K. C. Gupta, R. Garg, and I. J. Bahl, *Microstrip Lines and Slotlines,* Artech House, Dedham, MA, 1979.

Reference 1 provides formulas, tables, and graphs for selecting and designing waveguide, coaxial cable, and microstrip. Reference 2 is a classic book on transmission lines. References 3 and 4 cover striplines and microstrip.

EXERCISES

6.1. Complete the following table comparing transmission lines at 4 GHz. (Numbers refer to Figure 6.5.)

Transmission Line	CW Power Handling	Attenuation (dB/100 ft)
1. 0.141-in. coax		
3. $\frac{7}{8}$ in. heliax		
6. WR 187 waveguide		
13. Microstrip		

6.2. Complete the following table comparing transmission lines at 10 GHz. (Numbers refer to Figure 6.5.)

Transmission Line	CW Power Handling	Attenuation (dB/100 ft)
1. 0.141 in. coax		
7. WR 90 waveguide		
12. Ridged waveguide		
13. Microstrip		

6.3. Complete the following table comparing transmission lines at 30 GHz. (Numbers refer to Figure 6.5.)

Transmission Line	CW Power Handling	Attenuation (dB/100 ft)
1. 0.141 in. coax		
9. WR 280 waveguide		

6.4. Arrange the following transmission lines in order of decreasing wavelength at 10 GHz.

a. Teflon-filled coaxial line
b. Microstrip on alumina
c. Rectangular waveguide
d. Air-filled coaxial line

(*Hint:* You don't have to calculate the wavelength for all four cases to know which ones are larger and smaller.)

6.5. Complete the following table.

| Cable Type | RG 196 | RG 214 | Semirigid | |
			0.141 in. dia	0.085 in. dia
Cable OD (in.)				
Flexible or semirigid				
Impedance (ohms)				
Attenuation at 10 GHz (db/100 ft)				
Power-handling at 10 GHz (W)				

6.6. What waveguide would you use for microwave equipment operating at each of the following frequencies?

Frequency (GHz)	EIA Waveguide Designation
16	
1	
3	
10	
30	

6.7. Complete the following table.

EIA Waveguide Designation	WR 1500	WR 284	WR 137	WR 28	WR 10
Frequency range (GHz)					
Outer dimensions (in. × in.)					
Attenuation at highest frequency (dB/100 ft)					
Power rating (MW)					

6.8. What is the difference between stripline and microstrip?

6.9. Compare the dielectric constants of the following stripline materials.

Material	Dielectric Constant
Teflon fiberglass	
Ceramic-filled Teflon fiberglass (epsilam 10)	
Alumina	
Beryllia	

6.10 The free-space wavelength at 5 GHz is 6.0 cm. What is the wavelength in stripline (with dielectric on both sides of the conductor) with the following materials?

Material	Wavelength at 5 GHz
Teflon fiberglass	
Alumina	

6.11. What is the width of a 50-Ω stripline using 0.030-in.-thick Teflon fiberglass?

6.12. What is the width of a 50-Ω microstrip line using 0.030-in.-thick Teflon fiberglass?

6.13. What is the guide wavelength for the microstrip line of Exercise 6.12 at 5 GHz?

6.14. What is the width of a 50-Ω microstrip line using 0.025-in.-thick alumina?

6.15. What is the guide wavelength for the microstrip line of Exercise 6.14 at 5 GHz?

7

MICROWAVE SIGNAL
CONTROL COMPONENTS

Microwave signal control components "control" the frequency, power, and other characteristics of the microwave signal. The function, operation, and important performance specifications for terminations, directional couplers, combiners, isolators and circulators, filters, multiplexers, attenuators, switches, phase shifters, and detectors are discussed. Since many of these components use microwave semiconductors or ferrites, these devices are discussed first.

7.1 MICROWAVE SEMICONDUCTORS

Microwave signal control components use PN (or varactor), Schottky, and PIN semiconductor diodes. Semiconductors are useful as electronic devices because their electrical characteristics can be varied by "doping" the material. Doping is reviewed in Figure 7.1. A semiconductor has four electrons in its outer shell. The outer shell can hold as many as eight electrons, and the atoms of the semiconductor are bound in a crystal by the outer electrons. For example, in a pure silicon crystal, each atom shares the four electrons in its outer shell with its four neighbors, and each neighbor shares one electron with this atom; this sharing of outer shell electrons holds the crystal together. Thus, there are no free electrons that can move through the material under the effect of an applied electric field, so the semiconductor material is an insulator.

If the dopant level is less than 10^{-9} which is one part per billion, then the number of unbound electrons is negligible, and the semiconductor material is type I, an insulator. However, the conductivity of the material can be

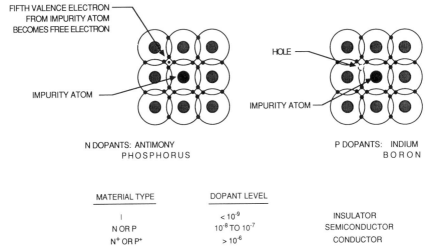

Figure 7.1 Semiconductor doping.

changed to any level simply by adding dopants, for example, antimony or phosphorus, which each have five electrons in their outer shell. When a dopant atom replaces a silicon atom in the crystal lattice, four of the outer shell electrons are shared with the neighboring silicon atoms to bind the crystal together, but the fifth electron is free to move through the material, so the material becomes a semiconductor. This type of doping makes N material, since the material has an excess negative charge.

On the other hand, P material has an excess positive charge. In this case, dopants such as indium or boron which each have only three electrons in their outer shell, are used. When the P dopant atom replaces a silicon atom, the three electrons are shared with the neighboring silicon atoms, but an electron is missing from the crystal lattice, and this missing electron is called a hole. An electron from a neighboring silicon atom fills this hole, but this leaves a corresponding hole in the atom that supplied the electron. As electrons fill the holes and leave other holes, the hole appears to "move" through the material, as a positive charge; hence, the designation P material.

The unique semiconductor action, particularly in regard to forming junctions between the N and P types of material, occurs at doping levels between 10^{-8}, which is one part in a hundred million, and 10^{-7}, which is one part in ten million. If dopants are added until the doping level exceeds 10^{-6}, which is one part in a million, then the material is called N^+ or P^+. The unique semiconductor junctions can no longer be formed when the material is so heavily doped, and N^+ and P^+ material are considered to be conductors. This type of doping makes silicon or gallium arsenide (GaAs), which in their pure form are insulators into conductors with a controlled level of conductivity. However, the material has no other special properties when doped, and appears a resistor with various resistance values. The unique characteristics or semiconductors, which make them useful for signal generation, amplification, and control, are achieved by forming junctions of differently doped semiconductors.

The basic junction is the *PN*, shown in Figure 7.2. The *PN* junction occurs when *P*-doped semiconductor material and *N*-doped semiconductor material are placed together. (The materials are not actually placed together, but the junctions are grown inside the material.) The electrons in the *N* material are attracted by the holes in the *P* material and move across the junction to fill the holes. The attraction is from electronic forces in the crystal lattice. As shown in Figure 7.2a, a depletion region is then formed between the *N* and the *P* materials, and this region is the *junction*. The junction no longer has electrons in the *N* material (because they have moved into the *P* material to fill the holes) nor holes in the *P* material (because they have been filled with electrons), so it is an insulator.

All of the electrons in the *N* material do not move over and fill the holes in the *P* material, because as the electrons leave the *N* region they leave the *N* region positively charged. Both *N* and *P* semiconductors are electrically neutral. The *N* material, because of its doping with atoms with five electrons in their outer shells, has free electrons that can move through the material, but the nucleus of each dopant atom has one more positive charge than do the silicon atoms, which balances out the extra electron in its outer shell. However, when the electron leaves the *N* material and enters the *P* material, it leaves the *N* material positively charge and correspondingly makes the *P* material negatively charged. The positively charged *N* material then exerts a force to draw electrons back into it. When this force balances the electron orbital forces, which pull the electrons into the holes, the flow of electrons into the holes stops. Consequently, the depletion region exists for only a

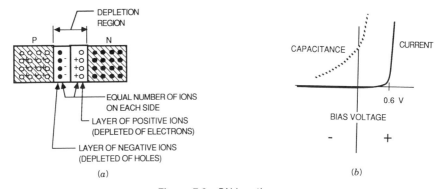

Figure 7.2 *PN* junctions.

small distance on either side of the junction, partly in the P material and partly in the N material.

The electrical characteristics of a PN junction are shown in Figure 7.2b. The current flowing through the junction is plotted as a function of the bias voltage applied to the junction. (The bias voltage is shown as the voltage applied to the P side relative to the N side.) With a negative bias, no current flows through the junction. The P side already has more electrons than it should, and connecting the negative terminal of the battery to the P side just makes the situation worse. Hence, the PN junction conducts only a small leakage current, which is usually assumed to be negligible. The diode is thus called *reverse-biased*. The PN junction then appears like a variable capacitance. As the negative bias is increased, the junction thickness becomes greater and the capacitance decreases.

When the diode is *forward-biased* the P material is connected to the positive terminal of the battery. However, current still will not flow until the forward bias is high enough to overcome the internal electronic forces. For silicon at room temperature this occurs when the external bias is 0.6 V. As shown in the figure, as the voltage increases above 0.6 V, the junction current increases rapidly.

At low frequencies, PN junctions are used as rectifier diodes for power supplies, for detectors, and as variable capacitors for tuning LC circuits.

Related semiconductor junctions used in microwave devices are compared in Figure 7.3. The PN junction is shown in Figure 7.3a. Figure 7.3b shows a Schottky junction, formed between a metallic conductor and an N-doped semiconductor. When the metal and N-doped material are placed together, electrons from the N material are attracted to the metal and leave an insulating junction in the N material. The characteristics of the Schottky junction are similar to those of the PN junction, except that the capacitance of the Schottky junction is reduced. By special fabrication techniques, the Schottky junction begins to conduct when the bias voltage just becomes positive.

An ohmic junction is shown in Figure 7.3c. The ohmic junction is used to make electrical connection to the N- or P-doped material. If a metal conductor is connected directly to N-doped semiconductor material, a Schottky junction is created. To make an electrical connection to the N material without any junction characteristics, a N^+ region, which is a heavily doped semiconductor, connected to the N region and the metal connected to the N^+ material. The ohmic junction has no unique characteristics; current flows equally well in either direction with no capacitance effects.

The PIN diode, shown in Figure 7.3d, is formed from P and N material with a thin layer of undoped or I type material between. The I material reduces the capacitance in the reversed-bias case, and the PIN diode is used as an electronically controlled attenuator, an electronic switch, a limiter, or in an electronic phase shifter.

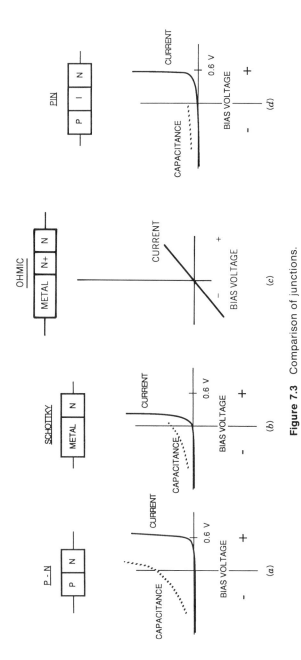

Figure 7.3 Comparison of junctions.

7.2 MICROWAVE FERRITES

The unique property of microwave ferrites is that the attenuation, propagation velocity, field configuration, and polarization of the microwave signals passing through them are affected by an applied external magnetic field. Ferrites are used in isolators, circulators, filters, and phase shifters.

Ferrites are iron oxide compounds. Although iron is a conductor, iron oxide is an insulator. Therefore, microwaves can pass through iron oxide ferrite materials. The iron atoms in the ferrite still have magnetic properties, and thus can be made to interact with the magnetic field of the microwave signal. Figure 7.4 explains the effect of microwave ferrites. The small sphere represents an electron spinning on its own axis. Since the electron has an electrical charge and is spinning, it has a magnetic moment, like a tiny magnet. Magnetic and nonmagnetic materials have spinning electrons, but the electrons in nonmagnetic materials occur in pairs and their magnetic effects cancel. A magnetic atom such as iron has unpaired electrons in its outer shell, so each atom has a net magnetic effect. When an external magnetic field is applied to a magnetic material, it tends to align the magnetic moments of the unpaired spinning electrons. However, since the electrons have mass and are spinning like a gyroscope, the magnetic moment of the electron does not align with the applied magnetic field: instead it rotates or precesses about it. The rate of precession depends on the strength of the applied field; at 1000 gauss, for example, the precession frequency is 2.8 GHz. (One Gauss is approximately the strength of the earth's magnetic field.) A field strength of several thousand gauss can be obtained with an electromagnet or a permanent magnet. The precession rate is proportional to the strength of the applied field; hence, doubling the magnetic field doubles the precession rate.

The properties of ferrites that make them useful for microwave devices are illustrated in Figures 7.5 and 7.6.

Figure 7.5*a* shows the attenuation in a ferrite-filled microwave transmission line at 10 GHz as a function of the magnetic field. No attenuation occurs at any magnetic field strength when the microwave signal is traveling down the transmission line in such a direction that the magnetic field of the microwave signal is oppositely aligned to the magnetic field in the ferrite material as shown by the dashed curve. When the microwave signal is traveling in the opposite direction, where the fields are aligned, there is no attenuation until the magnetic field is 3600 gauss which makes the precession frequency of the magnetic moments in the ferrite material equal to 10 GHz. The microwave signal is then attenuated (solid curve).

Even at magnetic fields above and below the attenuation condition, the microwave signal is affected by the ferrite, as shown in Figure 7.5*b*. The microwave signal travels through the ferrite at different velocities in opposite directions, depending on whether the magnetic field of the microwave signal is aligned with the magnetic field in the ferrite material. The difference in

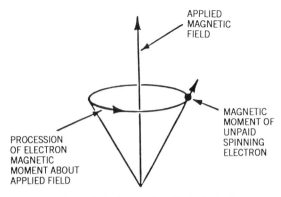

Figure 7.4 Magnetic effects in ferrite atoms.

velocity between one direction of propagation and the other is shown in Figure 7.5*b* as a function of the magnetic field. Note that it is impossible to show a difference in the propagation velocity at 3600 gauss, because the microwaves traveling in one direction are all absorbed. This velocity difference between propagation directions is used in Y-junction circulators.

Figure 7.6 shows how ferrites are used to provide frequency sensitive coupling between two transmission lines. The transmission lines are coupled together with two loops, at right angles to each other. Normally, there is no coupling because the loops are at right angles and their electromagnetic fields are at right angles. If a small ferrite sphere is placed in the center of the loops, there is still no coupling except at the magnetic resonant frequency. Figure 7.6 shows the coupling between the transmission lines as a function of frequency with two levels of magnetic field applied to the ferrite sphere in the center of the loops. If the magnetic field is 1800 gauss (solid curve), no coupling occurs between the transmission lines except at 5 GHz. If the

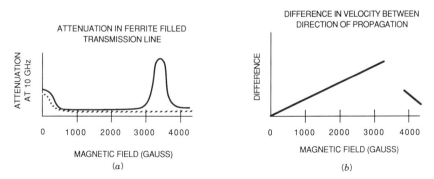

Figure 7.5 Microwave propagation in ferrite filled transmission lines.

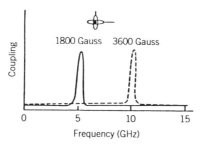

Figure 7.6 Coupling with a ferrite sphere.

magnetic field is 3600 gauss, then no coupling occurs between transmission lines except at 10 GHz (dotted curve). By varying the applied magnetic field, the coupling frequency can be varied. Ferrites are used in this way in electronically tuned filters.

Different ferrite materials are useful at microwave frequencies, including nickel ferrite, magnesium ferrite, and yttrium iron garnet (YIG). Because of its sharp frequency response, YIG is a popular material in ferrite devices. However, in applications where the ferrite device must operate over a wide frequency range (e.g., an isolator or a circulator), nickel ferrite is preferred.

7.3 TERMINATIONS

Terminations, or loads, are shown in Figure 7.7. They absorb all the power at the end of a transmission line and reflect none. Their purpose is to terminate a microwave equipment without allowing the power to escape into the surroundings or to be reflected back into the equipment. A termination for coaxial transmission lines is shown in Figure 7.7a, and a termination for microstrip is shown in Figure 7.7b. A low-power waveguide termination is shown in Figure 7.7c, and a high-power waveguide termination, air-cooled and capable of handling several hundred watts of power, is shown in Figure 7.7d. Terminations contain a tapered absorber, usually consisting of a carbon-impregnated dielectric material that absorbs the microwave power and reflects none of it.

The important specifications for terminations are their input SWR and their power-handling capability. These specifications are shown in Figure 7.7 for each termination.

7.4 DIRECTIONAL COUPLERS

Directional couplers sample the power traveling in one direction down a transmission line. As shown in Figure 7.8, input power P_i enters the direc-

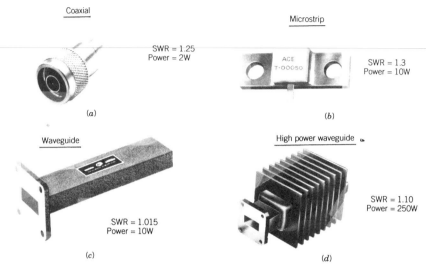

Coaxial

Microstrip

SWR = 1.25
Power = 2W

SWR = 1.3
Power = 10W

(a)

(b)

Waveguide

High power waveguide

SWR = 1.10
Power = 250W

SWR = 1.015
Power = 10W

(c)

(d)

Figure 7.7   Terminations.

tional coupler, a sample of it is coupled out of the directional coupler, as indicated by power P_c, and the rest passes into the output transmission line P_o.

The important specifications of a directional coupler are its coupling, insertion loss, isolation, and directivity. *Coupling* specifies how much of the input power is sampled. It is defined as 10 times the log of the coupled power divided by the incident power. (The absolute value is used because the log of the power ratio is negative.) Typical values of coupling are 3, 6, 10, 20, 30, 40, and 50 dB. However, any value of coupling can be obtained by properly designing the coupler.

The *insertion loss* specifies the output power relative to the input power. The output power is less than the input power for two reasons: (1) some of the input power is coupled into the coupling port and does not reach the output; (2) some of the power is absorbed inside the directional coupler. The insertion loss in dB is ten times the log of the output power divided by the input power, again taking the absolute value.

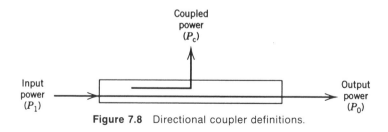

Coupled
power
(P_c)

Input
power
(P_1)

Output
power
(P_0)

Figure 7.8   Directional coupler definitions.

The directional coupler is supposed to sample power in only one direction, but since no coupler is perfect it does sample a small amount of power in the wrong direction. This wrong-direction sampling is called the *isolation* and is given in dB by ten times the log of the power at the coupling port divided by the power that is sent into the coupler in the wrong direction.

The *directivity* of the coupler is another way of specifying its performance in the wrong direction; and it is equal to the isolation in dB minus the coupling in dB. Therefore, the amount of power sampled in the wrong direction, which is the isolation, is equal to the directivity plus the coupling.

In addition to sampling the power traveling down a transmission line, directional couplers are used to attenuate the power by a known amount (by extracting the power through the coupled arm) and to measure the reflected power from a mismatch.

The principle of operation of directional couplers is shown in Figure 7.9. A waveguide directional coupler is shown but coax or stripline couplers operate in a similar fashion. The upper drawing shows that the directional coupler consists of two transmission lines connected through a series of holes: the larger the holes, the greater the coupling. The holes are spaced one quarter of a wavelength apart to achieve directional coupling properties. The directional properties are explained in the lower drawing. Incident power is assumed to be traveling in the main transmission line from left to right. Part of the power is coupled into the auxiliary transmission line at the first hole, and part at the second hole. These power samples are in phase because they have both traveled the same distance. Hence, they add and continue to

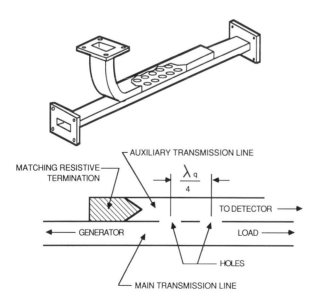

Figure 7.9 Directional coupler operation.

travel down the auxiliary transmission line from left to right until they appear at the output port of the coupler.

In contrast, power traveling down the transmission line from right to left has the same amount of power coupled at the second hole as at the first hole. However, when these samples combine in the auxiliary guide, they are 180° out of phase, because the power coupled through the first hole has traveled half a wavelength further than the power coupled through the second hole. Thus, they cancel, and no power appears at the coupled port.

Typical directional couplers are shown in Figure 7.10. A waveguide coupler is shown in Fig. 7.10a. A coaxial coupler is shown in Figure 7.10b, including a cutaway view in Figure 7.10d. A microstrip coupler is shown in Figure 7.10c. Coupling is controlled in the coaxial and microstrip couplers by the spacing between the main transmission line and the coupled line. Directional properties are obtained by making the coupling region one quarter of a wavelength long. The waveguide coupler can handle high power and has high directivity, but is limited in its frequency coverage. In contrast, coaxial and microstrip couplers cover broad frequency ranges, but have limited power-handling capability and poor directivity. The directivity specification is not critical when the coupler is being used for sampling microwave power, but it is extremely important when the coupler is used for a return loss measurement, to measure the small power reflected from a mismatch.

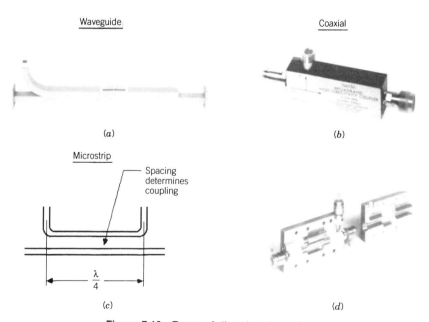

Waveguide

Coaxial

(a)

(b)

Microstrip

Spacing determines coupling

$\dfrac{\lambda}{4}$

(c)

(d)

Figure 7.10 Types of directional couplers.

7.5 COMBINERS

Combiners are used to combine two or more transmission lines into one transmission line. They can also be used to divide the microwave signal from one transmission line into two or more transmission lines.

The simplest way of combining or dividing microwave signals is to use a T connection of two transmission lines into another transmission line. However, this simple combiner provides no isolation between the two input lines, so the signal in one input line would get into the other input line and the combined line. The combiners shown in Figures 7.11 through 7.13 are used instead. One of the most widely used is the 3-dB quadrature (90°) hybrid. Its behavior is shown in Figure 7.11.

Figure 7.11a picts an input signal with a power level P entering the top left port of the hybrid. The hybrid splits the power into two equal parts; therefore, at each output port on the right, half of the input power appears. However, the ports are 90° out of phase. Because the power divides equally, the hybrid is called a 3-dB hybrid, and because the power is 90° out of phase in one output compared with the other one, the hybrid is called a quadrature hybrid.

Figure 7.11b shows two input signals applied to the hybrid. Half of signal 1, with no phase shift, and half of signal 2, with a 90° phase shift, appear in

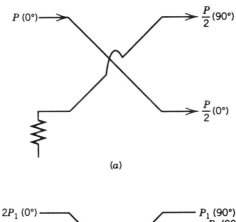

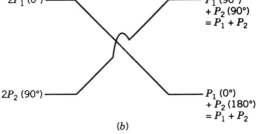

Figure 7.11 Coupling effects in a hybrid combiner.

the lower output arm. In the upper output arm, half of signal 1, with a 90° phase shift, and half of signal 2, with no phase shift, appear. If the two input signals are 90° out of phase, they have a 90° phase shift in the upper output arm and so add, but they are 180° out of phase in the lower arm and thus subtract.

Three designs of 3-dB quadrature hybrids are shown in Figure 7.12. Conventional labeling is shown in Figure 7.12 *a:* the input port is labeled 1, the coupled output port is labeled 2, the direct output port is labeled 3, and the isolated port on the input is labeled 4.

One type of 3-dB quadrature hybrid is simply a 3-dB directional coupler, as shown in Figure 7.12 *b.* The coupling lines must be very close if only 3-dB coupling is to be obtained, which makes this coupler difficult to fabricate.

The branch line coupler in Figure 7.12 *c* is much easier to fabricate. It consists of direct quarter-wave connections between the two main lines of the coupler. Note that the impedance of the various sections of main line and connecting lines must be carefully adjusted. The disadvantage of this coupler is its limited bandwidth.

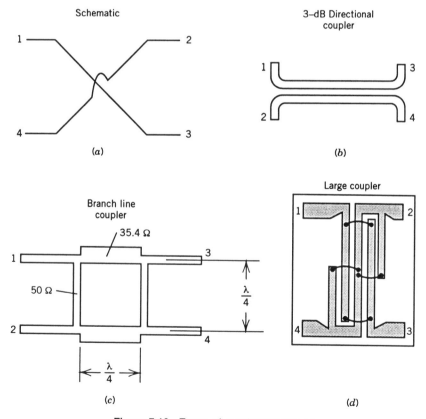

Figure 7.12 Types of quadrature hybrids.

The Lange coupler in Figure 7.12 d has good bandwidth (as high as 4 : 1) and is relatively easy to fabricate. Its coupling mechanism is similar to that of the 3-dB directional coupler, but it uses several sections of transmission lines for the coupling, so the spacing between individual lines can be greater and the coupler is therefore easier to fabricate. The Lange coupler requires wire bonds between the coupling arms.

Two other types of combiners are illustrated in Figure 7.13: the hybrid ring shown in Figure 7.13a and the magic T shown in Figure 7.13b. Both are 3 dB hybrids with a 180° phase shift between the signals in the output arms. The Wilkinson combiner shown in Figure 7.13c has 0° phase shift between the output signals.

The hybrid ring works as follows. The distance around the ring is chosen to be $1\frac{1}{2}$ wavelengths. A signal entering arm 1 divides into halves with each half traveling in opposite directions around the ring. Both halves reach arm 3 in phase, because each half has traveled the same distance, so the signals combine. In arm 2, one signal has traveled a quarter wavelength; the other signal, traveling in the opposite direction, has traveled 1 1/4 wavelength, so

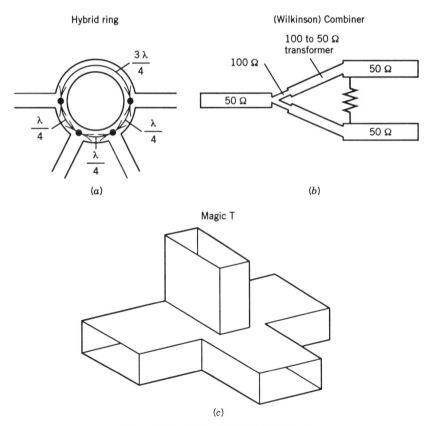

Figure 7.13 Other types of combiners.

they are also in phase and combine. Half of the input signal that reaches arm 4 travels through half a wavelength, and the other half of the signal reaching arm 4 travels through a full wavelength. Therefore, these signals are 180° out of phase and cancel. Consequently, the signal entering arm 1 divides; half of the signal exits arm 3, the other half exits arm 2, and these signals are 180° out of phase. None of the input signal exits arm 4. A similar analysis shows that none of the signal entering arm 4 appears at arm 1, but it divides equally between arms 2 and 3. Such 3-dB 180° hybrids are used in balanced modulators and balanced mixers, which are described later.

The in-phase (Wilkinson) power divider shown in Figure 17.3 has the advantages of in-phase power division and excellent amplitude balance between the output ports over a 2 : 1 or wider bandwidth. To avoid mismatches, the 50-Ω input line is connected to two parallel 100-Ω lines. Transformers are used to match between the 100-Ω lines and each 50-Ω output line. Isolation is achieved between ports by the terminating resistor. The resistor does not attenuate the output signals because the two signals are in phase in both lines at the point where the resistor is connected. However, any unequal mismatch or out-of-phase condition that would couple power from one line to the other is attentuated by the resistor. The disadvantage of the Wilkinson combiner for power applications is that the termination must be mounted inside the coupler, which limits its power-handling capability. The termination that dissipates the power in a 3-dB quadrature hybrid or hybrid ring can be outside the hybrid.

7.6   ISOLATORS AND CIRCULATORS

An isolator allows microwaves to pass in one direction but not the other. This isolating effect is achieved with ferrites. A typical microwave isolator is shown in Figure 7.14. A ferrite material is mounted inside a waveguide, and an external magnetic field, supplied by the C-shaped permanent magnet, magnetizes the ferrite.

If the external magnetic field and the magnetic field of the microwave signal are oppositely directed, the ferrite has no effect and the microwave

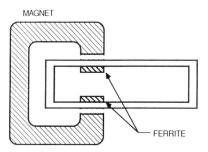

Figure 7.14   Isolators.

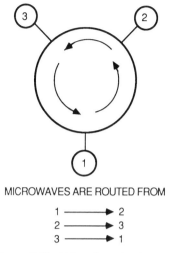

MICROWAVES ARE ROUTED FROM

1 ⟶ 2
2 ⟶ 3
3 ⟶ 1

Figure 7.15 Y-junction circulators.

signal is unattenuated. However, if the microwave signal travels in the opposite direction through the waveguide so that the magnetic fields are aligned, the microwave signal will be strongly attentuated.

The most important specifications for isolators are the isolation, which is the insertion loss in the reverse direction, and the forward insertion loss. The isolation should be high, and the forward insertion loss should be low. Typical values, are 20 dB for isolation and 0.5 dB for insertion loss.

Circulators as shown in (Figure 7.15) route microwave signals from one port of the device to another. For example, a microwave signal entering port 1 is directed out of the circulator at port 2. A signal entering port 2 is routed to leave the circulator at port 3 and does not get back into port 1. A signal entering port 3 does not get into port 2, but goes out through port 1.

The important specifications of a circulator are the insertion loss, which is the loss of signal as it travels in the direction that it is supposed to go, and the directivity, which is the loss in the signal as it travels in the wrong direction. Insertion loss is typically 0.5 dB, and directivity is 20 dB.

Typical Y-junction circulators are shown in Figure 7.16. A is a circulator for use with coaxial transmission lines. B is a circulator for use with waveguides. The inside details of each of these circulators are essentially the same, as shown in C. The circulator consists of

1. A ferrite disk, which is the black circular region in the center
2. Matching transitions between the ferrite disk and the transmission line connectors at the three ports
3. Biasing magnets, which are placed inside the package just above the ferrite disk.

A. COAXIAL

B. WAVEGUIDE

C. INSIDE VIEW

Figure 7.16   Typical Y-junction circulators.

The principle of operation of the Y-junction circulator is shown in Figure 7.17. The circulator uses the difference in propagation velocity in two directions in the ferrite (see Figure 7.5b). The magnetic field is adjusted so that the phase shift of the microwave signal is different in the clockwise direction through the circulator than it is in the counterclockwise direction. When the

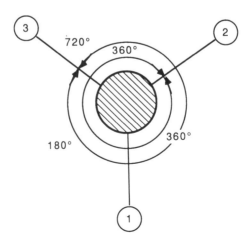

Figure 7.17   Y-junction circulator operation.

signal enters port 1, it divides in half; one part experiences a 360° phase shift when traveling counterclockwise from port 1 to port 2. (The dimensions of the circulator were chosen so that this 360° phase shift occurs.) The other half of the signal, traveling in the other direction, travels twice as far to reach port 2, but the ferrite is magnetically biased so that the microwave signal travels clockwise at twice the velocity as in the counterclockwise direction. Consequently, the signal traveling clockwise from port 1 to port 2 also experiences a 360° phase shift, and the two signals add in phase and the power exits through port 2.

The signal that travels clockwise from port 1 to port 3 experiences a 180° phase shift, since it travels only a third of the distance around the circulator and is traveling twice as fast in the clockwise direction as in the counterclockwise direction. The other half of the signal traveling in the counterclockwise direction from port 1 to port 3 experiences a 720° phase shift. The two signals reaching port 3 are 180° out of phase and cancel, so no power is at port 3.

Figure 7.18 shows two other circulator configurations. In Figure 7.18a the circulator is converted to an isolator by the addition of a termination on the third port. Microwave power passing in port 1 goes out port 2, and any power reflected from the output transmission line reenters the circulator at port 2 and leaves port 3 into the termination. The advantages of using the circulator as an isolator are that the Y-junction circulator is easier to fabricate in microstrip line than the isolator, and the power is absorbed outside the circulator in an external termination rather than absorbed in the ferrite itself, as with the isolator.

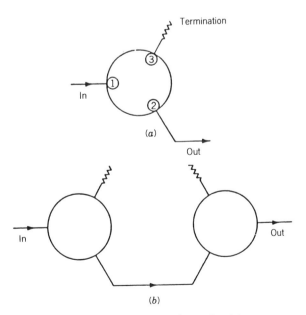

Figure 7.18 Isolators made from circulators.

Circulators may be connected in series, as shown in Figure 7.18*b*, to form an isolator with increased isolation. This increased isolation, however, is achieved at the expense of insertion loss. For example, a single circulator might have 0.5 dB of insertion loss and 20 dB of isolation. When two of these circulators are connected in series, the resulting isolator will have 40 dB isolation and 1 dB insertion loss.

7.7   FILTERS

Microwave filters pass a range of frequencies and reject other frequencies. A typical application of a filter is at the input of a microwave receiver. Microwave frequencies from many systems are picked up by the receiving antenna and could enter the receiver and cause interference. The filter passes only those frequencies in the assigned operating range of the system and rejects all others.

Filter performance is shown in Figure 7.19. The graph shows the attenuation of a microwave signal passing through the filter as a function of frequency. At frequencies below the filter passband, the attenuation is high and most of the microwave signal is attenuated. In a narrow frequency range about the center frequency of 10 GHz, ±10 MHz on either side of the center frequency, almost all of the signal passes through the filter. At frequencies above the passband, most of the signal is attenuated.

The important characteristics of filters are

- Passband type
- Out-of-band attenuation
- In-band insertion loss
- Selectivity
- Absorptive or reflective
- Power handling
- Tunability.

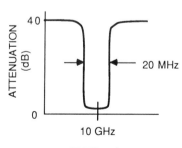

Figure 7.19   Filters.

Filters have four types of passbands. Figure 7.20a shows a low-pass filter, which passes all frequencies up to a certain frequency and then attenuates all higher frequencies. A high-pass filter, on the other hand, illustrated in Figure 7.20b, allows no signals to pass until a certain frequency is reached, and then passes all higher frequencies. A waveguide is an example of a high-pass filter. A bandpass filter, Figure 7.20c, passes signals only in a specified frequency band; above or below this band, the microwave signal is attenuated. In contrast, a bandstop filter, Figure 7.20d, passes microwave signals at almost all frequencies, except in a narrow range where it stops or prevents the signals from passing.

A filter should have a large out-of-band attenuation and a low in-band insertion loss. Selectivity defines the frequency range over which the filter characteristics change from passing the signal to blocking it. The smaller the frequency range of this change, the greater is the filter selectivity.

Some typical, fixed-tuned microwave filters are shown in Figure 7.21. The bandpass cavity filter in A is formed by a series of obstacles in the waveguide that divide it into resonant cavities. Each cavity is approximately a half wavelength long at the bandpass frequency. This type of filter is reflective, since power that cannot get through the series of cavities is reflected back down the transmission line toward the generator.

The low-pass, high-power waveguide filter in B is absorptive, in that it absorbs all the incident power in the stop band rather than reflecting it back into the input transmission line. The filter is a waveguide whose walls consist of an array of smaller waveguides, and for this reason the filter is termed "leaky wall." At passband frequencies, the small waveguides in the main

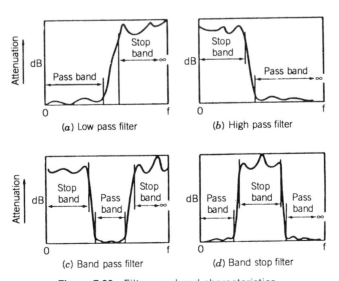

Figure 7.20 Filter passband characteristics.

A. BAND PASS CAVITY FILTER

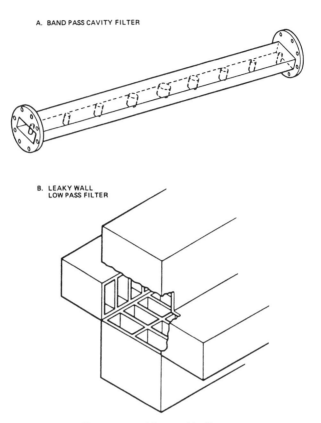

B. LEAKY WALL
LOW PASS FILTER

Figure 7.21 Waveguide filters.

waveguide wall are below cutoff and, therefore, too small to propagate microwaves, so they have no effect. At the upper edge of the filter's passband, where the frequency is high enough so that these auxiliary waveguides can propagate microwaves, the power is transferred from the main waveguide to the auxiliary waveguides, where it is absorbed by terminations at the outer end of the auxiliary waveguide. This leaky-wall filter can handle both high peak and high average power and reflects no power back toward the source.

Figure 7.22 shows stripline filters. These filters can be fabricated on either microstrip or stripline. To reduce losses, which improves the filter selectivity, they are usually fabricated on stripline with Teflon-fiberglass substrates. Figure 7.22a shows a low-pass filter, Figure 7.22b a high-pass filter, which is composed of inductive sections of stripline with lumped capacitors. A bandpass filters is shown in Figure 7.22c.

The waveguide and stripline filters in Figures 7.21 and 7.22 are fixed-tuned.

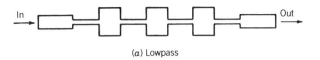

(a) Lowpass

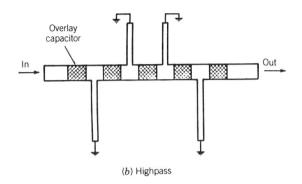

(b) Highpass

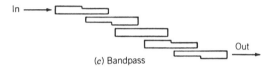

(c) Bandpass

Figure 7.22   Stripline filters.

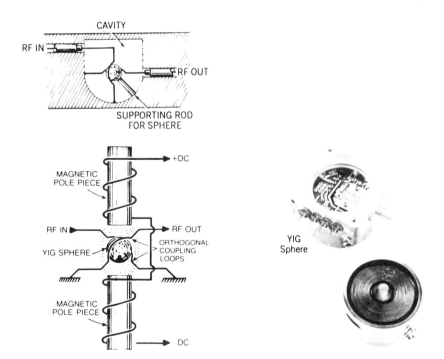

Figure 7.23   Electronically controlled YIG filter.

159

The YIG filter shown in Figure 7.23 can be electronically tuned over wide frequency ranges. As shown in the upper sketch of Figure 7.23, the YIG filter consists of a small 0.5 mm-diameter YIG sphere inside two coupling loops. As was shown in Figure 7.6, no coupling exists between the loops, so power from the transmission line feeding the loops is not coupled into the other transmission line, because the loops, and consequently their electromagnetic fields, are at right angles to each other. However, at the resonant frequency of the YIG material, which is controlled by the applied magnetic field, the microwave signal is coupled from one loop to the other by the action of the YIG itself. For example, when the magnetic field is 1800 gauss power is coupled in a narrow frequency range around 5 GHz. When the magnetic field is doubled to 3600 gauss, the microwave power is coupled when the input frequency is 10 GHz. By varying the external magnetic field with an electromagnet, the frequency of the YIG filter can be electronically changed over a wide frequency range, from 1 to 20 GHz. The YIG with its magnetic pole pieces, which apply the magnetic field to the YIG, is shown in the drawing. A complete YIG filter with its electromagnet is shown in the photo. By changing the current to the electromagnet coils, the frequency of the YIG filter can be electronically changed.

Fixed-tuned and electronically tuned filters allow the desired range of frequencies to be transmitted, and the unwanted frequencies to be reflected or absorbed. In contrast, a multiplexer, like a filter, allows the desired frequencies to be transmitted, but saves the undesired frequencies and routes them back into the transmission line for use in some other part of the equipment. Two types of multiplexers are shown in Figure 7.24.

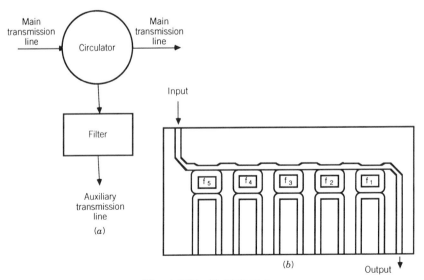

Figure 7.24 Multiplexers.

Figure 7.24 shows a multiplexer formed from a circulator and a filter. The microwave signal in the main transmission line enters arm 1 of the circulator and leaves arm 2. Arm 2 contains a reflective bandpass filter. Over the frequency range of the filter passband the microwave signal passes through the filter into the auxiliary transmission line. At all other frequencies above and below the passband, power is reflected back into the circulator and exits arm 3 to be transmitted down the main transmission line.

Figure 7.24*b* shows a multiplexer using an array of stripline couplers and stripline filters, with each pair removing a band of frequencies from the signal. The dimensions of the coupling resonators are different for each multiplexer, so each couples out a different frequency band. The advantage of the stripline multiplexer is that it can be formed on a single substrate with a single photoetching operation and does not need the ferrite circulator. Its disadvantage is that the filtering obtained by the stripline circuit does not have as good selectivity as is obtained by the all-metal waveguide-type filter.

7.8   ATTENUATORS

Attenuators are used to adjust the power level of microwave signals. Attenuators may be fixed, mechanically variable, or electronically variable. Fixed coaxial attenuators, also called pads, are shown in Figure 7.25. They consist

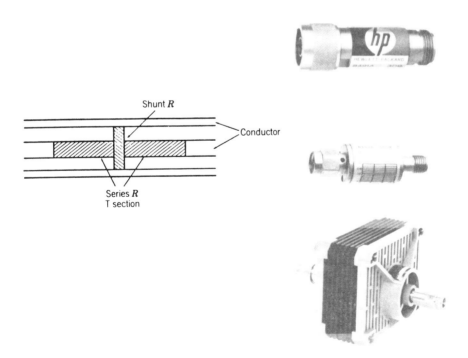

Figure 7.25   Fixed attenuator (Photos courtesy of Hewlett Packard).

of coaxial transmission lines, which have lossy material on a disk extending from the center to the outer conductor and along the center conductor. This lossy material forms a resistive T, which absorbs some of the microwave power without reflecting any of it.

Coaxial attenuators cover the frequency range from dc up to about 18 GHz, and they can have any value of attenuation. Typical values are 3, 6, 10, and 20 dB. The upper photograph shows an attenuator with N connectors. The center photograph shows an attenuator with SMA connectors. The lower photograph shows a high-power attenuator, which provides 30 dB of attenuation and can absorb 25 W. Note the cooling fins that allow the absorbed power to be dissipated into the surroundings.

Mechanically variable attenuators are shown in Figure 7.26. One means of obtaining a mechanically variable attenuator in waveguide is shown in the upper sketch. The attenuator consists of a vane of absorbing material inserted into the waveguide through a slot in the broad wall. The greater the penetration of the vane, which is controlled by the adjusting knob, the greater the attenuation, and the dial can be calibrated directly in dB.

Electronically variable attenuators are achieved with PIN diodes. The PIN diode is a PN junction with a layer of undoped or I material between the P and N regions to reduce the capacitance in the reversed-bias state.

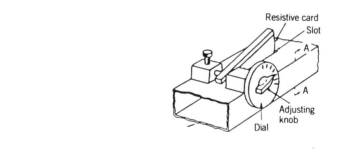

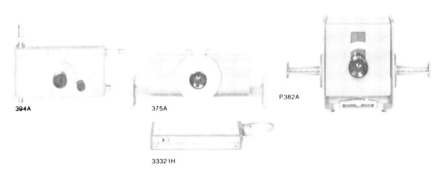

Figure 7.26 Mechanically variable attenuators (Photos courtesy of Hewlett Packard).

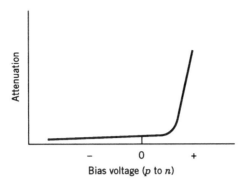

Figure 7.27 Electronically variable attenuator using a *PIN* diode.

Figure 7.27 shows the attenuation, when the diode is placed in a microwave transmission line, as a function of the bias voltage. With a negative or reverse bias, the diode provides no attenuation, since no current flows and it has very little capacitance. When it is forward-biased, the diode conducts and attenuates the microwave signal. Varying the positive-bias level varies the attenuation.

A reflective *PIN* diode attenuator is shown in Figure 7.28. It consists of

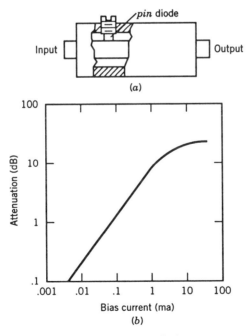

Figure 7.28 Reflective *PIN* diode attenuator.

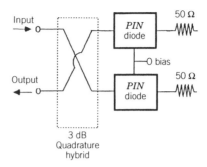

Figure 7.29   Matched *PIN* diode attenuator.

a *PIN* diode mounted in shunt across a coaxial transmission line. When the diode is reverse-biased, the *PIN* diode has very little effect on the microwave signal, and the signal travels through the assembly without attenuation. However, when the *PIN* diode is forward-biased, it looks like a resistor, and as the bias current increases, the resistance decreases. As this occurs, the *PIN* diode reflects more and more of the incident signal, so less of it leaves the output. Typical attenuation as a function of the forward-bias current is shown in Figure 7.28*b*. Since the forward-biased current varies so rapidly with forward-biased voltage, *PIN* diode characteristics are usually specified in terms of the forward-biased current rather than the forward-biased voltage. Note that as the bias current is varied from 0.01 to 10 mA, the attenuation varies from a fraction of a dB up to 20 dB. If additional attenuation is required, additional *PIN* diodes can be added to the attenuator assembly.

When the *PIN* diode is used in shunt across the transmission line, the power that does not get through the attenuator to the output is reflected back into the input line. This problem can be avoided by using a matched attenuator as shown in Figure 7.29. In this case, a 3-dB quadrature hybrid is used with two *PIN* diodes. The input signal divides between the diodes, and, as the bias level on the diodes is varied, varying but equal amounts of power are reflected from each diode. As this reflected power travels back through the hybrid, because of the phase relationships of the 3-dB quadrature hybrid the reflected powers add up at the lower port and cancel at the input port. Therefore, the input sees a perfect match, and the amount of power leaving the output is controlled by the *PIN* diodes. Note that the remaining power that is not reflected passes through the *PIN* diodes and is absorbed in the internal terminations.

7.9   SWITCHES

A switch directs microwave power from one transmission line to another, or turns the microwave power on and off. Switches can be mechanically or electronically actuated. Mechanically actuated switches connect and disconnect the transmission lines by mechanical means. Electronically actuated microwave switches use PIN diodes. With a reverse bias, the *PIN* diode has very little effect on the microwave signal. With forward bias, the *PIN* diode completely absorbs the microwaves. Hence, for use in a switch, the *PIN* diode control voltage is not continuously varied, as in an attenuator, but is switched from a reverse-biased level to a forward-biased level.

A *PIN* diode on-off switch is made by using the variable attenuators shown in Figures 7.28 and 7.29. However, rather than varying the control voltage continuously, the control voltage is switched between one level and another.

7.10 PHASE SHIFTERS

Microwave signals are characterized by amplitude and phase. The amplitude of microwave signals is controlled with attenuators. Their phase is controlled by phase shifters.

Phase shifters, like attenuators, can be mechanically or electronically adjustable. Figure 7.30*a* shows two mechanically controlled coaxial phase shifters, often called *line stretchers*, because the phase adjustment is accomplished by varying the length of the coaxial transmission line.

(*a*) MECHANICAL

(*b*) ELECTRONIC

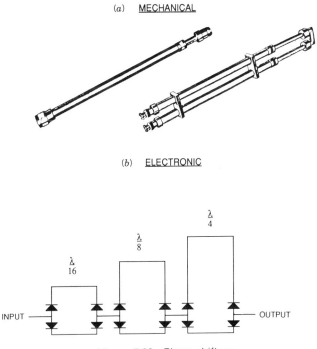

Figure 7.30 Phase shifters.

Figure 7.30*b* shows a three-section electronically adjustable phase shifter. Each section works as follows: As the microwave signal enters the section, it may take a short path or a long path. The length difference between the paths in the first section is a one sixteenth of a wavelength, or 22.5°. The length difference in the second section is one eight of a wavelength or 45°, and in the third section it is one quarter of a wavelength, or 90°. Depending on which path the signal takes, any phase shift of 22.5°, 45°, 67.5°, 90°, 112.5°, or 157.5° can be obtained. Each pair of diodes forms a single-pole, double-throw switch to route the microwave signal in one direction or the other.

Because the propagation velocity through a ferrite can be controlled by an external magnetic field, electronically controlled ferrite phase shifters can also be made.

7.11   DETECTORS

Components for detecting and measuring microwave power include thermistors, thermocouples, and Schottky diodes. Thermistors and thermocouples use the heating effect of microwaves to determine the microwave power level. They give extremely accurate power measurements, but cannot measure powers less than a microwatt and their response time is seconds. In contrast, the Schottky diode, which rectifies the microwave signal to get a dc voltage proportional to the microwave power, has a fast response time of nanoseconds and can measure powers as low as nanowatts. Schottky diodes are accurate only if carefully calibrated. Examples of microwave power meters using these sensors were shown in Chapter 3.

A thermistor consists of a small bead of semiconductor material placed in a transmission line. As this bead absorbs microwave power, its temperature rises and hence, its resistance changes. By measuring the thermistor resistance, its temperature can be determined and the microwave power can then be determined. This calculation is made by the microwave power meter.

A thermocouple consists of a small bead of dissimilar metals. The microwave power heats the thermocouple junction, which develops a voltage proportional to its temperature, which is in turn proportional to the microwave power. The power meter is used to obtain the power reading from the voltage reading. The power must be large enough to cause a significant temperature rise, and time must be allowed for the element to reach its stable temperature. This accounts for the slow response of the thermistor and thermocouple.

The application of a Schottky diode as a detector is shown in Figure 7.31*a*. The average value of an ac signal cannot be directly measured, since the signal is positive as much as it is negative. The value of the ac signal, whether large or small, averaged over one or many cycles is zero.

When the microwave signal is applied to a Schottky diode, the diode allows current to flow in one direction during the positive half of the cycle but not in the other direction during the negative half of the cycle. Conse-

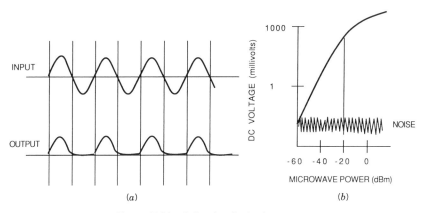

Figure 7.31 Schottky diode detector.

quently, the diode rectifies the current, and only half of each microwave cycle appears at the diode output. The average value of the rectified microwave signal is proportional to its amplitude, so the Schottky diode provides a dc output proportional to microwave power. The characteristics of a Schottky diode detector are shown in Figure 7.31*b,* where the dc voltage (mV) is shown as a function of the microwave power (dBm). The lowest power that can be measured with a Schottky diode, where the dc output voltage caused by the rectified microwave power is equal to the noise signal generated by the diode, is called the *tangential signal sensitivity* and is approximately -60 dBm. At microwave power levels from this value up to about -20 dBm, the dc voltage out of the diode is directly proportional to the microwave input power. This is called the square-law range, and many microwave instruments that use Schottky diodes have their calibration based on the assumption that the diode operates in the square law range. If the input power increases above the square-law range, the output continues to increase, but not directly with microwave power, so measurement errors can occur.

   In some measurement instruments, the Schottky diodes are operated in their nonlinear range by measuring the diode characteristics as shown in Figure 7.31*b* and adding corrections to the instrument readings to account for these nonlinearities. Schottky diodes are used as microwave detectors and for power measurement in network analyzers, spectrum analyzers, power meters, and other microwave measurement equipment and systems.

ANNOTATED BIBLIOGRAPHY

1. K. Chang, *Microwave and Optical Components* Vols. I, II, Wiley, New York, 1989.
2. T. Saad, *Microwave Engineers' Handbook,* Vol. I, II, Artech House, Dedham, MA, 1988.

3. G. Matthaei, L. Young, and E. M. T. Jones, *Microwave Filters, Impedance-Matching Networks, and Coupling Structures,* Artech House, Dedham, MA, 1980.
4. T. Laverghetti, *Solid State Microwave Devices,* Artech House, Dedham, MA, 1986.
5. W. S. Cheung and F. H. Levien, *Microwaves Made Simple: Principles and Applications,* Artech House, Dedham, MA, 1985.

Reference 1 provides design principles for signal control components. Reference 2 provides formulas, tables, and graphs for designing signal control components. Reference 3 provides detailed design information on couplers and filters. References 4 and 5 cover microwave semiconductor signal control components.

EXERCISES

7.1. Describe what each of the following signal control components does:

1. Termination (load)
2. Directional coupler
3. Combiner
4. Isolator
5. Circulator
6. Detector
7. Filter
8. Multiplexer
9. Fixed attenuator
10. Variable attenuator
11. Switch
12. Phase shifter

7.2. Complete the following table of directional coupler characteristics:

Input Power (dBm)	Coupled Power (dBm)	Output Power (dBm)	Coupling (dB)	Insertion Loss (dB)
0	−20	−1		
30	10	29		
50	10	48		
13			20	1
10			3	4
	−10		10	1

7.3 A power of 23 dBm is incident into a 10-dB directional coupler in the correct direction. What is the power at the coupled port?

7.4 An input signal is divided in a hybrid into two equal outputs. What will be the phase difference between the output signals if the following hybrids are used?

3-dB quadrature _____
3-dB 180° _____
Wilkinson combiner _____

7.5. A signal with an amplitude P_1 is applied to one of the input arms of a 3-dB quadrature hybrid. A signal P_2, which is 90° out of phase with P_1, is applied to the other input arm. What is the signal in each of the output arms of the 3-dB quadrature hybrid?

7.6. Complete the following table of isolator characteristics:

Input Power (dBm)	Transmitted Power in Forward Direction (dBm)	Transmitted Power in Reverse Direction (dBm)	Forward Insertion Loss (dB)	Isolation (dB)
10	9	−13		
25	24	5		
0			1	20
10			1	25
−10			1	20

7.7 Complete the following table of circulator characteristics:

Input Power in Arm 1 (dBm)	Output Power in Arm 2 (dBm)	Output Power in Arm 3 (dBm)	Insertion Loss (dB)	Directivity (dB)
0	−1	−20		
10	9	−13		

Input Power in Arm 1 (dBm)	Output Power in Arm 2 (dBm)	Output Power in Arm 3 (dBm)	Insertion Loss (dB)	Directivity (dB)
− 5			1	20
		− 10	1	25
	13		1	20

7.8. An isolator has an insertion loss of 1 dB and an isolation of 17 dB. If a 23-dBm signal is put through the isolator in the wrong direction, how much power is transmitted out of the isolator?

7.9. The isolator of Exercise 7.8 is placed in front of a short. Assuming the isolator has a perfect match, what is the return loss of the isolator and short combination?

7.10 A circulator has an insertion loss of 1 dB and a directivity of 23 dB. If a 5-dBm signal is put in arm 1 of the circulator, what is the output power from arm 2 (the correct direction of transmission) and arm 3?

7.11 A circulator has an insertion loss of 1 dB and a directivity of 21 dB. If a 0-dBm signal is put into the input port and port 2 is shorted, how much power leaks back into the input port?

7.12. Two circulators, which have directivities of 19 dB and 23 dB, respectively, are combined to form an isolator. What is the isolation of the isolator?

7.13. The following list refers to the filters in Figure 7.32.

1. What type of passband does filter 1 have?
2. What type of passband does filter 2 have?
3. What is the insertion loss of filter 1 at 4 GHz?
4. What is the insertion loss of filter 1 at 6 GHz?
5. What is the insertion loss of filter 1 at 10 GHz?
6. What is the insertion loss of filter 2 at 4 GHz?
7. What is the insertion loss of filter 2 at 6 GHz?
8. What is the insertion loss of filter 2 at 10 GHz?

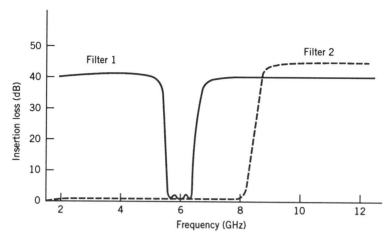

Figure 7.32 Filter characteristics for exercises.

7.14. What type of filter can be electronically tuned?

7.15. Three fixed attenuators with an insertion loss of 3, 6, and 10 dB are connected. What is the total insertion loss of the combination?

7.16. What is the power level of the smallest signal that can be detected above the noise by a Schottky diode?

7.17. What is the power level of the largest signal that will still be in the square-law range of a Schottky diode?

8

MICROWAVE SEMICONDUCTOR AMPLIFIERS

Microwave semiconductor amplifiers are used in all low power and medium power microwave equipment. High power tube amplifiers, used as the final output stage in transmitters, are covered in Chapter 12.

Amplifier performance characteristics are defined, and various microwave semiconductor devices used for amplifiers, including bipolar transistors, FETs, HEMTs, HBTs, and IMPATTS, are described. The most suitable types for particular frequency ranges and power levels are specified.

Microwave transistors, including bipolar transistors, FETs, HEMTs, and HBTs are discussed. Each transistor's operation and the unique fabrication techniques to make it work at microwave frequencies are described. Transistor packaging and mounting are then explained.

Since the performance of microwave transistors is best described by S-parameters, they are defined, equations for calculating transistor performance are given and sample calculations made.

After a transistor has been chosen for its performance characteristics, it must be biased for the correct operating current and voltage and matched into the transmission line. The methods for this, and for isolating the biasing and matching elements from each other, are given.

Medium power transistor amplifiers work only to about 100 GHz. Above this frequency, and for the highest possible output power above 20 GHz, IMPATT devices must be used. IMPATTS are most easily built as oscillators, and present problems when they are used as amplifiers, but there is no choice at high frequencies or at high power in the EHF range because transistors are not suitable. Operating principles of IMPATT devices are reviewed, and their use in reflection-type amplifiers is described.

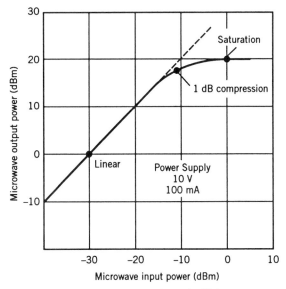

Figure 8.1 Power, gain and efficiency.

8.1 AMPLIFIER PERFORMANCE CHARACTERISTICS

The most important amplifier performance characteristics are

Power, gain, efficiency
Bandwidth
Phase shift
Harmonic and spurious power
Intermodulation products

Power, gain, and efficiency are defined in Figure 8.1. The graph shows the microwave power output of an amplifier as a function of the microwave power input at a single frequency. The output power is greater than the input power since the device is an amplifier. At low input power levels, marked "linear" on the graph, output power is directly proportional to input power. If input power doubles (increases by 3 dB), the output power also does. However, the output power cannot increase indefinitely, because it cannot be greater than the dc power input. The output power finally reaches a saturation value of 20 dBm (100 mW), and no more power can be obtained from the amplifier, no matter how great the input microwave power is. Gain is defined as

$$\text{Gain} = \frac{P_{\text{out}}}{P_{\text{in}}}$$

or

$$\text{Gain} \quad (\text{dB}) = P_{\text{out}} - P_{\text{in}} \quad (\text{dBm})$$

The gain continually changes, depending on the output power level. The three important values of gain are

Linear or small-signal gain
Saturation gain
Gain at 1 dB compression

In the linear gain region, the output power is directly proportional to the input power, so the gain is independent of output power level.

The saturation gain is the gain of the amplifier when the output power first reaches its saturation value.

The 1 dB compression point is defined as the point where the gain has decreased from its linear value by 1 dB.

Efficiency is defined as

$$\text{Efficiency} = \frac{P_{\text{out}}}{P_{\text{dc in}} + P_{\text{in}}}$$

The microwave input power is often negligible compared to the dc power. The maximum efficiency occurs when the microwave output power is a maximum, which occurs at saturation. The power supply power remains the same, independent of microwave output power, so the efficiency decreases from saturation.

Example 8.1   Using Figure 8.1, calculate the linear gain, saturation gain, gain at 1 dB compression, saturation efficiency, and efficiency at 1 dB compression.

Solution   From Figure 8.1, $P_{\text{in}} = -30$ dBm and $P_{\text{out}} = 0$ dBm (other values for P_{in} in the linear region could be used and the value of P_{out} read from the graph), so

$$\begin{aligned}
\text{Linear Gain} &= P_{\text{out}} - P_{\text{in}} \quad (\text{dBm}) \\
&= 0\,\text{dBm} - (-30\,\text{dBm}) \\
&= 30\,\text{dB}
\end{aligned}$$

The saturation output power occurs at 20 dBm, which corresponds to 0 dBm input power. Hence,

$$\text{Saturation Gain} = P_{\text{out}} - P_{\text{in}}$$
$$= 20\,\text{dBm} - 0\,\text{dBm}$$
$$= 20\,\text{dB}$$

The output power at the 1 dB compression point is 17 dBm, which corresponds to -12 dBm input (see the graph). Hence,

$$\text{Gain at 1-dB Compression} = P_{\text{out}} - P_{\text{in}}$$
$$= 17\,\text{dBm} - (-12\,\text{dBm})$$
$$= 29\,\text{dB}$$

Converting dBm to milliwatts, the saturated output power is 100 mW, and the microwave input power is 1 mW. The dc power is $(10\,\text{V})\,(100\,\text{mA}) = 1000$ mW. Hence,

$$\text{Saturation Efficiency} = \frac{P_{\text{out}}}{P_{\text{dc in}} + P_{\text{in}}}$$
$$= \frac{100\,\text{mW}}{1000\,\text{mW} + 1\,\text{mW}}$$
$$= 10\%$$

At the 1-dB compression point, the power has decreased by 3 dB, so the efficiency has decreased to half of its saturation value, or 5%. Note than at an input power level of -30 dBm, the output power is 0 dBm or 1 mW, and the efficiency is only 0.1%.

The bandwidth of a microwave amplifier is defined in Figure 8.2. Figure 8.1 showed the microwave output power as a function of microwave input power at a single frequency. Figure 8.2a shows the same type of curve at three frequencies, and at each frequency the amplifier has a different linear gain, saturation gain and saturation power.

The power at 1-dB compression and the linear gain (remember, the gain at 1-dB compression is 1 dB less than the linear gain, by definition) of this amplifier are shown in Figure 8.2b as a function of frequency. An amplifier's bandwidth is the frequency range over which its performance meets the specification requirements, which include gain and power. For power at 1-dB compression, the amplifier meets the requirement from f_1 to f_H, but at the higher frequency f_3 the power has dropped below the requirement. In contrast, the linear gain of the amplifier meets the specification requirement from f_L (which is above f_1) to f_3.

The useful frequency range of this amplifier would therefore be limited to a frequency of f_H by the power requirement and f_L by the gain requirement. The bandwidth is defined as

$$\text{BW} = \frac{f_H - f_L}{f_M}$$

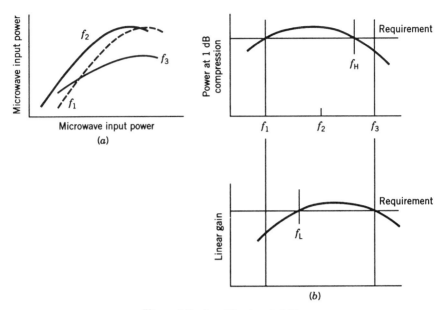

Figure 8.2   Amplifier bandwidth.

The center frequency f_m is defined as the mean between the high and low frequencies. For example, if f_H = 4 GHz and f_L = 2 GHz, then BW = 67%.

The phase of the output power as well as its amplitude is sometimes important, particularly when the outputs of two amplifiers must be combined, because their powers add only if they are in phase. Figure 8.3a shows the phase of the output power relative to the input power as a function of the input power level. This graph is of the amplifier in Figure 8.1, which reached its saturation output power at 0 dBm. The phase changes with input power level and, at saturation, has shifted by 50° from the phase in the linear range. The slope of this curve is the AM/PM coefficient, and indicates how much phase modulation appears on the amplifier output signal due to amplitude modulation on the input signal.

Figure 8.3b shows the phase shift of an amplifier as a function of frequency. Amplifiers usually consist of several amplifying devices connected to various lengths of microwave transmission lines and thus have a large total phase shift. Of interest for an amplifier is its phase variation compared with a reference coaxial cable that has the same number of wavelengths as the amplifier. As shown in Figure 8.3b, the phase increases with frequency as the wavelength becomes smaller, and what is of interest is the fine structure variation of the actual amplifier itself compared with the reference coax.

The discussion up to now has considered the output power of the amplifier at the same frequency as the input. The output is expected to be at the same

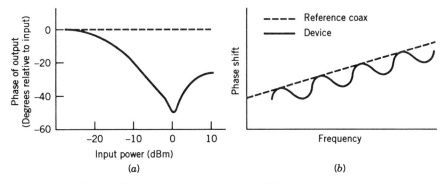

Figure 8.3 Phase of output power relative to input power.

frequency as the input, since the device is an amplifier. However, signals other than the amplified input signal appear in the amplifier output. These outputs are shown in Figure 8.4 and include harmonic, spurious, and white noise powers. The fundamental is the amplified input frequency. The harmonic is an integral multiple of the input frequency. (Only the second harmonic and the third harmonic are shown.) The harmonics' power levels are negligible in the linear region, but become large in the saturation region due to distortion of the output signal waveform as a result of the amplifier's nonlinear characteristics. Typically, the second harmonic is about 10 dB below the fundamental at saturation.

The spurious outputs are due to internal oscillation and can occur with no input signal at the amplifier. (They may be modified, however, by the signal.) They are not related to the input signal and are typically 40 dB below the fundamental. The white noise is at a very low power level and is only

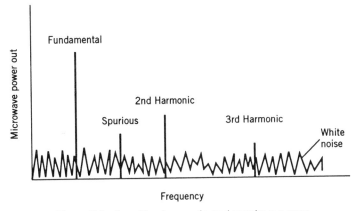

Figure 8.4 Amplifier harmonic and spurious power.

important if the amplifier input signal is at the picowatt level. At the microwatt level, the input signal is more than 60 dB greater than the noise, so the amplified noise is negligible compared with the amplified signal. Receivers work with input signal levels of about a picowatt, so the noise characteristics of receiver amplifiers are extremely important.

Intermodulation products are shown in Figure 8.5. So far the discussion has concerned an amplifier with only one input frequency. However, the microwave signal transmitted by a radar or a communication system has modulation sidebands that interact in the amplifier due to its nonlinear characteristics, to produce many other frequencies, called *intermodulation products*. With many sidebands in the input signal, the intermodulation products can be very complicated, so amplifiers are usually tested, as shown in Figure 8.5, with two equal input signals applied to the amplifier. As shown by the graph, the intermodulation products are negligible when the amplifier is operated in the linear range, but they become large near saturation because they are caused by the harmonic of one signal mixing with a higher-order harmonic of the other signal. Third-order intermodulation products are caused by the second harmonic of one signal mixing with the fundamental of the other. Fifth-order intermodulation products are caused by the third harmonic of one signal mixing with the second harmonic of the other, and so on. An infinite set of intermod products exists. For example, as shown in Figure 8.5, when the second harmonic of 3.2 GHz, which is at 6.4 GHz, mixes with the fundamental at 3 GHz, an intermod product at 3.4 GHz is formed. When the third harmonic of 3.2 GHz, which is at 9.6 GHz, mixes

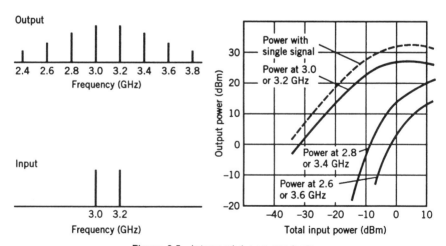

Figure 8.5 Intermodulation products.

with the second harmonic of 3 GHz, which is at 6.0 GHz, an intermod product at 3.6 GHz is formed.

The optimal output power level of an amplifier for high efficiency is at saturation, since the highest efficiency is obtained at this point. However, at saturation, large harmonics are formed, and the intermod products are only 10 dB below the fundamental. The optimal output power for low signal distortion is in the linear range, but amplifier efficiency is only a fraction of a percent. The best compromise, then, between low signal distortion and high efficiency, and the usual operating point for amplifiers, is the 1-dB compression point. There, the output power is approximately 3 dB below the saturation value and thus the efficiency is half the saturation value, but the intermod products and harmonics are more than 20 dB below the fundamental.

8.2 TYPES OF MICROWAVE SEMICONDUCTOR AMPLIFIERS

The output powers of microwave semiconductor devices that can be used for amplifiers are compared in Fig. 8.6 as a function of frequency.

The bipolar transistor is normally used at the low frequency end of the microwave band. Although the FET provides greater power, the bipolar transistor power output is adequate (>100 W at 1 GHz), and the device is less expensive than the FET. Bipolar transistors can be built to work up to 10 GHz, but is normally used to about 6 GHz.

FETs provide about double the power for a given frequency, or double the frequency for a given power, as compared to bipolar transistors, and they can be used up to about 30 GHz. At 10 GHz the FET can provide several watts of output power.

High electron mobility transistors (HEMTs) are similar to FETs but provide greater power and higher frequency response.

As frequency increases in the EHF band (30–300 GHz), the power from any transistor rapidly decreases, and IMPATT devices must be used.

The IMPATT actually uses transit time and so produces microwave power up to the upper end of the microwave band at 300 GHz. The problem with using IMPATTS is that they are diodes and have only one terminal. (They are easily used as oscillators, since the output power can be taken from this terminal.) As amplifiers, they must be used with a circulator, which even when fabricated on microstrip line, is many times larger than the IMPATT chip, so cascading amplifiers is difficult.

The dotted line in Figure 8.6 provides a convenient way of remembering the maximum power capability of microwave semiconductor amplifiers. It shows that the maximum power obtainable from any of the devices is 100 W at 1 GHz, 10 W at 10 GHz, and 1 W at 100 GHz. The product of the maximum power (watts) and the frequency (GHz) is therefore 100.

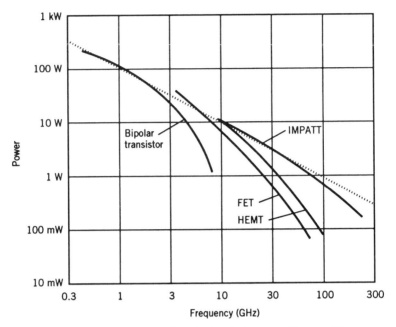

Figure 8.6   Microwave semiconductor devices used for amplifiers.

8.3   BIPOLAR TRANSISTORS

The operation of a bipolar transistor is shown in Figure 8.7. Figure 8.7*a* shows that the bipolar transistor consists of an *N*-doped semiconductor called the *emitter*, a *P*-doped semiconductor called the *base*, and an *N*-doped semiconductor called the *collector*. The transistor consists of two *PN* junctions: one between the emitter and the base, the other between the base and the collector. The base-collector junction is reverse-biased by a rather large voltage, 10 V in this example. The emitter-base junction is forward-biased at 0.7 V. The current-voltage characteristics of a *PN* junction are illustrated in the circle Figure 8.7a. When the junction is reverse-biased, no current flows. Current begins to flow when the junction is forward-biased about 0.6 V. When, for our example transistor, the emitter-base junction is biased at 0.6 V, 0.5 mA of electron current flows from the emitter to the base. When the junction is biased at 0.7 V, 1.0 mA flows; when the junction is biased at 0.8 V, 1.5 mA flows.

Normally, no current flows across the reverse-biased base-collector junction, because no electrons are available in the negatively biased *P* region of the base. However, in the bipolar transistor, the base region is made thin and electrons are available because they are emitted into the base from the emitter. As the electrons flow from the emitter across the emitter-base junc-

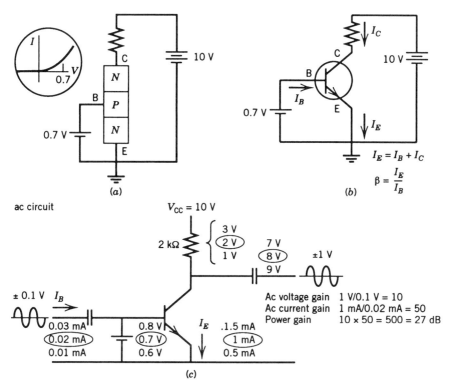

Figure 8.7 Bipolar transistor principles.

tion into the base, they see 0.7 V at the base and 10 V in the collector. Consequently, most of the electrons flow into the collector across the base-collector junction, and only about 2% of the electrons flow to the base. The emitter current thus divides, and the ratio of emitter current to base current is called the *beta* of the transistor: $\beta = \dfrac{I_E}{I_B}$

Figure 8.7*b* shows the schematic circuit of a bipolar transistor connected as a common emitter. Conventional current flow, opposite to electron current flow, is shown. The emitter current is shown to be equal to the sum of the base current plus the collector current.

Figure 8.7*c* shows the amplification of a ±0.1V ac signal by a bipolar transistor. The emitter-base junction is dc biased at 0.7 V. With no ac signal applied and an emitter-base bias of 0.7 V, 1 mA of current flows from the emitter into the base. Of this emitter current, 0.02 mA goes into the base and the rest (≈ 1 mA) goes into the collector. When the input ac signal to be amplified is added to the dc bias, the voltage between the emitter and the base is 0.8 V and 1.5 mA of current flows from the emitter into the base. The base current is 0.03 mA and the collector current is approximately

1.5 mA. When the input ac signal is negative, it subtracts from the dc bias; the emitter-base voltage is then 0.6 V, and 0.5 mA of current flows from the emitter into the base, most of which flows into the collector. As the electron current leaves the collector, it flows through the 2-kΩ collector resistor and develops a large voltage. In Figure 8.7, when the input ac voltage is zero, the collector current is 1 mA and the voltage drop across the collector resistor is 2 V. As the input ac signal is raised so that the base-emitter voltage is 0.8 V, the collector current is 1.5 mA and the voltage drop across the collector resistor is 3 V. When the emitter-base voltage is 0.6 V, the collector current is 0.5 mA and the drop across the collector resistor is 1 V. The voltage drop across the collector resistor subtracts from the supply voltage of 10 V. The output voltage is taken through a capacitor to remove the dc component. The ac output voltage is ±1 V around the 8-V dc value. The bipolar transistor has provided a voltage gain, amplifying the ± 0.1-V ac input signal to a ± 1-V output signal. The ac voltage gain is 1 V/0.1 V = 10.

There is also an ac current gain. The base current, which is the input current, varies over a 0.02 mA range, while the collector current, which is the output current, varies over a 1-mA range (0.5 to 1.5 mA). The current gain is the beta of the transistor, 50. The power gain is the voltage gain times the current gain: 10 × 50 = 500 = 27 dB.

The difference in operation between a microwave bipolar transistor and a low frequency bipolar transistor is shown in Figure 8.8. The operation is the same except for the degrading effects of transit time, internal capacitance and resistance, and external lead inductance that occur at microwave frequencies. Because the electrons take a finite time to travel across the emitter-base junction, through the base, and across the base-collector junction, the ac current is reduced at microwave frequencies, compared with low frequencies.

The significance of transit time is illustrated as follows. The velocity of microwaves, which is the same as the velocity of light, is 3×10^{11} millimeters per second. Electrons in a semiconductor travel approximately 1/3000 slower, or 10^8 millimeters per second. The distance that a microwave or an electron travels in one microwave cycle is the velocity divided by the frequency. Table 8.1 compares the distance traveled by a microwave and

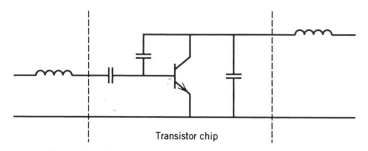

Transistor chip

Figure 8.8   Bipolar transistor at microwave frequencies.

Table 8.1 Microwave versus Electron Transit Times

Frequency (GHz)	Distance Traveled by Microwave (mm)	Distance Traveled by Electron	Required Electrode Spacing in Semiconductor (μ)
1	300	0.10 mm = 100 μ	10
10	30	0.01 mm = 10 μ	1
100	3	0.001 mm = 1 μ	0.1

1 mm = 1000 μ = 0.040 in.
0.025 mm = 25 μ = 0.001 in.
0.001 mm = 1 μ = 0.00004 in.

the distances traveled by an electron in one cycle at various frequencies. Also shown is the required electrode spacing in a microwave semiconductor, assuming that the electron must get from one electrode to the other in one tenth of a microwave cycle. This reasonable assumption gives an idea of what the required spacing inside a semiconductor must be at various frequencies. At 1 GHz, microwaves travel 300 millimeters in one cycle. In contrast, an electron travels only 0.1 millimeters or 100 microns. Consequently, the electrodes must be no more than 10 microns apart.

At 10 GHz, the distance traveled by a microwave in one cycle is 30 millimeters, the distance traveled by an electron is only 10 microns and the electrodes must be spaced 1 micron apart.

At 100 GHz, the microwaves travel only 3 millimeters in one cycle, and the electrons travel only 1 micron so the required spacings between the semiconductor elements is 0.1 micron.

A distance conversion table is shown at the bottom of Figure 8.9 to relate microns to millimeters to inches. As shown, 1 millimeter is 1000 microns or .040 inches. One thousandth of an inch is 25 microns, and finally 1 micron is 1000th of a millimeter of 40 millionths of an inch.

The wavelength of light is approximately 0.5 microns. This is the smallest dimension that can be fabricated in semiconductors by using optical microscopes and optical photoetching techniques. It thus represents a fabrication limit for microwave devices. Electron beam microscopes and electron beam lithography can be used to fabricate dimensions smaller than 0.5 micron, but these techniques are expensive.

Returning again to Figure 8.8, as the microwave signal enters the transistor, it must flow through the capacitance of the forward-biased emitter-base junction. This capacitance exists in low frequency transistors, too, but its reactance at low frequencies is negligible. Feedback capacitance between the collector and base is important at microwave frequencies, namely the capacitance of the reverse-biased base-collector junction. An output collector capacitance between the collector and emitter exists also. In addition, the bonding wires connecting the transistor chip to the input and output microwave transmission lines have inductance. This complicated circuit is best described by the measured transistor S-parameters, discussed later. The S-parameters are used for amplifier design.

The special fabrication techniques used in a microwave bipolar transistor to reduce the effects of transit time and internal capacitance and resistance will now be discussed. See Figure 8.9. The major time delays are the charging time of the emitter-base junction, the electron transit time through the base, and the electron transit time through the base-collector junction. The emitter-base charging time is minimized by using interdigital finger geometry for the emitter and the base as shown in part A of the figure. The interdigital finger design provides large areas for current to flow from the emitter to the base around the periphery of the fingers. At the same time, the design minimizes the emitter-base capacitance. This capacitance is determined by the area of the fingers, so they are made as narrow as fabrication allows, approximately 1 micron wide and 20 microns long. For comparison, a piece of paper is approximately 75 microns thick, so the length of the fingers on a bipolar transistor is approximately equal to one fourth the thickness of a piece of paper.

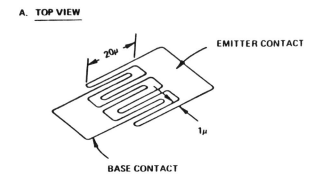

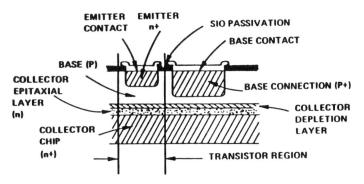

Figure 8.9 Microwave bipolar transistor.

The base transit time is reduced by making the base width as small as possible. Since no voltage is applied across the base, only the fabrication process limits the base width. The planar fabrication process, described later, permits a base width of a few tenths of a micron.

The transit time through the base-collector junction is controlled by the junction thickness. Its thickness cannot be arbitrarily reduced because it must stand off the reverse voltage applied to the transistor. To minimize transit time, the operating voltage of the transistor must be as low as possible, which in turn reduces transistor power capability.

Typical delay times for a bipolar transistor such as shown in Figure 8.9, are an emitter-base charging time of 3 picoseconds, a base transit time of 6 picoseconds, and a base-collector transit time of 14 picoseconds, for a total transit time of 25 picoseconds from the emitter to the collector. This may seem very small, but at 4 GHz one microwave cycle lasts only 250 picoseconds, so the total delay of a 4-GHz signal in passing through this transistor is one tenth of an microwave cycle.

To obtain the submicron dimensions required to reduce transit time, bipolar transistors are made using the planar process. Silicon is used for bipolar transistors, because of the ease with which the planar fabrication process can be carried out in silicon. The planar fabrication process is shown in Figure 8.10. The fabrication begins with a silicon wafer approximately 200 μm thick (step 1). This wafer forms the supporting substrate for the

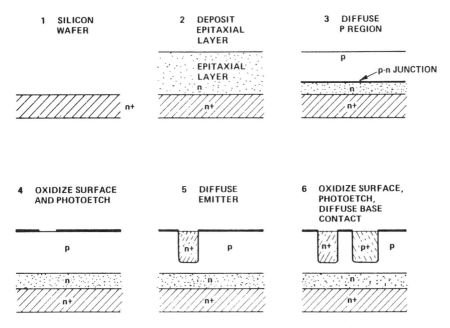

Figure 8.10 Fabrication of a microwave bipolar transistor.

transistor. A lightly doped N layer of silicon is then deposited on top of the substrate (step 2), and the various transistor elements are diffused into this epitaxial layer. The third step is to diffuse P dopants into the N epitaxial layer. Enough P dopants are put in to cancel the N doping and make the P base region of the transistor. The junction between the P material and the remaining N material forms the base-collector junction.

The fourth step is to oxidize the top surface of the transistor to form a silicon dioxide layer. The interdigital emitter contact is then photoetched through the oxide layer, and N impurities are diffused into this region to form the emitter (step 5). The final step is to etch the base contact fingers and diffuse additional P impurities to form the base connection.

The depth of the emitter extension into the base region controls the base width, a critical dimension that must be maintained to a few tenths of a micron to reduce the base transit time.

Figure 8.9B shows a cross section of the transistor made with the planar process. The active transistor region is between the emitter finger, through the base, to the collector. The interdigital base fingers, which are interleaved with the emitter fingers, simply provide an electrical contact to the base region.

The bipolar transistor in Figure 8.9, with only four base fingers, provides only a few milliwatts of output power. The operating voltage of this transistor is around 10 V. The base-collector junction, which must withstand this voltage, is made as thin as possible to reduce transit time effects. The only way, therefore, to obtain increased power is to use more interdigital fingers, and this approach is illustrated in Figure 8.11, which shows several bipolar transistor amplifiers connected to amplify a -10-dBm signal to $+30$ dBm, which is the 1-W level. The first stage in this chain is a small

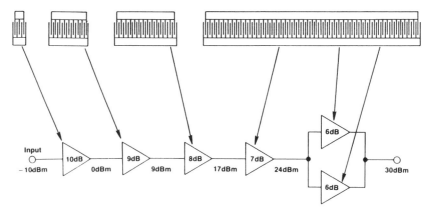

Figure 8.11 Transistor amplifier chain.

transistor with only four emitter fingers. This transistor provides 10 dB of gain, but its output power is 0 dBm. The next transistor contains 10 emitter fingers and provides 9 dB of gain, so it amplifies the signal to 9 dBm, almost 10 mW. The next transistor has 15 fingers and amplifies the signal to 17 dBm, which is 50 mW. The next transistor has 50 fingers and amplifies the signal to 250 mW. Additional fingers cannot be added arbitrarily, because it becomes increasingly difficult from a fabrication standpoint to make each finger perfect. Any flaw in the semiconductor material would cause one finger to short between the emitter and collector, ruining the transistor. The maximum number of fingers is 50 to several hundred. After that, the best approach is to make individual transistors and connect them in parallel, as shown in the output stage of Figure 8.11, where two transistors each with 50 fingers are paralleled to provide 1 W output power.

Using many fingers to obtain the high output power from a bipolar transistor causes several problems. The first, already mentioned, is to fabricate many fingers without one of them shorting out. The second problem is that of uneven current flow into the different fingers. This occurs if the base width varies from finger to finger: some fingers will draw more current than others, which causes hot spots. To eliminate this problem, most bipolar power transistors use a ballast resistor. The resistance is formed by making part of the finger interconnection of resistive material. Thus, even if a high current flows in one finger, a voltage drop develops across the ballast resistor and the current flow into the finger is reduced.

The third problem with multiple fingers is matching, in order to get the microwave signal into the transistor. A 1-W transistor operating at 10 V and a few hundred milliamps has a resistance of just a few ohms, and must be matched into a 50-Ω transmission line.

The lower the microwave frequency, the less severe are the transit time effects, because the microwave cycle is longer and the electrons have more time to move through the transistor. At low microwave frequencies, the base-collector junction thickness can be made large and a higher voltage can be applied, so more power can be obtained at low microwave frequencies than at high microwave frequencies. This effect was clearly illustrated in Figure 8.6, where the performance of bipolar transistors as a function of frequency was shown. Above 10 GHz it is not possible to fabricate silicon bipolar transistors with the small sizes required to reduce transit time. The silicon bipolar transistor is usually used only to about 6 GHz. At higher frequencies, the FET and HEMT are used.

The transistor chips shown in Figure 8.11 are only a few hundred microns wide, and the internal dimensions are only a fraction of a micron. These very tiny transistor chips must be mounted into microwave transmission lines so that the microwave signal to be amplified can be applied to the transistor, and the amplified output power from the transistor can be taken out into an external transmission line. The packaging and mounting of the transistor chips are discussed later.

8.4 FIELD-EFFECT TRANSISTORS

The operation of a field effect transistor (FET) is shown in Figure 8.12. The FET is fabricated on a semi-insulating substrate, which serves as the transistor support. An epitaxial layer of N-doped semiconductor material is deposited on top of the substrate, and the FET is built into this layer. The FET has a source, a gate, and a drain. The source is at one end of the transistor, and the drain is at the other. A positive voltage is connected to the drain, and electrons are drawn from the source to the drain.

The gate is placed between the source and drain on the top surface of the epitaxial layer. Microwave FETs are made with a metal-to-semiconductor junction (also called a Schottky junction) at the gate. For this reason the microwave FET is often called a MESFET, meaning that the gate is a metal-to-semiconductor junction.

The semiconductor material used for FETs is gallium arsenide (GaAs). Silicon could be used, but electrons travel approximately twice as fast in gallium arsenide as in silicon, so better high frequency performance is obtained. For this reason the FET performs better than the bipolar transistor, which is made of silicon. Likewise, gallium arsenide could have been used for

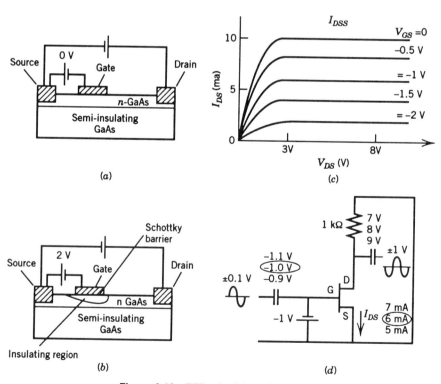

Figure 8.12 FET principles of operation.

bipolar transistors, but silicon bipolar transistors is less expensive. (Special gallium arsenide bipolar transistors are described in the next section.) In a FET *PN* junctions do not have to be grown through the material, and all the electrodes are mounted on the top surface. Because microwave FETs are made with gallium arsenide, they are sometimes called GaAsFETs. They are also sometimes called GaAsMESFETs because they are made with gallium arsenide and use a metal-to-semiconductor junction for their gate.

The most commonly used transistor for computer circuitry is a MOSFET. Computer circuitry operates around 10 MHz rather than microwave frequencies. The MOSFET uses a metal-to-oxide to semiconductor junction for its gate. The layer of oxide reduces the power consumption, which is important for putting thousands of FET computer gates on a single chip, but the added capacitance of the oxide layer greatly reduces the frequency response. Because frequency response is not a critical parameter, MOSFETs use silicon because fabrication is easier.

In Figure 8.12*a*, the gate is shown operated at the same voltage as the source; that is, the source-to-gate voltage is zero. In this case electrons move through the entire thickness of the epitaxial layer and the FET draws the maximum current, called the saturated drain-source current, I_{DSS}.

As shown in Figure 8.12*b*, as the gate voltage is made negative (a microwave FET is always operated with its gate voltage negative with respect to its source voltage), a reversed-biased Schottky junction is formed around the gate. Electrons are drawn out of the semiconductor into the gate electrode from a region around the gate. This region becomes an insulator, and the electron flow from the source to the drain is therefore impeded, since electrons can no longer flow through the Schottky barrier. As the negative voltage on the gate is increased, the size of the insulating barrier region increases, and the current flow from the source to the drain is further reduced. Figure 8.12*c* shows the source-to-drain current I_{DS} at various gate-to-source voltages. If the gate voltage is negative enough, the insulating region around the gate can be made to extend across the entire epitaxial layer and completely cut off the current flow.

The FET achieves amplification because a small voltage applied to the gate controls a large amount of current flowing through the transistor, and this current can be used to generate a large voltage in the output circuit.

Figure 8.12*d* shows a FET used as a low frequency amplifier. A large dc voltage of 10 V is applied between the source and the drain, and the drain is positive with respect to the source. The gate is biased at a dc voltage of −1 V, and from Figure 8.12*c*, which shows I_{DS} as a function of the gate voltage, 6 mA of current flows through the FET and through the 1-kΩ drain resistor. Conventional current flow is shown, with the current flowing from the drain to the source, which is opposite to electron current flow. Significantly, in a FET none of the transistor current flows into the gate circuit, because the gate junction is reverse-biased. This is in contrast to a bipolar transistor, where a small fraction of the emitter current flows into the base.

Figure 8.12d shows an input ac signal with an amplitude of ±0.1 V applied to the transistor. When added to the gate bias voltage, this voltage changes the gate voltage from −1.0 V to −1.1 V or −0.9 V. From the current-versus-gate-voltage characteristic of the transistor, this changes the current flowing through the transistor to from 6 mA to 5 mA or 7 mA. As this current flows through the 1-kΩ resistor, the voltage drop across the resistor varies from 5 to 7 V. The output voltage of the transistor amplifier, taken from the output capacitor, then varies ±1 V as the input voltage varies ±0.1 V, so the voltage gain of the amplifier is 10. The gate draws no current, so the current gain of a FET at low frequencies is very high, a major advantage. The FET therefore has a very high input resistance. However, at microwave frequencies the capacitance between the gate and the source has an appreciable reactance, and the current gain of the transistor is no longer infinite.

The microwave FET has the same problems as the microwave bipolar transistor. Microwave performance is degraded because of transit time effects, internal capacitance and resistance, and external lead inductance. The FET, like the bipolar transistor, is best described by its experimentally measured S-parameters, which take all these effects into account.

The fabrication details of a microwave FET, which minimize the adverse effects of transit time and the internal capacitance and resistance, are shown in Figure 8.13.

The gate length determines the transit time of the FET. Typical gate lengths are 0.5 micron with source to drain spacings of two to three times the gate length. Note that, as defined in Figure 8.13, the gate length is the short dimension of the gate, and the width of the gate is the long dimension. The gate width determines the power capability of the FET, and, for the FET shown in Figure 8.13a, is 25–50 microns, roughly the length of the base and emitter fingers of a bipolar transistor. With this gate width, the FET provides about 1 mW of power. For higher power levels, parallel gates must be used.

Figure 8.13c,d show the top views of power FETs. The increased power is obtained by using multiple sources, gates, and drains. The problem with the FET is that all three electrodes are on the top surface of the substrate, so some sort of bridging connection must be used. (Although bipolar transistors have the base and emitter electrodes on the top surface, the collector is on the lower surface.) As shown in Figure 8.13c, the interconnection of the three electrodes, all on the same surface, is accomplished by forming the source and the drain electrodes first as interdigital fingers. After this is done, the surface is covered with an insulating layer and the gate electrode is deposited over it. A special technique called *bridging* is used to etch away the insulating layer where the gate connection passes over the source electrode to reduce capacitance.

Another technique shown in Figure 8.13d uses wire bonds to connect a common gate connection to the individual gates. Since the gate lengths are less than a micron, the fabrication of a wire-bonded high power FET is extremely difficult.

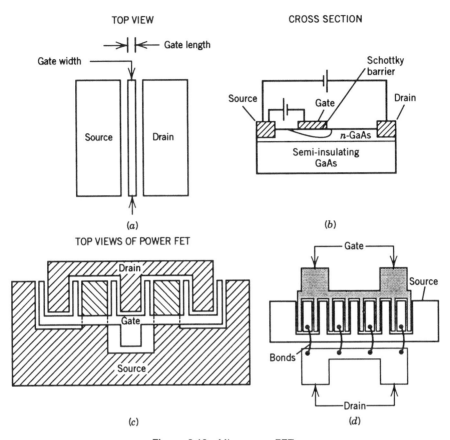

Figure 8.13 Microwave FETs.

The total gate width for a 1-W FET must be several thousand microns. This requires about 40 gates, each of which is 25 to 50 microns long.

8.5 HEMTs

The high frequency performance of FETs and bipolar transistors can be improved with special fabrication techniques and materials. As shown in Figure 8.6, an improvement over FET performance is obtained by using the HEMT, which stands for "High Electron Mobility Transistor". A comparison of the FET and HEMT is shown in Figure 8.14a,b. The FET as shown in Figure 8.14a, is built on a semi-insulating Gallium Arsenide substrate, and an N-doped layer of Gallium Arsenide is deposited on it. An electrical connection is made at each end of the N-doped layer to form the source and drain electrodes. To achieve an electrical connection, a layer of N^+-doped material must be between the metal electrode and the N-doped semiconductor. The gate is a metal-to-semiconductor junction and is deposited directly

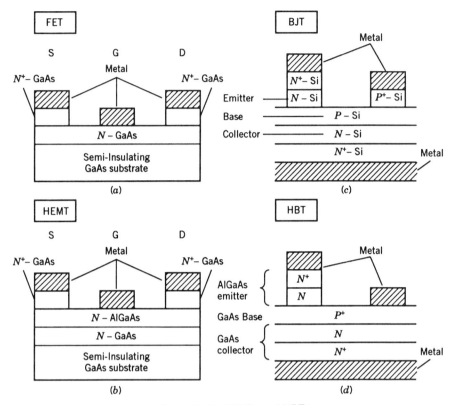

Figure 8.14 HEMTs and HBTs.

on the N-doped Gallium Arsenide with no N^+ intermediate layer. The electrons flow through the N-doped Gallium Arsenide channel from the source to the drain, and the thickness of the channel is controlled by the negative voltage applied to the gate.

Like the FET, the HEMT is built on a semi-insulating Gallium Arsenide substrate. Unlike the FET, however, the HEMT consists of multiple layers of different materials, including lightly N-doped Aluminum Gallium Arsenide and N-doped Gallium Arsenide. This layer of different semiconductor materials is called a *heterojunction*. The source, drain, and gate electrodes are the same as in the FET.

The difference between the FET and the HEMT is that in the FET the electrons travel in the N-doped Gallium Arsenide channel, whereas in the HEMT the electrons travel in the heterojunction layer between the two materials. The electrons suffer fewer collisions with the ionized doping atoms, which results in a higher electron velocity. This higher velocity of electron flow in the heterojunction channel from the source to the drain gives the HEMT its better high frequency performance, and its name. Some HEMTs have an even more complex arrangement of layers of semiconductor materials, including layers of gallium arsenide, aluminum gallium arsenide, indium phosphide and indium gallium arsenide. The purpse of all these layers is to increase the velocity of the electrons in the channel from the source to the drain, which allows the HEMT to operate up to almost 100 GHz, as shown in Figure 8.6, whereas the standard FET can only operate to about 30 GHz.

A greatly improved bipolar transistor called the heterojunction bipolar transistor (HBT) is compared to the standard Si bipolar transistor in Figure 8.14c,d. The standard silicon BJT is shown in Figure 8.14c. (This is the same drawing as Figure 8.9.) In the standard silicon bipolar junction transistor, the emitter is made of N-doped silicon, the base is P-doped silicon, and the collector is N-doped silicon.

The HBT is shown in Figure 8.14d. Its emitter is made of N-doped Aluminum Gallium Arsenide. The N^+ layer allows the emitter electrodes to be connected to the N-doped Aluminum Gallium Arsenide. The base is made of P^+-doped Gallium Arsenide, and the collector is made of N-doped Gallium Arsenide. Because different materials are used for the emitter and base, the transistor is called a *heterojunction* bipolar transistor. The use of the emitter-base heterojunction allows the base to be heavily doped to reduce its resistance, which significantly reduces the charging time of the base-collector junction. The heavy doping and the use of Gallium Arsenide instead of silicon allows the HBT to operate into the EHF band and provides higher efficiency than either the FET or HEMT.

8.6 TRANSISTOR PACKAGING AND MOUNTING

The tiny microwave bipolar and FET chips must be connected into microwave transmission lines so that the microwave input signal to be amplified can be put into the transistor and the amplified microwave signal can be taken out. The chips may be connected directly by wire bonds to input and output microstrip lines, or they may be mounted in a package and the package can then be mounted in a microstrip or other type of transmission line.

Figure 8.15 shows transistor packaging. The transistor chip is in the upper left. The chip dimensions are 350 microns by 350 microns, so the chip itself is about as wide as five thicknesses of a piece of paper. Remember that the internal dimensions of the transistor are a fraction of a micron.

A drawing of the transistor chip mounted inside the transistor package is shown in the upper right. Bonding wires connect the transistor electrodes to the microstrip lines of the package. The drawing shows that the bonding pad areas are approximately 40 μ square, which is about half the thickness of a piece of paper. The leads extending from the package are 0.020-in.-wide ribbons and are designed for direct connection to a 0.025 in. microstrip line. A photograph of the packaged transistor mounted in a microstrip test fixture is shown at the bottom of Figure 8.15. The source electrode (or the emitter

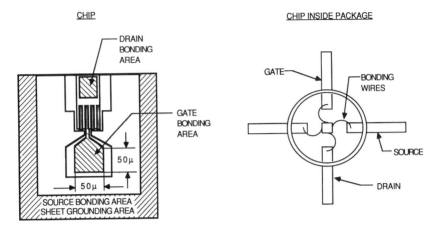

Figure 8.15   Transistor packaging.

electrode if a bipolar transistor is being packaged) is connected to the package itself. When the package is mounted in the microstrip circuit, the source is connected to the microstrip ground plane. The gate electrode (or the base electrode in a bipolar transistor) is connected to the input microstrip line, and the drain electrode (or the collector electrode in a bipolar transistor) is connected to the output microstrip line.

8.7   S-PARAMETERS

The microwave performance of a transistor is specified by its S-parameters. As shown in Figure 8.8, the equivalent circuit of a microwave transistor is very complicated, compared with a low frequency transistor, due to the internal resistances, capacitances and lead inductances that have significant reactances at microwave frequencies. Designing a transistor amplifier with this complex equivalent circuit is most easily done by using the transistor's measured S-parameters, which completely characterize the effects of all the elements of the transistor's equivalent circuit. S-parameters were defined in Chapter 4. The transistor is treated as a two-port microwave component. By definition,

a_1 = electric field of the microwave signal entering the transistor input
b_1 = electric field of the microwave signal leaving the transistor input
b_2 = electric field of the microwave signal leaving the transistor output
a_2 = electric field of the microwave signal entering the transistor output

Then,

$$S_{11} = \frac{b_1}{a_1} \text{ with } a_2 = 0 \qquad S_{12} = \frac{b_1}{a_2} \text{ with } a_1 = 0$$

$$S_{21} = \frac{b_2}{a_1} \text{ with} \qquad\qquad a_2 = 0 \qquad S_{22} = \frac{b_2}{a_2} \text{ with} a_1 = 0$$

and

$$\text{Input return loss} \approx 20 \log \mid S_{11} \mid$$
$$\text{Transistor gain} \approx 20 \log \mid S_{21} \mid$$
$$\text{Isolation} \approx 20 \log \mid S_{12} \mid$$
$$\text{Output return loss} \approx 20 \log \mid S_{22} \mid$$

S-parameters can be used to characterize any microwave device, as discussed in Chapter 4, but they are particularly useful for characterizing microwave semiconductor amplifiers.

Since b_1 and a_1 are electric fields, their ratio is a reflection coefficient. The input reflection coefficient of the transistor is approximately S_{11}. Therefore the input return loss is approximately 20 log $S_{11.}$

S_{21} is related to transistor gain. It tells how much electric field leaves the transistor relative to the total electric field that goes into the transistor.

S_{12} specifies the electric field leaving the transistor input when no signal is applied to the input and a signal is applied to the output. This means the transistor is passing a signal in the wrong direction (the transistor does this, due to its internal capacitance), so S_{12} specifies the transistor isolation. This leakage signal modifies the input and output matching requirements and may cause the amplifier to oscillate.

S_{22} is similar to S_{11}, just looking in the other direction into the transistor. It is approximately equal to the output reflection coefficient.

Figure 8.16 shows the S-parameters of a FET, taken from its data sheet, tabulated and plotted. Each S-parameter has an amplitude and a phase. S-parameters depend on frequency and on the transistor's operating voltage and current. S_{11} and S_{22}, which are related to reflection coefficients, are plotted on a Smith chart. S_{21} and S_{12} are plotted in polar form, showing their amplitude and phase. Note that $S_{21} > 1$, as it should be since the transistor is an amplifier. $S_{12} \ll 1$, as it should be since only a minimum signal should be coming in the backward direction through the transistor. Note the behavior of S_{11}, as shown on the Smith chart, as a function of frequency. S_{11} lies along a constant-resistance circle with a capacitive reactance component that decreases with frequency, ultimately becoming inductive at frequencies

S—MAGN AND ANGLES:

V$_{DS}$ = 3V, I$_{DS}$ = 10mA

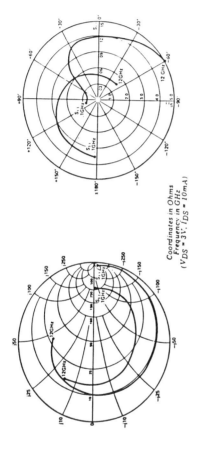

FREQUENCY (MHz)	S11		S21		S12		S22	
100	1.00	−4	3.43	176	.02	109	.69	−2
200	1.00	−6	3.47	175	.01	77	.69	−3
500	1.00	−16	3.49	164	.01	80	.68	−9
1000	.98	−31	3.39	154	.03	73	.68	−20
2000	.92	−60	3.25	125	.06	53	.66	−39
4000	.75	−117	2.56	79	.12	15	.54	−78
6000	.67	−160	2.02	43	.11	−6	.50	−110
8000	.60	166	1.64	10	.12	−21	.50	−146
10000	.57	127	1.55	−18	.14	−38	.51	178
12000	.55	86	1.50	−52	.15	−60	.55	137

Coordinates in Ohms
Frequency in GHz
($V_{DS} = 3V$, $I_{DS} = 10mA$)

Figure 8.16 S-parameters of a low power transistor.

above 8 GHz. The input circuit of the transistor could be represented as a resistance, inductance, and capacitance in series. At low frequencies the capacitance has a large reactance. As frequency increases, the reactance decreases until around 8 GHz it is canceled by the effect of the lead inductance.

The linear gain of a transistor can be calculated exactly from its S-parameters and a knowledge of the input and output transmission line impedances. The calculation is complicated because both the amplitude and phase of the S-parameters and the transmission line mismatches must be considered, so the calculation usually requires the use of a CAD program.

However, an estimate of the limits of amplifier gain can easily be made using simplified formulas that envolve only the amplitudes of S_{11}, S_{12}, S_{21}, and S_{22}. These simplified formulas are as follows:

$$\text{Unmatched Gain} = |S_{21}|^2$$
$$\text{Maximum Available Gain (MAG)} = \frac{|S_{21}|^2}{(1 - |S_{11}|^2)(1 - |S_{22}|^2)}$$
$$\text{Maximum Stable Gain (MSG)} = \frac{|S_{21}|}{|S_{12}|}$$

If the transistor is not matched to the transmission line, some of the input power will not get in, and some of the amplified power will not get out, so the gain will be low. The Unmatched Gain is the lower limit on the amplifier gain.

The Maximum Available Gain will be achieved if the transistor is perfectly matched, and so represents the upper limit on amplifier gain.

The actual gain will be between the Unmatched Gain and the Maximum Available Gain, depending on the exact impedance of the input and output match. Note that the amplitude and phase of these impedances, as well as the S-parameters, vary with frequency.

Because some power leaks through the transistor in the reverse direction, as specified by S_{12}, the transistor may become unstable and oscillate as the match is optimized to achieve the Maximum Available Gain. The Maximum Stable Gain defines the maximum gain that can be achieved without oscillation. If the Maximum Stable Gain is less than the Maximum Available Gain, then the Maximum Available Gain cannot be achieved without oscillation, and the maximum gain that can be achieved is the Maximum Stable Gain. If the Maximum Stable Gain is greater than the Maximum Available Gain, then the maximum gain that can be achieved is the Maximum Available Gain.

Example 8.2 Using the amplitude of the S-parameters at 6 GHz from Figure 8.16, find the MAG, MSG, and the unmatched gain.

Solution From Figure 8.16, at 6 GHz we find

$$|S_{11}| = 0.60 \qquad |S_{12}| = 0.10 \qquad |S_{21}| = 2.34 \qquad |S_{22}| = 0.39$$

Therefore,

$$\text{Maximum Available Gain} = \frac{\mid S_{21}\mid^{2}}{(1 - \mid S_{11}\mid^{2})(1 - \mid S_{22}\mid^{2})}$$

$$= \frac{(2.34)^{2}}{(1 - 0.60^{2})(1 - 0.39^{2})} = 10\,\text{dB}$$

$$\text{Maximum Stable Gain} = \frac{\mid S_{21}\mid}{\mid S_{12}\mid} = \frac{2.34}{0.60} = 13.7\,\text{dB}$$

$$\text{Unmatched Gain} = \mid S_{21}\mid^{2} = 2.34^{2} = 7.38\,\text{dB}$$

Since the Maximum Stable Gain is greater than the Maximum Available Gain, the Maximum Available Gain of 10 dB be achieved without oscillations occuring.

8.8 TRANSISTOR BIASING AND MATCHING

After a transistor is selected for an amplifier, it must be biased for the proper operating voltage and current, and then matched. Biasing, matching, and isolating the biasing and matching elements from each other will be described in this section.

The performance of a microwave transistor depends on the voltage applied to the transistor and on its collector or drain current. Figure 8.17 shows the variation of transistor performance with operating voltage. The curves show S_{21} as a function of collector current I_C with collector-emitter voltage V_{CE} as a parameter.

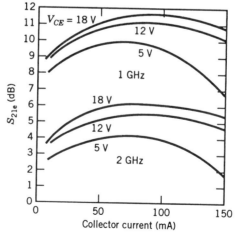

Figure 8.17 S_{21} versus collector current for a microwave transistor.

Bipolar transistor biasing is shown in Figure 8.18. Biasing means obtaining the correct values of collector emitter voltage, V_{CE}, and base emitter voltage V_{BE} from the power supply to obtain the desired value if I_C. Two power supplies could be used, one to supply V_{CE} and one to supply V_{BE}. However, it is preferable to use one power supply and obtain the correct voltages from the biasing resistors, particularly when a chain of transistors is used for a multistage amplifier, to permit all transistors to operate from a common supply voltage.

The major biasing problem is the extreme sensitivity of transistor performance to operating temperature. The current I_C is shown as a function of

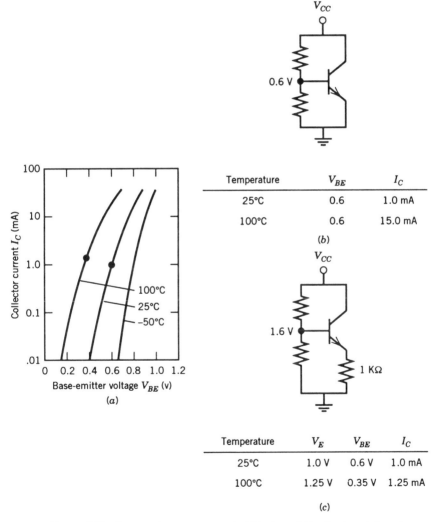

Temperature	V_{BE}	I_C
25°C	0.6	1.0 mA
100°C	0.6	15.0 mA

(b)

Temperature	V_E	V_{BE}	I_C
25°C	1.0 V	0.6 V	1.0 mA
100°C	1.25 V	0.35 V	1.25 mA

(c)

Figure 8.18 Temperature compensated biasing of bipolar transistors.

V_{BE} in Figure 8.18a, with the transistor operating temperature as a parameter. Note the extreme change in I_c with temperature. Transistor temperature depends not only on the ambient temperature but also on the power dissipated by the transistor and the transistor cooling.

A simple biasing scheme, where V_{BE} is obtained from the power supply by a resistive divider network, as shown in Figure 8.18b, is not satisfactory. If V_{BE} is fixed at 0.6 V, I_C is the desired 1 mA when the transistor temperature is 25°C. However, according to the graph, when the transistor temperature increases to 100°C, with $V_{BE} = 0.6$ V, I_C rises to 15 mA, a 15-fold increase! The same problem would occur if the resistive divider network, which supplies 0.6 V for V_{BE}, were replaced by a separate bias supply.

A common method of biasing bipolar transistors is shown in Figure 8.18c. An emitter feedback resistor is used in conjunction with the resistive divider network. The values of the divider resistors are selected so $V_B = 1.6$ V. The emitter resistor is chosen so that, with $I_E = 1$ mA, as desired, the voltage drop across the emitter resistor is 1.0 V. Then $V_{BE} = 0.6$ V, as required, when the transistor is operated, as shown in the table, at 25°C. As transistor temperature rises, I_C increases, but the voltage drop across the emitter resistor also increases, so that V_{BE} decreases, which reduces I_C. As shown in the table (derived from the transistor current characteristics shown in the graph), when transistor temperature increases to 100°C, the stable operating point occurs for $I_C = 1.25$ mA. Then $V_E = 1.25$ V, and, since the voltage divider network has fixed the base voltage at 1.6 V, $V_{BE} = 0.35$ V, which gives the 1.25 mA when the transistor temperature was 100°C. Therefore, the emitter feedback resistor has greatly reduced the variation of I_C with temperature.

The characteristics of a FET are not nearly as temperature-sensitive, so fixed bias voltages can be used. The simplest way to bias a FET is to use two power supplies, one for V_{DS} and the other for V_{GS}. Another common method for biasing a microwave FET is to use a source resistor. The source resistor has the advantage of providing feedback to stabilize the FET performance and requires only one power supply.

Once the transistor has been properly biased, it must be matched to the microstrip transmission line. Note that S_{11} and S_{22}, which specify the mismatch on the input and output respectively, depend on the biasing conditions. Matching is done from the Smith chart, as shown in Chapter 5.

After biasing and matching, the biasing and matching circuits must be isolated.

This problem is illustrated in Figure 8.19a, using an FET as an example. To bias the FET, the gate must be connected to ground, the source connected to ground through the source resistor, and the drain connected to the supply voltage through the drain resistor. All of these are dc connections, but are made to the same gate, source, and drain leads that are connected to the matching circuits.

The source must be connected to the microstrip ground plane, the gate must be connected to the input microstrip line, which contains the input

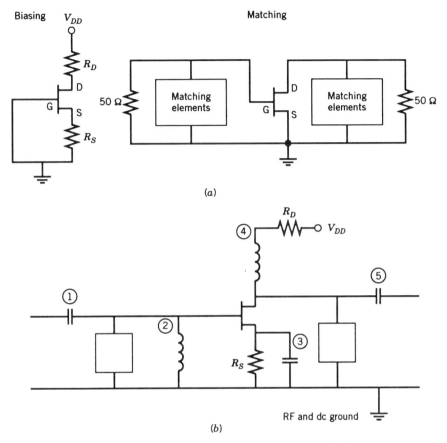

Figure 8.19 Isolating the biasing and matching circuits.

matching elements, and the drain must be connected to the output microstrip line, which contains the output matching elements. Consequently, each electrode must be simultaneously connected to an RF circuit and a dc circuit, and the two circuits must not interfere. For example, the drain electrode is connected to the output microstrip line, which is at dc ground potential, and to the positive power supply through the drain resistor. The microwave signal must not leak from the output line through the drain resistor, and the output microstrip line must not short out the power supply.

The required isolation between the biasing and the matching circuits is accomplished with RF chokes (which pass dc and block the RF) and coupling capacitors (which pass RF and block dc). Figure 8.19b shows the FET with its biasing and matching circuits and the appropriate chokes and capacitors to provide the isolation. Input coupling capacitor 1 allows the microwave signal to enter the transistor gate but prevents the input microstrip line, which is at dc ground, from shorting out the gate bias voltage. The gate must be connected to dc ground, but RF must not leak through this ground connection. Therefore, this connection is made through RF choke 2. The source is connected to RF ground through coupling capacitor 3, which allows the source to be at RF ground but allows the biasing source resistor to be used between the source and dc ground.

The drain is connected to the drain resistor through RF choke 4, which is connected to the positive supply voltage. The choke prevents the microwave signal from being shorted out by the drain resistor and the power supply. Coupling capacitor 5 allows the microwave signal to pass into the output microstrip line, but prevents the output microstrip line, which is at dc ground, from shorting out the drain voltage.

An example of combining the biasing and matching circuitry is shown in Figure 8.20. The source electrode is connected to the ground plane, which serves as the RF ground and the dc ground, through a bypass capacitor. The source electrode thus is dc-insulated from ground, but the bypass capacitor allows the microwave signal to go directly to ground so that the source is at RF ground. A bypass capacitor is not required on the input since the dc voltage on the gate is at ground. The gate dc connection is made through an RF choke, which is actually a small coil of wire. The source resistor is shown

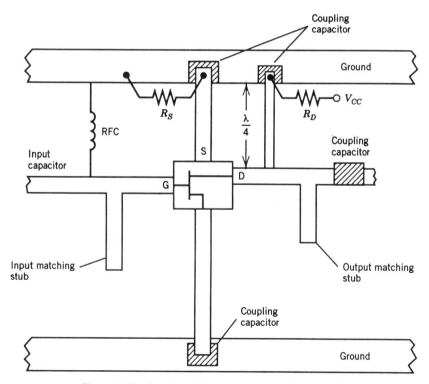

Figure 8.20 Combined biasing and matching circuits.

between the dc ground and the source. The drain must be connected to the supply voltage through an RF choke to the drain resistor, which is in turn connected to the power supply. A small RF choke coil could be used to connect the drain electrode to the drain resistor, or a quarter-wave section of transmission line could be used. The transmission line is shorted at its output end through a bypass capacitor to the microwave ground. At its other end, where it connects to the drain electrode, it presents an open circuit (a quarter-wave section of transmission line, if grounded at one end, is open at the other end). This technique is easier to fabricate but it operates over only a narrow frequency range. The drain resistor is connected to the RF grounded end of the quarter-wave line, which, although at RF ground, is insulated from ground by the bypass capacitor. The choke line is designed to have a high impedance by making its width narrow to further reduce the possibility of eakage of the microwave signal down the line. Finally, a bypass capacitor is added in the output microstrip line to allow an output microwave signal to pass through but to provide dc isolation between the dc grounded output line and the drain resistor.

8.9 IMPATT AMPLIFIERS

As shown in Figure 8.6 IMPATT amplifiers must be used at the high end of the microwave band because microwave transistors do not work well above 30 GHz due to transit time limitations. The IMPATT actually uses transit time effects to generate microwaves instead of trying to fight these effects, as is done with bipolar and field-effect transistors. Figure 8.21 shows a single-stage IMPATT amplifier, consisting of a circulator connected to an IMPATT diode which is mounted in a resonant circuit or cavity. Since the IMPATT is a single-port device, a circulator must be used to separate the input and output powers. The microwave power to be amplified enters one arm of the circulator and is routed into the IMPATT diode and cavity. The incoming

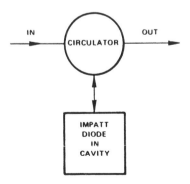

Figure 8.21 IMPATT amplifier.

microwave signal is amplified, as will be explained shortly, and leaves the cavity from the same port that it entered. It is then routed by the circulator into the output transmission line.

IMPATT diode operation is shown in Figure 8.22. Figure 8.22a shows the fabrication details of the IMPATT. It consists of a PN junction between the P^+ and the N regions, a drift region of I material, and an N^+ connection. The I region thickness is controlled so that the electron transit time through it will be half a microwave cycle at the selected operational frequency. The electron velocity in both silicon and gallium arsenide at the high electric fields at which the IMPATT operates is 10^7 centimeters per second (10^{11} microns per second). For operation at 10 GHz, the I region should be 5 microns thick; it will take an electron half of an RF cycle at 10 GHz to move through the 5 micron wide I region. The PN junction width is made as small as possible (about 1 micron). When a reverse bias of 85 V is applied across the IMPATT, the electric field gradient looks like Figure 8.22b. About 35 V of the 85 V is across the PN junction, and since the junction region is 1 micron wide the voltage gradient is 35 volts per micron (350 kV/cm), which is the avalanche breakdown voltage of a reverse biased PN junction. The remaining 50 V appears across the 5 micron I region, so the voltage gradient there is 100 kV/cm, below the avalanche voltage.

The interaction process is shown in sketches of Figure 8.22c, which show the electric field profile across the IMPATT at four times during an RF cycle. The voltage across the IMPATT and the current through it are shown in Figure 8.22d. When the microwave voltage across the IMPATT is zero, the electric field across the PN junction is slightly less than 350 kV/cm; hence, the junction does not avalanche. This voltage gradient is adjusted by the bias, and if a microwave signal is not added to the diode, the PN junction is below the avalanche voltage and the diode draws no current. When the microwave signal to be amplified is added, it raises the entire voltage gradient curve (point 2), and the voltage across the PN junction, now equal to the bias voltage plus the microwave input voltage, exceeds the critical value of 350 kV/cm. The junction therefore breaks down, and an avalanche of holes and electrons are formed.

The holes pass immediately into the P^+ region and appear as current at the output terminals of the IMPATT. The hole current is in phase with the applied voltage, so the hole current makes the IMPATT behave simply as a passive resistor. The electrons formed during the avalanche do not appear immediately at the output terminals of the IMPATT because they must travel through the I region to get to the N^+ output terminal. At time 3, when the microwave voltage has again returned to zero, these electrons have traveled halfway through the i region. At time 4, when the microwave voltage across the IMPATT has reached its maximum in the negative direction, the electrons reach the N^+ region and appear at the output terminal.

This electron current is shown by the dashed curve of Figure 8.22d and it exits the IMPATT 180° out of phase with the applied voltage. This condition

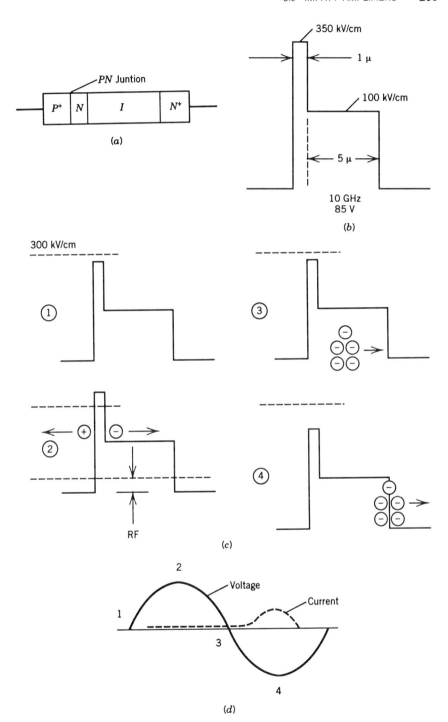

Figure 8.22 IMPATT operating principles.

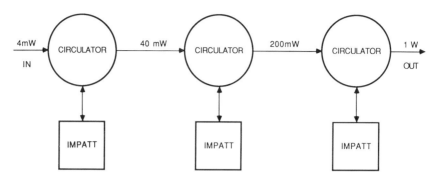

Figure 8.23 Multistage IMPATT amplifier.

is exactly what is required for amplification, and the IMPATT looks like a negative resistance. In this condition and when put into a resonant circuit, the IMPATT will generate microwave power.

An IMPATT diode has two possible modes of operation as an amplifier. The mode just described is the negative-resistance mode, where the input signal entering the IMPATT through the circulator is actually amplified due to the negative-resistance phenomena in the IMPATT. The amplified signal passes out of the diode through the same port at which the input signal entered and, because of the circulator, passes into the output line. The second operational mode is called injection-locked. The IMPATT is biased so that it is oscillating all the time, but the frequency is locked to the input frequency so that the power leaving the IMPATT is at the same frequency as the input. The negative-resistance mode provides the optimal bandwidth, which can be as high as 20%. In contrast, the bandwidth of an injection-locked amplifier is only a few percent. However, the injection-locked mode provides two or three times greater efficiency that the negative-resistance mode.

A multistage IMPATT amplifier is shown in Figure 8.23. Microwave power at the 4-mW level enters the input of the amplifier and passes through the first circulator into the first negative-resistance amplifier stage. The signal is amplified by 10 dB in the first stage, and the 40-mW signal passes through the second circulator into an IMPATT amplifier operating in the negative-resistance mode. The signal is amplified by 7 dB and goes into the output stage, which is operated in the injection-locked mode to obtain high efficiency. The last stage also has 7-dB gain, so 1 W of output power is obtained. This amplifier therefore has 24-dB gain.

ANNOTATED BIBLIOGRAPHY

1. K. Chang, *Microwave and Optical Components*, Vol II, Wiley, New York, 1989.
2. J. V. DiLorenzo and D. D. Khandelwal, *GaAs FET Principles and Technology*, Artech House, Dedham, MA, 1982.
3. R. A. Soares, *GaAs MESFET Circuit Design*, Artech House, Dedham, MA, 1988.

4. F. Ali, A. Gupta, and I. Bahl, *HEMTS and HBTs: Devices, Fabrication and Circuits,* Artech House, Dedham, MA, 1991.

5. G. D. Vendelin, *Design of Amplifiers and Oscillator by the S Parameter Method,* Wiley, New York, 1981, pp. 40–131.

References 1 through 4 represent complete design and fabrication details on bipolar transistors, FETs, HEMTs, and HBTs. Reference 5 covers designing microwave amplifiers using *S*-parameters.

EXERCISES

8.1. Figure 8.24 shows the microwave output power as a function of microwave input power for a bipolar transistor amplifier at 2 GHz. The dc input power is 100 mA at 10 V. Determine the following:

a. The output power at an input power of -30 dBm
b. Gain at an input power of -30 dBm
c. Efficiency at an input power of -30 dBm
d. Output power at saturation
e. Gain at saturation
f. Efficiency at saturation
g. Output power at 1-dB compression
h. Gain at 1-dB compression
i. Efficiency at 1-dB compression

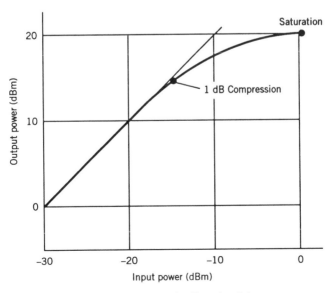

Figure 8.24 Data for Exercise 8.1.

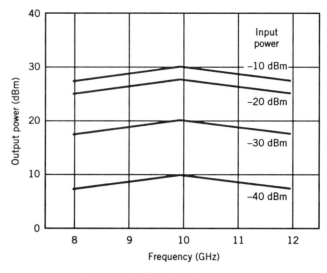

Figure 8.25   Data for Exercise 8.2.

8.2. Figure 8.25 shows the swept output power of a FET amplifier at several input power levels. Using this data, plot transfer curves for this amplifier at 10 GHz and 12 GHz.

8.3. Figure 8.26 shows the phase of the output power from a bipolar transistor amplifier (relative to the input power) as a function of input power level.

a. What is the phase change when the input power level is changed from − 10 to 0 dBm?

b. What is the AM-PM coefficient at − 5 dBm?

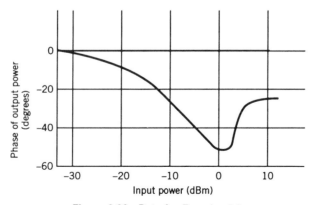

Figure 8.26   Data for Exercise 8.3.

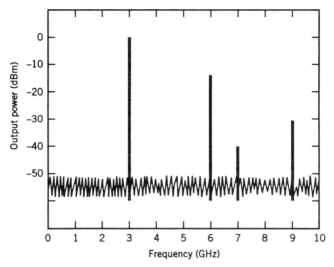

Figure 8.27   Data for Exercise 8.4.

8.4. Figure 8.27 shows the output of a bipolar transistor amplifier with an input signal at 3 GHz, as measured on a spectrum analyzer. What are the power levels and the frequencies of the following components in the output.

Output Component	Frequency (GHz)	Power (dBm)
Fundamental		
Second harmonic		
Third harmonic		
Spurious		

8.5. Figure 8.28 shows the fundamental ouput power and the second harmonic output power of a FET amplifier as a function of the fundamental input power at 8 GHz. The dc input power is 100 mA at 10 V. Determine the following:

a. The harmonic frequency
b. Fundamental power at saturation
c. Second harmonic power at fundamental saturation
d. Separation between fundamental and second harmonics at fundamental saturation
e. Second harmonic power at second harmonic saturation

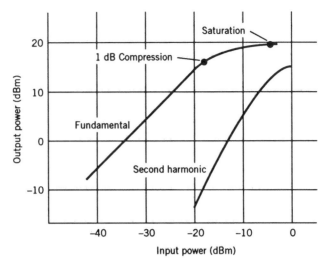

Figure 8.28   Data for Exercise 8.5.

f. Fundamental power at 1-dB compression
g. Second harmonic at 1-dB compression of the fundamental
h. Fundamental to second harmonic separation at 1-dB compression
i. Fundamental efficiency at 1-dB compression
j. Linear gain of the fundamental

8.6. Figure 8.29 shows the intermodulation products of a FET amplifier operated with input signals at 3.0 and 3.2 GHz as a function of the total power of the two input signals:

a. What is the fundamental power at saturation
b. What is the separation between the fundamental and the third-order intermod at fundamental saturation?
c. What is the fundamental power at 1-dB compression?
d. What is the separation between the fundamental and the third-order intermod at 1-dB compression of the fundamental?

8.7. For the intermodulation data of Figure 8.29:

a. What are the frequencies of the two third-order intermodulation products?
b. What are the frequencies of the two fifth-order intermodulation products?

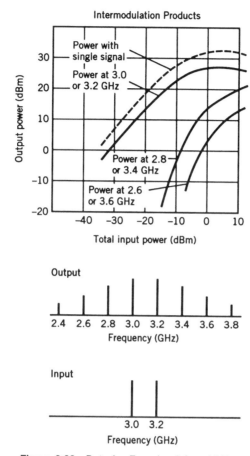

Figure 8.29   Data for Exercise 8.6 and 8.7.

8.8. What would be the best microwave semiconductor device for each of the following amplifiers?

a. A 10-W amplifier at 1 GHz
b. A 0.5-W amplifier at 20 GHz
c. A 0.5-W amplifier at 100 Ghz

8.9. What *S*-parameter approximately represents the following:

a. Input reflection coefficient
b. Transistor gain
c. Isolation
d. Output reflection coefficient

S-parameters for a bipolar transistor amplifier, taken from the manufacturer's data sheet, are shown below. Use them for Exercise 8.10 through 8.12.

Freq	11		21		12		22	
500.00	.571	− 131.4	12.403	107.9	.032	43.7	.629	− 38.2
1000.00	.592	− 168.4	7.034	80.8	.040	35.0	.494	− 47.9
2000.00	.600	156.4	3.637	46.8	.057	27.8	.459	− 68.2
3000.00	.598	131.6	2.466	17.4	.075	17.7	.476	− 92.6
4000.00	.598	109.0	1.864	− 10.4	.095	4.8	.502	− 114.8
5000.00	.587	88.1	1.510	− 35.4	.117	− 9.6	.540	− 140.9
6000.00	.560	67.9	1.254	− 60.6	.135	− 24.5	.607	− 165.1
7000.00	.488	44.2	1.063	− 86.1	.156	− 42.5	.690	174.4
8000.00	.407	10.6	.932	− 111.6	.181	− 61.1	.749	160.6

8.10. Plot S_{11} and S_{22} on a Smith chart at the following frequencies.

a. 500 MHz   b. 1 GHz   c. 2 GHz   d. 4 GHz

8.11. Calculate the following gains at 1 GHz using S-parameters.

a. Unmatched transistor gain
b. Unilateral gain with matched transistor
c. Maximum stable gain

8.12. Calculate the following gains at 4 GHz using S-parameters.

a. Unmatched transistor gain
b. Unilateral gain with matched transistor
c. Maximum stable gain

9

MICROWAVE OSCILLATORS

The purpose of a microwave oscillator is to generate a microwave signal. This chapter begins with a discussion of how an oscillator works. An oscillator consists of two parts: a resonator to control the frequency of the microwave signal, and an active device to generate the power. The various types of resonators are compared. The active devices and how they can be made to oscilate using feedback or negative resistance are discussed.

Oscillator performance requirements is then described. The most important performance requirements are related to frequency—including frequency stability, frequency tuning, and phase noise. The various types of oscillators are discussed, including fixed and mechanically tuned oscillators, IMPATT and Gunn oscillators, electronically tuned oscillators such as YTOs and VCOs, harmonic multipliers, upconverters, and phase-locked oscillators.

9.1 OSCILLATOR PRINCIPLES

A microwave oscillator consists of two parts:

1. A resonator to control the frequency, and
2. An active device to generate the microwave power.

A typical microwave resonator is shown in Figure 9.1a. The resonator is a half-wavelength long section of waveguide connected by small coupling holes to the input and output transmission lines.

A graph of the power transmitted through the resonator as a function of frequency is shown in Figure 9.1b. When the microwave signal frequency

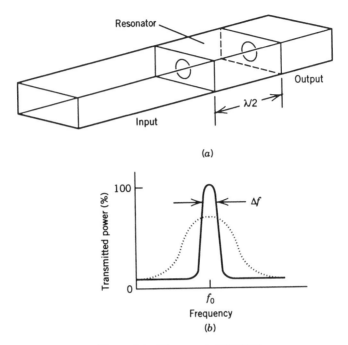

Figure 9.1 Microwave resonator.

is above or below the cavity resonant frequency, f_0, all of the power is reflected from the cavity. Power is transmitted through the cavity only in a narrow frequency range around f_0. The most power is transmitted at f_0, when a half wavelength of the microwave field exactly fits into the cavity.

The frequency f_0 of microwave resonators, can be fixed, or mechanically changed by altering the cavity dimensions, or electronically changed.

The quality of the resonator is the "sharpness" of its frequency response, as defined by its Q, where

$$Q = \frac{f_0}{\Delta f}$$

where the resonator bandwidth Δf is the range between the frequencies when the response has changed from its resonant value by 3 dB. Therefore, these frequencies are called the half-power frequencies. The solid curve shows the response of a high-Q resonator, which has a sharp resonance. The dotted curve shows the response of a low-Q resonator.

Microwave resonators that can be used for oscillators are shown in Figure 9.2. A popular resonator used with bipolar and FET oscillators is the lumped inductor and capacitor (A), which can be used up to about 20 GHz. At these very high frequencies, the inductor consists of short lengths of bond wires that electrically connect the transistor chip to the circuit.

A. LUMPED INDUCTOR AND CAPACITOR

D. MICROSTRIP

B. RECTANGULAR METAL CAVITY

E. DIELECTRIC RESONATOR

C. COAXIAL METAL CAVITY

F. SAW

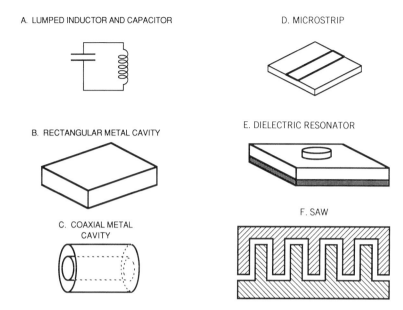

Figure 9.2 Resonator types.

The rectangular metal cavity (B) provides the highest Q of any resonator, and therefore provides the most stable frequency. However, it is large, especially at the low end of the microwave frequency band, and not compatible with microstrip mounting of the active devices. The metal coaxial cavity (C) has almost as good a Q as the metal rectangular cavity, but is much smaller since a coaxial transmission line does not need a half wavelength width as the rectangular cavity does. The coaxial cavity usually has a gap between its center conductor and one end of the cavity, which can be mechanically adjusted for tuning the cavity. The active device is located in the gap, where the electric field of the microwave signal is high.

The microstrip resonator (D) is easy to fabricate, but is has a low Q, and oscillators using it have poor frequency stability.

The dielectric resonator (E) has almost as good a Q as the metal rectangular or coaxial cavities, and its size is compatible with microstrip dimensions in the microwave band above about 5 GHz.

The surface acoustic wave (SAW) resonator (F) provides a small resonator at the low end of the microwave band below 1 GHz, with a much higher Q than could be obtained with stripline resonators. It uses lithium niobate, a piezoelectric material. The microwave signal sets up mechanical vibrations in the piezoelectric material at the microwave frequency, and the resonance is controlled by mechanical vibrations of the material.

The resonators used with an oscillator must often be mechanically or electronically tuned to change the oscillator frequency. Figure 9.3a shows

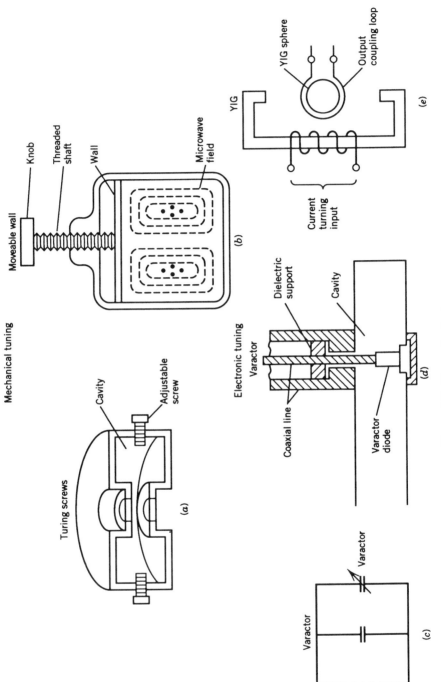

Figure 9.3 Resonator tuning.

216

screws extending into the cavity to slightly change its dimensions, to change the oscillator frequency a few percent. Figure 9.3*b* shows a moving wall, which can make a large change in the cavity dimensions and, consequently, in the oscillator frequency. It is also possible to mechanically change the frequency of a dielectric or microstrip resonator by mechanically moving the ground plane above the resonator.

Figure 9.3*c* and 9.3*d* show electronic tuning of a lumped inductance and capacitance circuit or a cavity using a varactor diode. The varactor diode is an electronically tuned capacitance; changing the diode voltage changes its capacitance, and consequently the frequency of the resonator. Figure 9.3*e* shows a YIG resonator, whose frequency can be changed over a wide range by adjusting the external magnetic field applied to the ferrite YIG material. Both types of tuning are discussed in more detail later.

The second part of the oscillator is the active device. Microwave semiconductors used as the active oscillator devices are shown in Figure 9.4 which compares their power capability as a function of frequency. The bipolar transistor is normally used at the low end of the microwave band, the FET in the center of the band, and the HEMT, IMPATT, and Gunn devices at the high end of the microwave band. Also shown is the varactor multiplier, which multiplies the frequency of a low frequency transistor oscillator. The varactor multiplier is discussed later.

The active device must be put into a condition to oscillate. As shown in Figure 9.5 external feedback or negative resistance is used to accomplish this. In a feedback-type oscillator, shown in Figure 9.5*a*, the active device is connected as an amplifier and a sample of the output power is fed back through the resonator to the input. Thermal noise at the amplifier input (which always exists in any microwave device) starts the oscillation. The input noise is amplified by the amplifier and fed back through the resonator to the input to be reamplified. Only amplified noise at the frequency that can pass through the resonator is amplified. Only a fraction of the output power is required to be fed back to the input to maintain the oscillation, and most of the power from the active device is delivered to the load.

A negative-resistance oscillator can take the two forms shown in Figure 9.5*b,c*. Figure 9.5*b* is a one-port active device, such as an IMPATT or Gunn. By properly biasing, these devices can be put into a negative-resistance condition, in which as current increases, voltage decreases. The oscillation requirement of a negative-resistance device at some frequency is that the device impedance be equal and opposite to the load impedance ($-R_D = R_L$). This impedance matching is achieved by the matching elements placed between the device output and the load. The condition $R_L = -R_D$ means that the device can provide power to the load. The condition $X_D = X_L$ determines the oscillation frequency.

Figure 9.5*c* shows a transistor as a negative-resistance oscillator. The transistor is matched to the load, and a feedback means is provided either externally or internally in the transistor to cause it to oscillate. The transistor

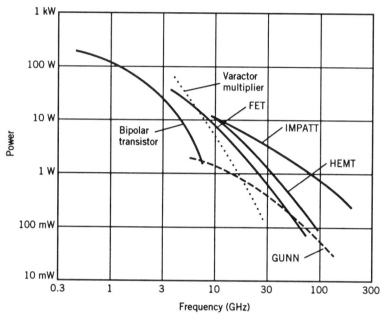

Figure 9.4   Microwave semiconductor devices used for oscillators.

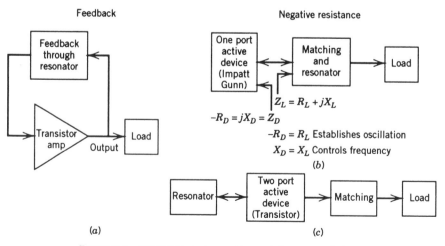

Figure 9.5   Feedback and negative resistance oscillators.

input impedance is then matched to the resonator impedance. At the frequency at which the reactances are equal, the device will oscillate.

9.2 OSCILLATOR PERFORMANCE REQUIREMENTS

Oscillator performance requirements are

1. Frequency
2. Frequency stability
   a. Frequency pushing (frequency change with power supply voltage or current)
   b. Frequency pulling (frequency change with load mismatch)
   c. Temperature
   d. Shock and vibration
3. Tuning
   a. Mechanical
   b. Electronic
      i. Tuning sensitivity (MHz/V or MHz/mA)
      ii. Linearity
      iii. Settling time
      iv. Post-tuning drift
4. Phase noise
5. Power and efficiency

The most important oscillator requirement is frequency, since the microwave oscillator is generating the microwave signal and controlling the frequency. The frequency must be stable. Frequency stability is defined by frequency pushing, frequency pulling, and variation of the oscillator frequency with temperature, shock and vibration. *Frequency pushing* is defined as the frequency change per volt of voltage change or milliamp of current change. *Frequency pulling* is the frequency change with the load mismatch—as the mismatch of the load is changed, the frequency of the oscillator may change.

As temperature is changed, the dimensions of the resonator often change, and thus the oscillator frequency changes. The specification on frequency stability with temperature is the change in frequency per degree of temperature change.

As the oscillator is shocked or vibrated, the dimensions of its resonator may change, which also causes frequency changes.

The oscillator may be tuned mechanically or electronically. Mechanical tuning is done by changing some dimension of the resonator. Requirements for electronic tuning include tuning sensitivity, linearity, settling time, and posttuning drift. The tuning sensitivity is specified as the frequency change in megahertz per volt (or megahertz per milliamp) as the voltage or current of the tuning control is varied.

Tuning linearity refers to the constancy of the tuning sensitivity across the tuning range. If the tuning voltage is changed by the same amount, the frequency change should be the same at any frequency across the tuning range of the oscillator. Some electronically tuned oscillators, particularly those with varactor tuning, are more sensitive at one end of their tuning range than the other.

Settling time and posttuning drift are defined in Figure 9.6. If the frequency of an oscillator is to be changed by changing the frequency control signal, the oscillator should respond immediately. Settling time and posttuning drift measure the time it takes for the oscillator to respond to the frequency control signal. Figure 9.6a shows the frequency control signal as a function of time. At time 0, the tuning voltage is changed and the oscillator is supposed to change frequency. The actual response of the oscillator as a function of time is shown in Figure 9.6b. When the frequency control signal is changed, the oscillator takes time to reach its final frequency. There may even be an overshoot where the oscillator frequency reaches the desired frequency,

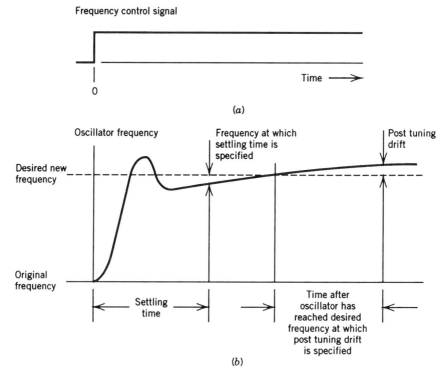

Figure 9.6 Settling time and post tuning drift.

goes above this value, below this value, and finally reaches the desired frequency. The settling time (usually microseconds) is the time that it takes the oscillator to come within a specified percentage of the desired frequency.

After the oscillator reaches its desired frequency, the frequency may drift as the oscillator heats up. The amount of drift is called posttuning drift.

One of the most important characteristics of oscillators is their phase noise. Figure 9.7 shows the amplitude of the oscillator power as a function of frequency over a small frequency range around the carrier. A graph of power from a perfect oscillator as a function of frequency should be a single line, meaning that the oscillator is generating its power at a single frequency. However, due to noise on the power supply voltage, the oscillator frequency is actually being frequency-modulated over a very small range around the main oscillator frequency. In Figure 9.7 the frequency range is only a few kilohertz away from the carrier, which is at a microwave frequency of, for example, 10 GHz. Consequently, phase noise frequency deviations are a few kilohertz out of 10 GHz, or about 1 part in a million. The oscillator power is usually at the carrier frequency, occasionally at other frequencies. Consequently, the oscillator output can be considered as a frequency spectrum. Phase noise is defined as the power in a 1-Hz bandwidth at a frequency f_m from the carrier. The phase noise is specified at a frequency away from the main oscillation frequency, called the carrier, and measured in dB below the carrier power. The phase noise of microwave oscillators is -60 to -120 dBc, which is 60 to 120 dB below the carrier.

It is not usually possible to measure the phase noise in only a 1-Hz band. Phase noise is usually measured over a larger bandwidth, such as 100 Hz or 1 kHz, and the measured values are converted to the power that would exist in a 1-Hz bandwidth.

The small oscillator frequency variation of 1 part in a million may not seem very important. However, this frequency variation is in the same range as the information frequency that is modulated onto the microwave carrier in a

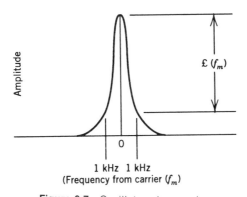

Figure 9.7   Oscillator phase noise.

communication or radar system. The importance of phase noise is shown in Figure 9.8*a* which shows the phase noise problem in a Doppler radar. The radar is tracking an airplane flying over a mountain. The airplane has a cross section of 100 square meters, which is the effective area of the airplane that reflects the microwave signal back to the radar. In contrast, the mountain, if it is 1 mile wide by $\frac{1}{3}$ mile high, has a 1 million square meter cross section. The reflecting area of the mountain is ten thousand times, or 40 dB, greater than that of the airplane. The radar signal reflected from the mountain is 40 dB bigger than the signal from the airplane, so the airplane would not normally be seen because its radar return is masked by the mountain. However, because the airplane is moving, the microwave frequency of the reflected signal is shifted by the motion of the airplane. Therefore, the echo from the airplane is at a different frequency and can be distinguished, although small, from the large echo off the mountain. However, if the oscillator in the radar transmitter has a large phase noise component, as shown in Figure 9.8*b*, which is only 40 dB below the main signal, the noise sideband reflected from the mountain will be as large as the main signal reflected from the airplane, and both appear to be coming from moving targets because they are shifted away from the carrier frequency. The airplane signal is shifted in frequency because the airplane is moving. The mountain signal is shifted away from the main frequency because it was shifted before transmission, it being a noise sideband of the transmitted signal. As shown in Figure 9.8*c*, the reflected noise sideband from the mountain and the reflected main signal

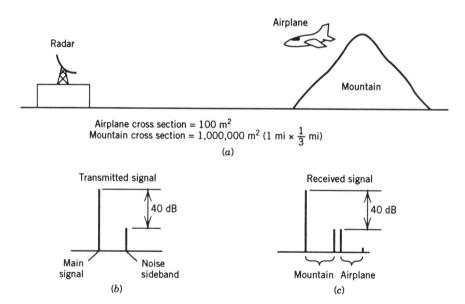

Figure 9.8 Significance of oscillator phase noise.

from the airplane appear to have the same frequency and are of the same amplitude, so the noise sideband causes a false signal. The solution to this problem is to reduce the noise sideband level by choosing an oscillator for this radar transmitter that has low phase noise.

9.3 FIXED-TUNED AND MECHANICALLY TUNED OSCILLATORS

Fixed-tuned oscillators consist of a resonator and an active device. The oscillation frequency is set by the resonator. Changing the resonator dimensions changes the oscillation frequency. Active devices used in fixed- or mechanically tuned oscillators include bipolar and field-effect transistors up to about 30 GHz, and IMPATT and Gunn devices from 8 to several hundred GHz.

Transistor oscillators using feedback and lumped inductance and capacitance resonators are shown in Figure 9.9. These oscillators are like their low frequency counterparts, except that very small values of inductance and capacitance must be used to make them resonate at microwave frequencies.

The basic transistor oscillator circuit is shown on Figure 9.9a. A common base configuration is often used for oscillators, although common emitter configurations can also be used. The transistor output is matched to the load, and the transistor input is matched. Feedback is supplied from the output to the input, and the feedback circuit contains the resonator, which controls the oscillation frequency.

The three types of transistor LC oscillators are the Colpitts, Hartley and Clapp shown in Figure 9.9b. These oscillators differ only in the way that the feedback is applied. In the Colpitts oscillator the feedback is supplied by the capacitive divider formed by C_1 and C_2. In the Hartley oscillator, the feedback is supplied by the inductive divider formed by L_1 and L_2. The Clapp oscillator is similar to the Colpitts, but uses an additional capacitor that allows the oscillator to be tuned in frequency without changing the amount of feedback.

In all three oscillators, the frequency is determined by the inductance and capacitance, from the standard LC resonance formula

$$f_{osc} = \frac{1}{2\pi\sqrt{LC}}$$

Transistor oscillators using dielectric resonators are shown in Figure 9.10. Dielectric resonators are most useful in the middle of the microwave band, where FETs would be the appropriate active device. In Figure 9.10a the dielectric resonator is shown in the feedback path, and it controls the frequency of oscillation. In Figure 9.10b the transistor is analyzed as a negative-resistance oscillator. The dielectric resonator is placed in the output, and the impedance of the load and resonator combination must be made equal to the transistor impedance.

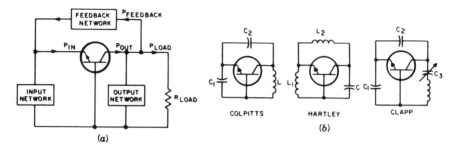

Figure 9.9 Transistor feedback oscillators with LC resonators.

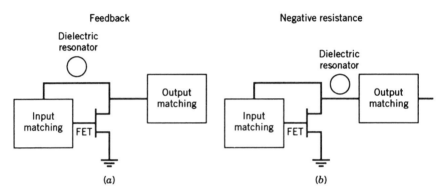

Figure 9.10 Transistor oscillators with dielectric resonators.

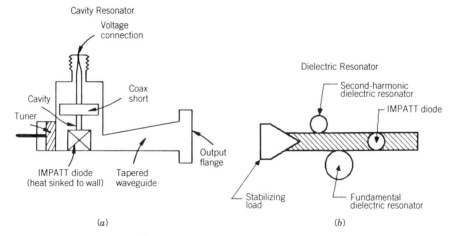

Figure 9.11 IMPATT oscillators.

IMPATT oscillators are shown in Figure 9.11. IMPATTs are used above about 30 GHz, since transistors do not provide adequate power above this frequency. In Figure 9.11a the IMPATT is shown in a metal cavity resonator. Mechanical tuning is provided by moving one wall of the cavity. The cavity is matched by the tapered waveguide to a standard external waveguide. The IMPATT is mounted in the cavity, and a voltage connection with an RF choke is provided. When the voltage is adjusted so that the voltage gradient across the *PN* junction of the IMPATT exceeds 350 kV/cm, the IMPATT junction avalanches and the IMPATT has negative resistance. Oscillations occur at the frequency of the mechanically tuned metal cavity.

A fixed-tuned IMPATT with dielectric resonators in a microstrip configuration is shown in Figure 9.11b. The resonant frequency is determined by the larger dielectric resonator, which is coupled to the stripline in which the IMPATT is mounted. The smaller dielectric resonator is at the harmonic frequency and controls the harmonic power to reduce its magnitude.

The transferred electron (Gunn) oscillator is an extremely simple oscillator. It consists of a single piece of gallium arsenide and contains no junctions. Microwaves are generated from the unique "electron transfer" effect of gallium arsenide. This effect is illustrated in Figure 9.12. The electron velocity in gallium arsenide and in silicon is plotted as a function of the electric field applied across the material. At field strengths above 15 kV/cm (1.5 V/μ), the electron velocity is the same in each material. At very low fields, however, the velocity is much greater in GaAs than in silicon. At zero electric field the electrons have no velocity. As the electric field is increased, which means that the voltage across the material is increased, the electrons travel faster and faster in GaAs. When the electric field reaches 5 kV/cm, the electrons have gained enough energy to be "transferred" from one energy-momentum state in the GaAs lattice to another. In the second energy-momentum state, they have the same velocity as in Silicon. This is where the name *transferred electron device* comes from. From 5 to 10 kV/cm the electron velocity is

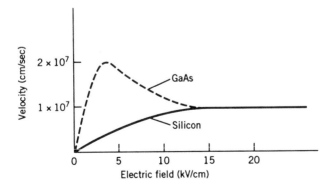

Figure 9.12 Transferred electron device (Gunn device) operation.

decreasing from a high value in one state to a low value in the second state; hence, as the voltage gradient increases, the current decreases. Alternatively, as the voltage gradient decreases, the current increases. This means that as the voltage goes up the current goes down, the negative-resistance condition. Simply by placing a piece of gallium arsenide material into a resonator and applying the correct voltage to it, so that the electric field is in the 7-kV/cm range, an oscillator is made. Its frequency is controlled by the resonator.

9.4   ELECTRONICALLY TUNED OSCILLATORS

Electronically tuned oscillators require a means of electronically changing the oscillator's resonator frequency, with either a varactor diode or a YIG.

The characteristics of a varactor diode are shown in Figure 9.13. A varactor diode is a reverse-biased *PN* junction and serves as an electronically variable capacitance. Schematics of the varactor are shown in Figure 9.13a at 0-V bias and at 20-V reverse bias. With 0-V bias, the electrons from the *N* material fill the holes in the *P* material, and a depletion layer is formed, partly in the *P* region and partly in the *N* region. The depletion layer is void of current carriers, since the free electrons in the *N* region have filled the holes in the *P* region. Consequently, no current flows through the *PN* junction, which consists of two conducting regions (the remaining *N* and *P* materials) separated by an insulator. Thus, the *PN* junction appears as a capacitance.

As the reverse bias is increased (a greater negative voltage is applied to the *P* material), the depletion layer increases, extending further into the *N* and *P* regions. Hence, the spacing between the capacitor plates, which are the *N* and *P* materials, increases and the capacitance decreases. Therefore, by varying the reverse-bias voltage, the capacitance of the *PN* junction can be controlled. The Figure 9.13b shows the capacitance in picofarads as a

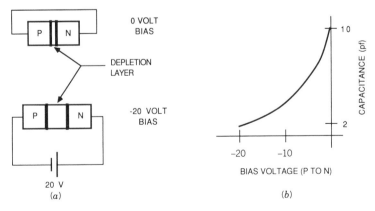

Figure 9.13   Varactor operation.

function of bias voltage. The capacitance changes over a 5:1 range as the tuning voltage is changed from 0 to −20 V.

Two examples of varactor-tuned oscillators are shown in Figure 9.14. They are commonly called voltage-controlled oscillators (VCOs). Figure 9.14a is a common base bipolar Colpitts oscillator. Other circuits, like the Hartley or Clapp, with either bipolar or field-effect transistors could be used. The transistor is matched to the load. One capacitor of the divider circuit is the varactor, whose voltage can be changed by applying a tuning voltage through an RF choke and a bypass conductor.

A varactor-tuned Gunn oscillator is shown in Figure 9.14b. The Gunn device and varactor are mounted in a metal cavity resonator. The cavity frequency can be changed by varying the varactor capacitance. However, a frequency change of only a few percent can be obtained because the microwave field is distributed in the cavity, and a capacitor in one part of the cavity has only a small effect on the cavity frequency.

The typical capacitance variation of a varactor diode is shown as a function of the reverse-bias voltage in Figure 9.14c. Note that the capacitance variation with voltage is nonlinear, with the capacitance changing more at low values of bias voltage than at high values. This leads to a nonlinear frequency versus voltage tuning curve for the oscillator. If linearity is required, a compensating network must be used to modify the tuning voltage before it is applied to the varactor.

YIG-tuned oscillators (YTOs) are shown in Figure 9.15. A transistor oscillator is shown in Figure 9.15a. A common gate FET is matched to the load

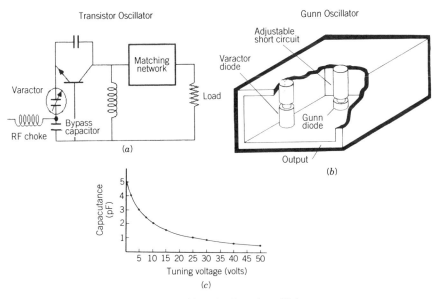

Figure 9.14 Varactor-tuned oscillators.

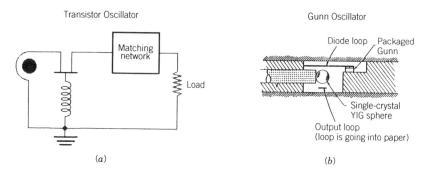

Figure 9.15 YIG-tuned oscillators.

and biased to the negative-resistance state. It then oscillates at the resonant frequency of the YIG.

A YIG-tuned Gunn oscillator is shown in Figure 9.15b. One coupling loop from the YIG is connected to the Gunn device. The other coupling loop connected to the output transmission line. The Gunn device is biased into its negative-resistance mode, and the oscillation frequency is controlled by the resonant frequency of the YIG, which is electronically controlled by varying the applied magnetic field. Since the YIG can be tuned over a wide frequency range, a broadband electronically tuned oscillator can be obtained this way.

A comparison of varactor and YIG-tuned oscillators is given in Table 9.1. The varactor-tuned oscillator can be rapidly tuned, since the varactor capacitance can be changed in microseconds. However, the varactor can be tuned only over a 2:1 frequency range when used in a lumped LC circuit, or a few percent when used in a resonant cavity. Another disadvantage is the varactor's nonlinear tuning with control voltage.

YIG tuning is linear with control current and has a wide range, over 1–20 GHz. However, the tuning rate is milliseconds. This slow tuning rate occurs because it is difficult to rapidly change the current through electromagnet coils, due to their inductance.

Table 9.1 Comparison of varactor and YIG-tuned oscillators

Tuning	Advantages	Disadvantages
Varactor	Fast tuning (μs)	Limited tuning range (2:1 in lumped circuits; 10% in cavities)
YIG	Wide tuning range (1–20 GHz) Linear tuning with control current	Slow tuning (ms)

9.5 HARMONIC MULTIPLIERS

A harmonic multiplier which uses a step-recovery varactor is shown in Figure 9.16. The output of this frequency multiplier is 1 mW at 9 GHz.

The frequency multiplication process begins with a 1-GHz transistor oscillator, which provides 10 mW of power. This power is at the wrong frequency and is more than required, but, because of the low frequency, it is easy to obtain. Waveforms and the corresponding frequency spectra at each stage of the multiplier are shown. The output of the transistor oscillator is a single frequency at 1 GHz, as shown by its waveform or the single line of its frequency spectrum.

The single-frequency microwave signal at 1 GHz is then applied to the step-recovery varactor multiplier. During the positive half of the microwave cycle, current flows from one side of the junction to the other side. As the voltage reverses and current can no longer flow, the charges stored on one side of the junction flow back in a very short time to the other side, so a current pulse is formed, as shown by the second waveform. This current spike generates harmonic frequencies, as shown by the frequency spectrum. The spiked waveform from the step-recovery varactor multiplier is composed of a series of harmonically related frequencies, at the fundamental of 1 GHz and up through the ninth harmonic and above. The ninth harmonic of 1 GHz is 9 GHz, the desired frequency. A filter removes all unwanted harmonics, leaving only the 9 GHz signal.

In many microwave applications, it is easier to get a high frequency signal, like 9 GHz signal in the example, by using harmonic multiplication of a lower

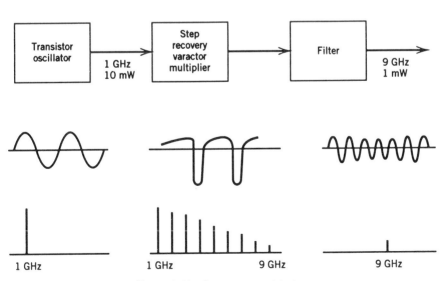

Figure 9.16 Frequency multiplier

frequency signal than it is to generate the desired signal directly with a high frequency oscillator.

The frequency multiplier in Figure 9.16 is also commonly used to obtain a stable harmonic of a quartz crystal controlled oscillator, to be used as a frequency reference in the microwave band. The standard frequency reference in electronic equipment is a quartz crystal oscillator. The oscillator frequency is controlled by the atomic dimensions of the quartz and provides a stable frequency from 1 to 100 MHz, the available resonant frequencies of quartz crystals. These resonant frequencies are well below the microwave band, but a stable microwave reference can be obtained by multiplying the stable frequency from the quartz crystal oscillator into the microwave band with a harmonic multiplier. The resulting harmonic will not have enough power for most microwave equipment but it can be used to "lock" the frequency of a higher power microwave oscillator, as described in the next section.

9.6 PHASE-LOCKED OSCILLATORS

The phase noise characteristics of some microwave oscillators are compared in Figure 9.17. (Recall that phase noise is the noise power in a 1-Hz band relative to the oscillator power at the carrier frequency, at frequencies slightly removed from the carrier.) Figure 9.17 shows the phase noise of 5-GHz microwave oscillators 1–10 kHz from the carrier.

The oscillators formed with bipolar and field-effect transistors and a metal cavity, a YIG resonator, and a dielectric resonator are called *fundamental* oscillators, since they generate microwave power at the resonant frequency of their cavity (5 GHz, here). From a phase noise standpoint, the best oscillator is the bipolar transistor in a metal cavity; the worst is the bipolar transistor in a YIG resonator, because of the noise current in the YIG tuning supply. The bipolar transistor in a dielectric resonator has greater phase noise than the same bipolar transistor in an all-metal cavity, because of the lower Q of the dielectric resonator. The FET in a cavity has higher phase noise than a bipolar in a cavity, because of the noise characteristic of a FET.

The requirements for achieving low phase noise in a fundamental oscillator, are (1) a high-Q resonator and (2) an active device that has low phase noise. The bipolar transistor in a metal cavity resonator best meets these requirements.

The phase noise of a low frequency quartz crystal controlled oscillator at 10 MHz and 100 MHz is also shown in Figure 9.17. Their phase noise is extremely low, but their output is not at microwave frequencies. However, with a harmonic multiplier, their stable frequency can be multiplied into the microwave band to be used as a reference signal, which can then control the microwave oscillator frequency. How this is accomplished will be discussed shortly, but if this is done, the phase noise of microwave oscillators can be

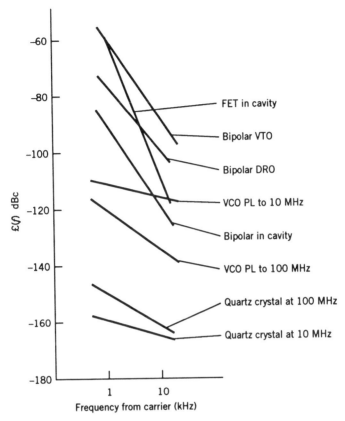

Figure 9.17 Phase noise of 5-GHz oscillators.

greatly reduced. Examples are shown in Figure 9.17: a VCO phase-locked to a harmonic of a 10-MHz quartz crystal oscillator and a VCO phase-locked to the harmonic of a 100-MHz quartz crystal oscillator. The harmonic multiplication increases the phase noise by the square of the multiplication ratio. In Figure 9.17 the microwave frequency is 5 GHz, so multiplying from a 10-MHz quartz crystal oscillator to a 5 GHz is a multiplication ratio of 500 times, or 27 dB. Thus, the phase noise is increased by the square of this, or 54 dB. Multiplying from a quartz crystal reference at 100 MHz up to 5 GHz is a multiplication of 50 times, or 17 dB, so the phase noise increases by 34 dB. Consequently, even though the phase noise of the 100-MHz quartz crystal oscillator is greater than that of the 10-MHz oscillator, the multiplication ratio is less, so lower noise is obtained by phase-locking the VCO at 5 GHz to the 100-MHz reference.

Figure 9.18 shows two methods of locking microwave oscillators to reduce their phase noise. The simplest design shown in Figure 9.18a uses a frequency discriminator and does not require a harmonically generated reference signal.

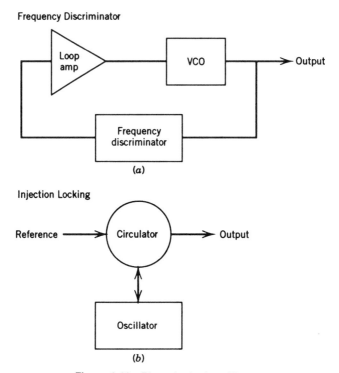

Figure 9.18 Phase locked oscillators.

Microwave power is generated by a voltage controlled oscillator, and a varactor diode electronically controls the oscillation frequency. A sample of the output is taken and passed through a frequency discriminator, which is a resonant circuit operated off resonance so that as frequency increases the discriminator output increases, and as frequency decreases, the discriminator output decreases. Thus, the discriminator provides a voltage signal proportional to the microwave frequency. The output from the frequency discriminator is applied to a low frequency amplifier and filter, and the resulting signal is used to control the frequency of the microwave VCO. If the VCO frequency departs from its desired value, an error signal is generated by the discriminator, which is sent back to readjust the VCO frequency. Although the discriminator method of controlling the microwave frequency and phase noise is the simplest, it is also the least effective.

The second technique for reducing phase noise, is by injection-locking the oscillator to the harmonically generated frequency reference, as shown in Figure 9.18b. A fundamental oscillator generates the microwave power. The reference signal, generated from the harmonic frequency multiplication of the quartz crystal oscillator frequency is applied through a circulator to the oscillator and locks the oscillator onto the reference frequency. The range over which the reference frequency can hold the oscillator frequency in lock depends on the ratio of the locking power to the fundamental oscillator power; the lower this ratio, the lower the locking range, and the greater the phase noise.

The most effective method of controlling the microwave oscillator frequency and reducing its phase noise is phase locking, which is explained in Figure 9.19. With phase locking, the microwave power is generated by a voltage controlled oscillator. A sample of the microwave signal is taken from the VCO output by a directional coupler and sent to a phase detector, where it is compared with a reference signal, usually the harmonic multiplied signal from a quartz crystal oscillator. The characteristics of the phase detector are shown in the graph. A phase detector generates a dc voltage, whose level is proportional to the phase difference between the two microwave signals. If the two signals are 180° out of phase, the output of the phase detector is zero. If the signals are not exactly 180° out of phase, an error voltage is developed, which is sent from the phase detector to the loop filter and amplifier. This error signal is amplified and applied to the VCO. The frequency range over which the phase-locked loop operates is determined by the loop filter bandwidth. The loop filter also determines the settling time, when the operating frequency of the oscillator is changed.

Figure 9.20 shows two phase-locked oscillators. Figure 9.20*a* is similar to Figure 9.19. A sample of the microwave power from the VCO is taken through the directional coupler and compared in a phase detector to a harmonic of the crystal-controlled oscillator. An output harmonic multiplier is used. It is often simpler to generate a microwave signal at the low end of the microwave band from the VCO with a coaxial cavity and a bipolar

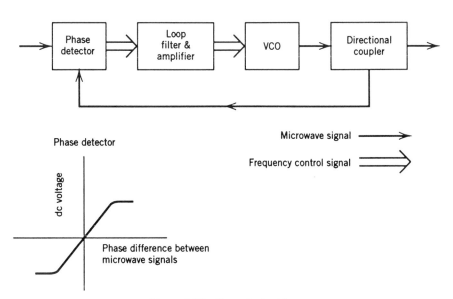

Figure 9.19 Phase-locked loop.

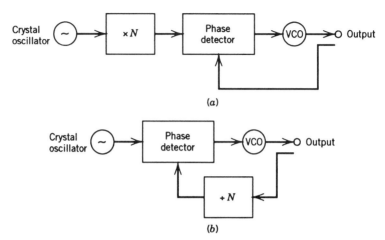

Figure 9.20 Phase-locked oscillators.

transistor, and to multiply the stable microwave frequency to the higher desired microwave frequency with a microwave varactor multiplier.

An alternative phase locked oscillator is shown in the lower sketch. Again the power is generated by a microwave VCO; a sample is taken and divided down in frequency by a digital dividing circuit (also called a scaling circuit) to the frequency of the crystal oscillator. The phase comparison is accomplished at the frequency of the crystal oscillator. Digital dividing circuits only work up to the lower part of the microwave band, so the VCO must operate there.

Various combinations of the techniques in Figure 9.20 are often used, such as multiplying the crystal reference frequency to a higher intermediate frequency while dividing the microwave signal to this lower intermediate frequency, and making the phase comparison at the intermediate frequency.

Figure 9.21 shows the phase noise of fundamental and phase-locked oscillators at frequencies up to 10 MHz from the carrier. A fundamental microwave oscillator using a bipolar transistor in a metal cavity to generate a microwave signal at 10 GHz is shown, along with a quartz crystal oscillator at 10 MHz. Ultimately both of these oscillators reach a noise floor of − 160 dBc. At frequencies close to the carrier, the quartz crystal oscillator has much less phase noise than the fundamental microwave oscillator. However, when the reference frequency is multiplied to 10 GHz, the phase noise is increased by the multiplication process. At about 50 kHz the phase noise of the fundamental oscillator is less than the frequency-multiplied signal from the 10-MHz crystal oscillator. The filter bandwidth of the phase-locked loop can be adjusted so that the phase locking is only effective up to 50 kHz. Therefore, the Phase Locked Loop (PLL) has the phase noise characteristics of the multiplied crystal reference up to 50 kHz and the phase noise characteristics of the fundamental oscillator at higher frequencies, as shown by the dotted curve.

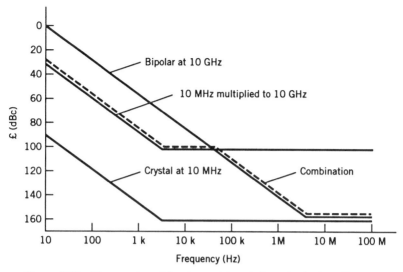

Figure 9.21 Phase noise of fundamental and phase-locked oscillators.

9.7 UP-CONVERTERS

Up-converters are used to add a lower frequency to a microwave frequency. In Figure 9.22 a 10-MHz Intermediate Frequency is added to the 6-GHz microwave frequency to provide an up-converted microwave frequency of 6.01 GHz. The frequency conversion is accomplished with varactor diodes, which are nonlinear capacitors. The microwave signal is applied to one arm of a hybrid. The varactor diodes are mounted in two of the other arms, and the Intermediate Frequency signal is applied to the diodes. The signals mix in the diodes, and the original microwave signal, the sum of the microwave signal and the IF signal, and the difference between the microwave signal

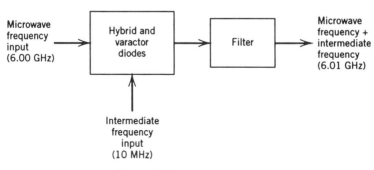

Figure 9.22 Up-converter.

and the IF signal all appear at the other output of the hybrid. The original input and the microwave input minus the intermediate frequency are filtered out, and at the output of the filter the microwave frequency shifted up by the Intermediate Frequency appears.

ANNOTATED BIBLIOGRAPHY

1. A. Sweet, *MIC and MMIC Amplifier and Oscillator Circuit Design,* Artech House, Dedham, MA, 1990.
2. G. D. Vendelin, *Design of Amplifiers and Oscillators by the S Parameter Method,* Wiley, New York, 1981, pp. 132–183.

Reference 1 covers microwave oscillator design and fabrication techniques. Reference 2 covers designing microwave oscillators.

EXERCISES

9.1. What is the purpose of an oscillator?

9.2. Name the two parts of an oscillator and the purpose of each part.

9.3. List six types of resonators used for microwave oscillators?

9.4. What are the four frequency stability specifications of microwave oscillators?

9.5. What is frequency pushing?

9.6. What is frequency pulling?

9.7. What are the four specifications for electronic tuning of a microwave oscillator?

9.8. What is settling time?

9.9. What is posttuning drift?

9.10. From Figure 9.17, what is the phase noise of each of the following oscillators at 10 KHz from the carrier?

Oscillator	Phase Noise (dBc in 1-Hz band)
FET in cavity	_____
Bipolar with DRO	_____
Bipolar in cavity	_____
VCO phased-locked to harmonic of 100-MHz crystal oscillator	_____

9.11. From Figure 9.17, what is the phase noise of a bipolar transistor in a metal cavity in a 100-Hz band at 10 kHz from the carrier?

9.12. What is the advantage of a varactor-tuned oscillator compared with a YIG-tuned oscillator?

9.13. What are the two advantages of a YIG-tuned oscillator compared with a varactor-tuned oscillator?

9.14. Name two uses of harmonic multipliers?

9.15. Draw a block diagram of a phased-locked loop.

10

LOW-NOISE RECEIVERS

The significance of low-noise receivers and the requirements placed on them by the microwave system are discussed. Next, the sources of noise—noise entering the receiver through the antenna and noise generated in the receiver itself—are identified.

Various units are used for measuring noise: noise power density, noise temperature, and noise figure. Each is defined, and formulas for converting among them are given.

Finally, each type of low-noise microwave receiver, namely bipolar and field-effect transistors, HEMTs, mixers, and paramps are discussed, including their operation and performance specifications.

10.1 THE SIGNIFICANCE OF LOW-NOISE RECEIVERS

Figure 10.1 shows a microwave communication system. This system transmits an electrical signal carrying information (multiplexed telephone calls, television programs, or data) from one location to the other in the form of amplitude, frequency, or phase modulation of the microwave signal. The microwave signal is therefore called the *carrier*.

The modulated microwave signal is transmitted from the transmitter antenna and spreads through space. Consider a satellite communication system that transmits 2 W of power from the satellite. The satellite antenna is designed to spread this power over the entire surface of the United States. The satellite signal is received by a 10-ft diameter antenna. Only a fraction of the transmitted power, equal to the area of the antenna divided by the area

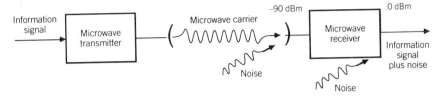

Figure 10.1 Microwave system.

of the United States, is received, so the received power is very small (about 1 pW (−90 dBm). The receiver must amplify this weak signal to the 1-mW (0-dBm) level, so it must have a 90-dB gain.

As shown in Figure 10.1, noise enters the receiving antenna in addition to signal, and additional noise is generated in the receiver itself. After amplification, the microwave signal must be 20 dB greater than the noise.

This required 20-dB carrier-to-noise ratio is illustrated in Figure 10.2. The upper sketch shows the microwave carrier waveform as a function of time, and the middle sketch shows the noise waveform. The noise is generated by the random motion of electrons in the environment into which the antenna looks and in the electronic components of the receiver. Noise occurs at all frequencies. A noise waveform is shown in Figure 10.2, but it actually varies continually because it is random in nature. This type of noise is called *white* because it has approximately the same amplitude at all frequencies. At the bottom of Figure 10.2 the noise adds to the signal and therefore distorts it.

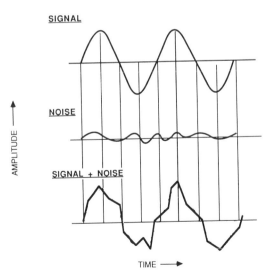

Figure 10.2 Signal and noise in a microwave receiver.

The example in Figure 10.2 is for a 20-dB signal-to-noise ratio, where the signal power is 20 dB, or 100 times, greater than the noise power. Since power is proportional to voltage squared, the signal voltage is 10 times greater than the noise voltage.

A block diagram of a microwave receiver is shown in Figure 10.3. The weak received signal enters the antenna and passes through a microwave filter. Signals at many frequencies (from other microwave systems) are picked up by the receiving antenna, but the filter limits the frequencies entering the receiver to those of the system for which the receiver is being used: interference is eliminated. The weak received signal is then amplified by a microwave amplifier (a bipolar transistor, a FET, a HEMT, or a paramp). This amplifier could have one stage or several. It could even be a combination of various amplifiers types.

The amplifier signal then enters a mixer, where the signal frequency is reduced to an Intermediate Frequency (IF) below the microwave band, where the signal can be more easily amplified. A typical intermediate frequency is 70 MHz. The signal is then easily amplified in an IF amplifier, which consists of several stages of low frequency transistors.

The amplified signal at the IF frequency then leaves the receiver and enters a demodulator, which removes the information signal from the carrier.

The requirements for a low-noise microwave receiver are shown at the bottom of Figure 10.3. The amplitude of the microwave signal as received by the antenna is about 1 pW (10^{-12} W or -90 dBm). The receiver must amplify this signal to 1 mW (10^{-3} W or 0 dBm). The low-noise receiver must therefore have a gain of 90 dB.

If the only function of a low-noise receiver were to amplify the received signal, its design would be very simple. However, the amplification must occur with a minimum of noise added to the signal.

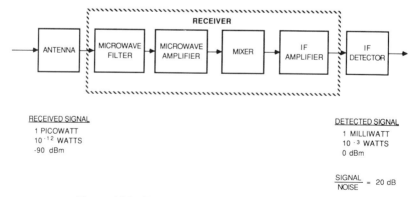

Figure 10.3 Block diagram of a low noise receiver.

10.2 SOURCES OF NOISE

Figure 10.4 shows the same receiver block diagram as Figure 10.3, but the sources of noise are included. The microwave signal enters the antenna but noise also enters. The antenna noise comes from the environment into which the antenna is looking. All the objects in the environment are generating noise because of the random motion of electrons in them, and the radiated microwave noise from these objects enters the antenna with the desired microwave signal. The incoming signal and the antenna noise pass through a microwave filter, where the signal is reduced by the insertion loss of the filter. The filter adds no noise, but its loss decreases the already weak microwave signal.

The filter decreases the microwave signal, and it might be expected that the filter would decrease the noise also. Although the filter does attenuate the noise entering it, the filter itself generates noise. Hence, effectively as much noise leaves the filter as enters it.

Both the attenuated signal and the antenna noise are amplified by the microwave amplifier, which adds noise. The signal passes through the mixer and IF amplifier, both of which add noise. All of these noise sources must be considered when determining the total signal to noise ratio at the receiver output. The most important noise sources, however, are at the receiver input—the antenna and the first amplifier. For example, if a microwave amplifier is used, the signal may be raised by 20 dB in the amplifier before it reaches the mixer. The mixer's noise contribution is therefore not very significant, because the signal entering the mixer is much larger than the signal entering the amplifier. If a microwave amplifier were not used, and the signal went directly into the mixer, the mixer noise would be very important.

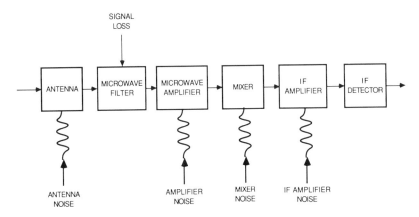

Figure 10.4 Sources of noise in a low noise receiver.

The magnitude of the antenna noise is shown in Figure 10.5, and the magnitudes of the amplifier and mixer noise are shown in Figure 10.6. Both figures show noise power density in picowatts per megahertz as a function of frequency. As shown on the right side of Figure 10.5, noise exists at all frequencies in the microwave band. However, only the small amount of noise in the receiver frequency band needs to be considered, because the rest is eliminated by the filter. Therefore, microwave noise for the antenna and receiver is specified in terms of *noise power density,* that is, picowatts of noise over a 1 MHz bandwidth. If the bandwidth of the microwave system is 2 MHz, the noise is twice as great as in Figures 10.5 and 10.6. If the bandwidth is 10 MHz, the noise is 10 times as great; if the bandwidth is 0.5 MHz, the noise is only half as great, and so on.

According to the antenna noise curves of Figure 10.5, the amount of noise entering the antenna depends on where the antenna is pointed. If the antenna is pointed toward the earth, it picks up noise from everything in the earth's environment. In most microwave systems the antenna is pointed toward the earth. The noise power density in this case is exactly the same as if the antenna had been replaced by a microwave termination operated at room temperature, which is the temperature of the earth. The antenna noise curve for an antenna pointed toward the earth applies to point-to-point microwave relay systems, where the receiving antenna looks along the surface of the earth at the transmitter, and to an airborne radar or electronic warfare systems, where the receiving antenna looks down toward targets located on the earth's surface. The same antenna noise curve also applies to a surface-based radar looking up at an aircraft. Since the radar beam is so wide, it looks along the surface of the earth while looking up at the aircraft. In other words, the altitude of the aircraft is so small relative to the antenna beam that the radar is still looking along the surface of the earth.

In all these cases, the noise power density of the antenna is 0.004 pW/MHz. If the system has a 25-MHz bandwidth, the antenna noise is 0.004 pW/MHz times 25 MHz, or 0.1 pW. This is the noise entering the receiver through the antenna. If the received microwave signal were 1 pW, then the signal-to-noise ratio is only 10 dB, which is unsatisfactory. The only way to make the system work satisfactorily, since nothing can be done about the antenna noise, is to increase transmitter power or the size of the receiving antenna, to increase the received signal. The amount of antenna noise is independent of the receiving antenna size. However, the amount of the received microwave signal is directly proportional to antenna size.

As Figure 10.5 shows, a significantly lower antenna noise can be obtained if the antenna is pointed at the cool sky. This situation applies only in satellite communication systems and radio astronomy. At the lower end of the microwave range, the antenna noise begins to increase even when the antenna is looking at the cool sky, due to cosmic noise sources. At the high end of the microwave range, the noise coming into the antenna also increases, due to atmospheric scattering. In the middle of the microwave

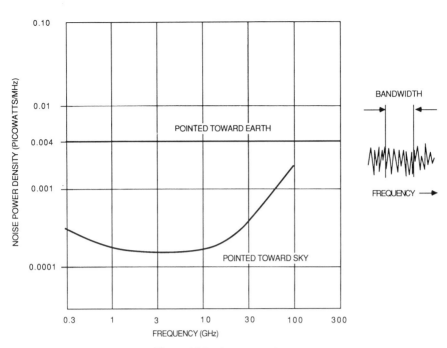

Figure 10.5 Antenna noise.

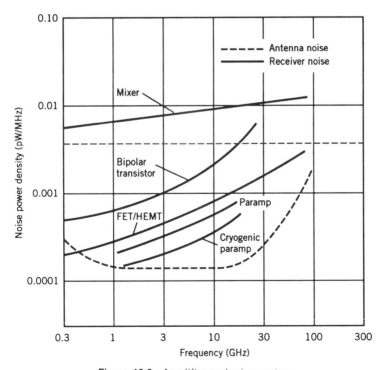

Figure 10.6 Amplifier and mixer noise.

range (1–10 GHz) the noise coming into the antenna is extremely low, being just slightly more than 0.0001 pW/MHz.

The noise power produced by amplifiers and mixers is shown in Figure 10.6. The scales are the same as those of Figure 10.5, with noise power density (pW/MHz) shown as a function of microwave frequency. The antenna noise is the dashed line. This antenna noise reference shows that it doesn't do much good to have a low-noise amplifier in a system when the antenna is looking at the surface of the earth, since the antenna noise is the overriding noise source, and the money spent on the low-noise amplifier has been wasted.

Figure 10.6 shows that the mixer noise is several times greater than the antenna noise when the antenna is looking at the earth. For example, at 10 GHz the noise power density of a mixer is 0.01 pW/MHz, and in a system with a 25-MHz bandwidth the mixer noise is 0.25 pW. A received microwave signal of 25 pW is therefore necessary to obtain the required 20-dB signal-to-noise ratio.

At frequencies up to about 6 GHz, the noise of the bipolar transistor amplifier is even lower than the antenna noise. The FET has even lower noise, and its noise remains below that of the earth-directed antenna to over 10 GHz. The parametric amplifier (paramp) has a still lower noise level, and its noise doesn't reach that of the earth-directed antenna until 30 GHz is reached. Cooling the parametric amplifier reduces its noise even more. Around 3 GHz the noise of a cooled paramp is not much greater than the noise of an antenna looking into the cool sky.

The total noise in a microwave system is the sum of the antenna and receiver noise. Both can be estimated from Figure 10.6 and the magnitude of the received microwave signal to achieve the required 20-dB signal-to-noise ratio can be determined.

Example 10.1 A satellite downlink has a frequency of 4 GHz, a bandwidth of 40 MHz, and uses a cryogenic paramp receiver. Find the power of the received signal to maintain a SNR of 20 dB.

Solution From Figure 10.5 the antenna noise power density is 0.00015 pW/MHz. Therefore,

$$
\begin{aligned}
\text{Antenna Noise} &= \text{Noise Power Density} \times \text{Bandwidth} \\
&= 0.00015 \, \text{pW/MHz} \times 40 \, \text{MHz} \\
&= 0.006 \, \text{pW}
\end{aligned}
$$

From Figure 10.6 the receiver noise power density is 0.0003 pW/MHz. Therefore,

$$
\begin{aligned}
\text{Receiver Noise} &= \text{Noise Power Density} \times \text{Bandwidth} \\
&= 0.0003 \, \text{pW/MHz} \times 40 \, \text{MHz} \\
&= 0.012 \, \text{pW}
\end{aligned}
$$

Then

$$\text{Total Noise} = \text{Antenna Noise} + \text{Receiver Noise}$$
$$= 0.006 + 0.012$$
$$= 0.018 \text{ pW}$$

Since the signal-to-noise ratio must be 20 dB for satisfactory operation of the satellite system, the received signal must be 100 times greater than the noise:

$$P_{\text{RX}} = 100 \times 0.018 \text{ pW}$$
$$= 1.8 \text{ pW}$$

10.3 NOISE UNITS

The following quantities are used to specify the noise entering the antenna and generated in the receiver:

Noise power density in pW/MHz
Noise power density in dBm/MHz
Noise temperature in °Kelvin
Noise figure in dB

These noise units are compared in Figure 10.7. which is the same as Figure 10.6, with noise unit scales added. For example, the noise of the antenna looking along the surface of the earth is 0.004 pW/MHz, which is equivalent to -114dBm/MHz, or a noise temperature of 290°K, or a noise figure of 3 dB.

The noise power output of a passive termination is directly proportional to the temperature of the termination. A passive termination is a microwave load not connected to an external source of power. The random motion of the electrons in the load generates microwave noise, and the amount of noise power depends on the termination temperature. If the temperature is doubled, the noise power output is doubled; if the temperature is reduced to one fourth, the noise output is reduced to one fourth. Noise power is related to the temperature of the termination by

$$N = kTB = 1.38 \times 10^{-14} \, TB$$

where

$$k = \text{constant} = 1.38 \times 10^{-14} \, \text{mW/°K-mHz}$$
$$N = \text{noise power (mW)}$$
$$T = \text{noise temperature (°K)}$$
$$B = \text{bandwidth (MHz)}$$

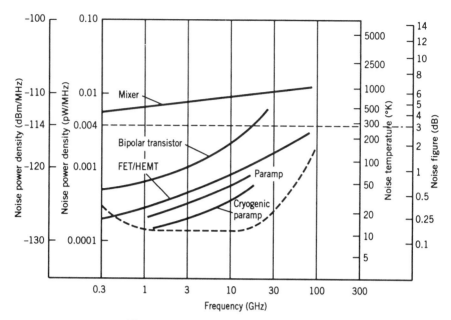

Figure 10.7   Comparison of noise units.

The total noise is directly proportional to the bandwidth. The temperature used for the calculation of noise power is the temperature measured on the Kelvin scale. The Kelvin temperature scale and the Celsius temperature scale are compared in Figure 10.8. At 0°K (-273°C), all electron motion ceases, so no noise is generated.

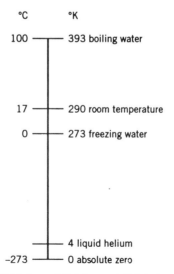

Figure 10.8   Comparison of Celsius and Kelvin temperature scales.

Example 10.2 Calculate the noise power for (a) a termination temperature of 290° K and a bandwidth of 1 MHz and (b) a temperature of 145 °K and a 10 MHz bandwidth.

Solution

(a) $\quad\quad\quad N = kTB$
$\quad\quad\quad\quad = (1.38 \times 10^{-14}\,\text{mW/°K·MHz})(290°\text{K})(1\,\text{MHz})$
$\quad\quad\quad\quad = 4 \times 10^{-12}\,\text{mW}$
$\quad\quad\quad\quad = -114\,\text{dBm}$

(b) $\quad\quad\quad N = kTB$
$\quad\quad\quad\quad = (1.38 \times 10^{-14}\,\text{mW/°K·MHz})(145°\text{K})(10\,\text{MHz})$
$\quad\quad\quad\quad = 20 \times 10^{-12}\,\text{mW}$
$\quad\quad\quad\quad = -107\,\text{dBm}$

Example 10.3 Redo Example 10.2(b) using dBs only.

Solution From Example 10.2(a) the noise for 290°K at 1 MHz is -114 dBm. In (b), $T = 145°$K, so the noise (directly proportional to temperature) must also be half as great as in (a), or -3 dB. Since the bandwidth is 10 times greater, the noise is increased by 10 dB. Hence,

$$N = -114\,\text{dBm} - 3\,\text{dB} + 10\,\text{dB}$$
$$= -107\,\text{dBm}$$

The noise power output of a passive termination depends on its actual physical temperature. The *noise temperature* of an amplifier or mixer is the temperature that a termination would have if it were putting out as much noise as the amplifier. The noise temperature of an amplifier has nothing to do with its actual physical temperature. It is simply a way of specifying how much noise the amplifier is putting out. It is commonly used in microwave system design because it greatly simplifies the calculations.

Example 10.4 From Figure 10.7 the noise power of a bipolar transistor at 3 GHz is 0.001 pW for a 1-MHz bandwidth. What is the noise temperature?

Solution

$$T = \frac{N}{kB}$$
$$= \frac{10^{-12}\,\text{mW}}{(1.38 \times 10^{-14}\,\text{mW/°K·MHz})(1\,\text{MHz})}$$
$$= 72.5°\text{K}$$

Note that 72.5°K is *not* the physical temperature of the transistor; the transistor is putting out as much noise as a passive termination cooled to 72.5°K.

Example 10.5 From Figure 10.7 the noise power of a mixer at 20 GHz is 0.01 pW for a 1 MHz bandwidth. What is the noise temperature?

Solution

$$T = \frac{N}{kB}$$
$$= \frac{10^{-11}\,\text{mW}}{(1.38 \times 10^{-14}\,\text{mW/°K·MHz})(1\,\text{MHz})}$$
$$= 725°\text{K}$$

Again this is *not* the actual temperature of the mixer—if it were, the mixer would melt!

Example 10.6 A satellite downlink is operating at 10 GHz with a 40-MHz bandwidth. It has a cryogenic paramp receiver. What received power is needed to achieve a signal-to-noise of 20 dB?

Solution From Figure 10.7,

$$T_{\text{antenna}} = 10°\text{K}$$
$$T_{\text{paramp}} = 20°\text{K}$$

Therefore,

$$T_{\text{total}} = 10°\text{K} + 20°\text{K} = 30°\text{K}$$

so the noise power is

$$N = kTB$$
$$= (1.38 \times 10^{-14}\,\text{mW/°K·MHz})(30°\text{K})(40\,\text{MHz})$$
$$\approx -108\,\text{dBm}$$

Therefore,

$$\text{Received Power} \approx -108\,\text{dBm} + 20\,\text{dB} \approx -88\,\text{dBm}$$

The *noise figure* (NF) is the difference between the signal-to-noise ratio (SNR) entering an amplifier and the signal-to-noise ratio (SNR) leaving the amplifier (see Figure 10.9):

$$\text{SNR}_{\text{in}}\,(\text{dB}) = \text{Input signal}\,(\text{dBm}) - \text{Input noise}\,(\text{dBm})$$
$$\text{SNR}_{\text{out}}\,(\text{dB}) = \text{Output signal}\,(\text{dBm}) - \text{Output noise}\,(\text{dBM})$$
$$\text{NF}\,(\text{dB}) = \text{SNR}_{\text{in}}\,(\text{dB}) - \text{SNR}_{\text{out}}\quad(\text{dB})$$

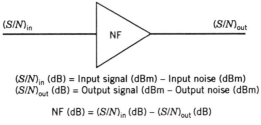

$(S/N)_{in}$ (dB) = Input signal (dBm) – Input noise (dBm)
$(S/N)_{out}$ (dB) = Output signal (dBm – Output noise (dBm)

NF (dB) = $(S/N)_{in}$ (dB) – $(S/N)_{out}$ (dB)

Figure 10.9 Noise figure.

The noise figure specifies how much noise is added by the amplifier. It is the only noise unit that can be directly measured.

Example 10.7   An amplifier has a gain of 40 dB and a noise figure of 3 dB. The input signal (S_{in}) is −80 dBm, and the bandwidth is 10 MHz. Find SNR_{in} and SNR_{out}.

Solution   The noise generated by the input signal source depends on its temperature, which for most applications is 290 K (room temperature). In a 1-MHz band, the noise power is −114 dBm. In a 10-MHz bandwidth, noise power is 10 dB greater, so here $N = -104$ dBm. Hence,

$$SNR_{in} = -80 \text{ dBm} - (-104 \text{ dBm}) = 24 \text{ dB}$$

Now,

$$\begin{aligned} S_{out} &= S_{in} + \text{Gain} \\ &= -80 \text{ dBm} + 40 \text{ dB} \\ &= -40 \text{ dBm} \end{aligned}$$

and

$$\begin{aligned} N_{out} &= N_{in} + \text{NF} + \text{Gain} \\ &= -104 \text{ dBm} + 3 \text{ dB} + 40 \text{ dB} \\ &= -61 \text{ dBm} \end{aligned}$$

Hence,

$$SNR_{out} = -40 \text{ dBm} - (-61 \text{ dBm}) = 21 \text{ dB}$$

By definition,

$$\begin{aligned} \text{NF} &= SNR_{in} - SNR_{out} \\ 3 \text{ dB} &= 24 \text{ dB} - 21 \text{ dB} \end{aligned}$$

which checks.

10.4   MIXERS

Every low-noise microwave receiver uses a mixer to shift the received carrier and its modulation sidebands out of the microwave band to a lower frequency, where it is easier and more economical to achieve the required 90 dB gain.

A block diagram of a mixer is shown in Figure 10.10. The microwave signal is at 10 GHz, and as shown at the upper left has a 10-MHz bandwidth, which means that the signal and its modulation sidebands are in the frequency range from 9.995 to 10.005 GHz. The modulation sidebands carry the information being transmitted by the microwave carrier. Microwave signals at other frequencies may have entered the antenna along with the desired signal, but these other signals have been filtered out before the signal reached the mixer. The microwave signal enters a combiner that contains Schottky mixer diodes (see Chapter 7).

A local oscillator signal, generated in the receiver, also enters the combiner. It is offset in frequency from the input microwave signal by the center frequency of the IF amplifier. In this example, 70 MHz has been chosen since it is a commonly used intermediate frequency. The local oscillator frequency is thus at 10.07 GHz.

The input microwave signal enters one arm of the combiner, and the local oscillator signal enters through another arm. These signals are combined and appear at the other arms of the combiner, but they are still at separate frequencies. The mixer diodes are placed in these other arms and mix the input and local oscillator signals. The frequencies leaving the mixer include the original microwave input signal, the local oscillator signal, a frequency that is the sum of the microwave and local oscillator signals (in this example it is 20.07 GHz), and the intermediate frequency, which is the difference between the microwave and local oscillator signal (in this example it is 0.07 GHz, or 70 MHz). All signals except the IF are filtered out, and the IF signal is then amplified by the 70-MHz IF amplifier. As noted, it is much easier to design a 70-MHz amplifier than a 10-GHz amplifier, which is why the carrier signal is converted from 10 GHz to 70 MHz. In the upper right of Figure 10.11 the 10-MHz bandwidth containing the modulation sidebands is shifted into the 70-MHz band. Therefore, if the original microwave frequency had been 10 GHz, the IF would be 70 MHz. The sideband of the modulated microwave carrier at 9.995 GHz would be shifted to 75 MHz, and the sideband of the microwave carrier at 10.005 GHz would be shifted to 65 MHz, and so on.

Mixer action is explained in Figure 10.11. The local oscillator (LO), the received microwave signal (RF), and the output of the mixer (IF) are plotted as functions of time during an interval of 10 nanoseconds. The LO amplitude is normally at least 10 times greater than the amplitude of the RF and IF output signals, but to save space the amplitudes are not drawn in this way in Figure 10.11.

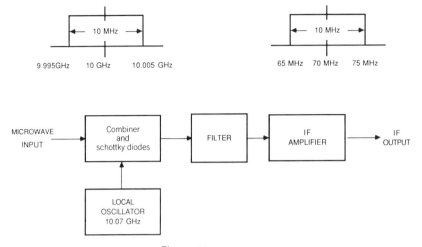

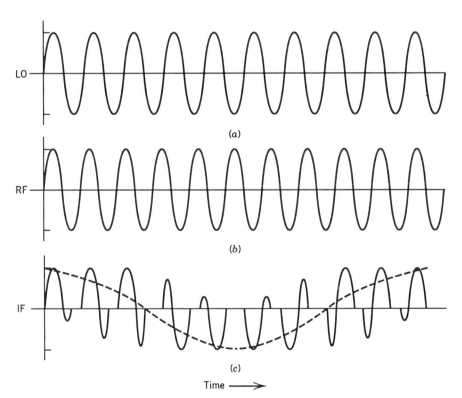

Figure 10.10   Mixer.

Figure 10.11   Mixer action.

The LO frequency is 1.0 GHz, so there are 10 LO cycles during a 10 nanosecond time interval. The RF frequency is 1.1 GHz, so there are 11 RF cycles during the 10 nanosecond time interval.

The LO signal is large, and when it is applied to the mixer diodes it biases the diodes off during its negative cycle and turns the diodes on during its positive cycle. Therefore, the RF signal can pass through the diodes when the LO is positive, but it cannot pass through the diodes when the LO is negative. The result is shown in curve C, where the RF signal only appears at times when LO signal is positive. This curve is the output of the mixer and is a complicated waveform containing many frequency components. However, as shown by the dashed envelope, one frequency is 1 cycle in the 10 nanosecond time period, which is 0.1 GHz. This frequency is the difference between the 1.0-GHz LO and 1.1-GHz RF signals. The output from the mixer is passed through a low-pass filter that removes the high frequency components and leaves only the 0.1-GHz IF.

Different types of mixers each use a different type of combiner to achieve the desired performance. Mixer choice depends on the requirements for handling the different signals that result from the mixing process. These signals are shown in Figure 10.12 and 10.13. Figure 10.12 shows a low-level single entering the mixer. In (a) the mixer input consists of a LO signal at 10.1 GHz and a microwave signal at 10 GHz. The signal is much smaller than the LO. The mixer output is four signals: the original microwave signal at 10 GHz, the LO signal at 10.1 GHz, the sum of the LO and microwave signals at 20.1 GHz, and the desired IF signal at 0.1 GHz (in this example, an IF of 0.1 GHz, or 100 MHz has been selected). This case was discussed in the previous section. At the mixer output the signal, the LO, and the sum frequencies are filtered out so that only the desired difference frequency goes to the IF amplifier.

In (a), the LO is offset in frequency above the signal, but in (b) the LO can also be offset below the signal. In (b) the LO is at 9.9 GHz, and the signal is, as before, at 10 GHz. The difference between these two signals is still at 0.1 GHz, so the IF is the same as before.

Figure 10.13c shows the problem of the "image" frequency. In this case, the signal is at 10 GHz and the LO is at 10.1 GHz, but an additional unwanted signal at 10.2 GHz enters the mixer (shown by the dashed line marked I for image). The signal and LO mix, and the difference frequency is at the desired IF of 0.1 GHz. The LO and image also mix to form a difference frequency (shown dashed), which also occurs at the IF of 0.1 GHz. Consequently, the downshifted image signal appears in the mixer output and interferes with the desired signal. If the receiver using the mixer covers only a narrow frequency range, the image frequency would be eliminated by the input microwave filter of the receiver. However, in many broadband receivers covering a wide frequency range, the image passes right through the filter and interferes with the desired signal. (Special mixers that cancel the image signal by proper phasing of the inputs to the mixer will be discussed shortly.)

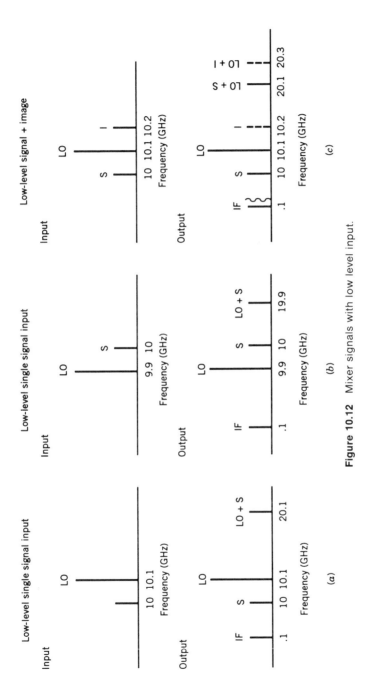

Figure 10.12 Mixer signals with low level input.

253

This situation is even worse at high input signal levels, as shown in Figure 10.13*a,* which shows a single input signal at a high level. As the high-level signal and the LO signal interact with the diodes, the desired IF signal at 0.1 GHz is generated along with harmonics of the signal at 20 GHz and of the LO at 20.2 GHz. When these harmonic signals mix, they form an IF at 0.2 GHz. This signal is called *spurious,* since it should not be present. This spurious signal (*a*) is called a *single-tone intermodulation product,* since it is produced by the interaction of the harmonic of a single input signal with the LO.

The problem is even more complicated if two input signals enter the mixer, which is the usual case, since the input microwave signal contains modulation sidebands. As shown in Figure 10.13*b,* these signals are labeled S_1 and S_2 and are at 9.95 and 10.00 GHz, with a 10.10 GHz LO. Both input microwave signals generate harmonics in the mixing process, and the harmonic of one signal mixes with the fundamental of the other to produce intermodulation sidebands. These intermodulation sidebands are the same type discussed in Chapter 8 for microwave amplifiers. The intermodulation sidebands then mix with the LO and shift the intermodulation sidebands down to the IF. Hence, the IF contains not only the two input signals, shifted down to the IF at 0.1 and 0.15 GHz, but also third-order intermodulation sidebands at 0.05 and 0.20 GHz and other high-order intermodulation sidebands that are not shown. These spurious signals are called *two-tone intermodulation products* because they are produced by the two input signals.

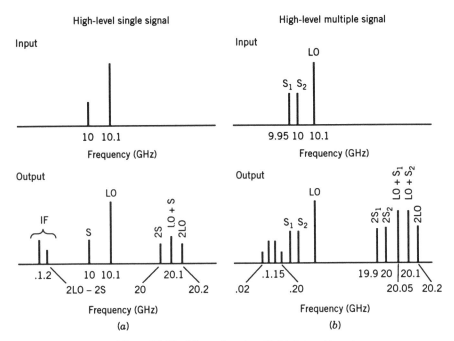

Figure 10.13   Mixer signals with high level input.

Table 10.1 Mixer Specifications

Conversion loss	Ratio of IF power to microwave signal input power
Noise figure	Ratio of signal to noise at mixer input to signal to noise at mixer output
SWR at LO and RF inputs	Mismatch at LO and signal ports
LO/RF isolation	Isolation between LO and RF ports
Spurious rejection	Suppression of single-tone and two-tone intermodulation frequencies
Harmonic suppression	Suppression of LO and signal harmonics
Dynamic range	Range of input power over which mixer provides required performance

Figure 10.13 looks complicated enough, but it has been simplified by considering only one type of interfering signal at a time. In an actual mixer all the different interfering signals are present. They can be minimized by special mixer designs, which will be described shortly.

Mixer specifications are shown in Table 10.1. An important specification is conversion loss, which is the ratio of the IF output power to the microwave signal input power. Some of the input microwave signal is converted in the mixer to the sum frequency, some of the input power remains at its original frequency, and part of the input microwave signal power is converted to the IF. If one third of the input signal is converted to each of these output frequencies, the IF output power is only one third of the input signal power, so the conversion loss of the mixer is 5 dB.

The definition of the noise figure for a mixer is the usual definition for noise figure: the ratio of the signal to noise at the mixer input relative to the signal to noise at the mixer output. The overall noise figure of a mixer is approximately equal to its conversion loss plus the noise figure of the IF amplifier.

Another important mixer specification is the SWR at the LO and RF ports. If these ports are mismatched, not all of the LO and not all of the signal can enter the mixer.

The LO-to-RF isolation specifies how much of the LO signal will leak out through the RF port. Note that the LO signal leaking into the RF port can be transmitted out the system antenna and can interfere with other systems.

The specification for spurious rejection describes the suppression of the single-tone and two-tone intermodulation products. Harmonic suppression refers to the suppression of the local oscillator and signal harmonics.

The dynamic range of a mixer is illustrated in Figure 10.14, which shows the IF mixer power as a function of the microwave signal power. The weakest signal that can be detected by the mixer is determined by the noise level of the mixer, which is

$$-114 \text{ dBm} + 10 \log \text{BW} + \text{NF}$$

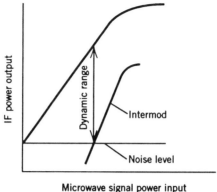

Figure 10.14 Mixer output as a function of microwave signal input.

The largest signal that can be handled by the mixer is determined by the acceptable level of intermodulation products, and is often specified by the requirement that the intermod products be below the noise level.

Unfortunately, no mixer meets all specifications of Table 10.1 in the optimal way. Figure 10.15 compares the most common types of mixers.

Schematics of some mixer types are shown. The major difference between all types is the combiner used. The single-ended mixer is the simplest type, but its performance is poor in most specifications. The combiner used in the single-ended mixer is a directional coupler, with the RF input being applied to the main arm of the coupler and the LO input entering through the coupled port. The combined signals are then mixed in a single diode, and a low-pass filter filters out the LO, the signal, and the sum frequency, allowing only the IF output to go to the IF amplifier.

The most common type is the balanced mixer. A 3-dB hybrid is used to supply the input microwave and LO power to two mixer diodes. A 90° or a 180° hybrid can be used, and different performance trade-offs are accomplished, depending on which hybrid is used.

The double-balanced mixer is conceptually two balanced mixers. It is normally made with a four-diode array, and in the lower part of the microwave band it actually consists of miniature wirewound transformers. Note that the double-balanced mixer with its four diodes has the best spurious and harmonic suppression of any mixer type.

The image rejection mixer actually separates out the signal and the image frequencies, so they appear at different output ports. This mixer consists of two balanced 90° mixers. The RF input signal is fed through a 90° hybrid to each mixer and the LO is fed through an in-phase power divider to the two mixers. The IF output from each of the balanced mixers is then combined in a 3-dB hybrid. When the phase relations between all of the mixer signals are considered, the signal frequency shifted down to IF appears at one port of

MIXER TYPE	CONVERSION LOSS	VSWR LO,RF	LO/RF ISOLATION	SPURIOUS REJECTION	HARMONIC SUPPRESSION
SINGLE-ENDED	GOOD	GOOD, POOR	FAIR	POOR	POOR
BALANCED (90°)	GOOD	GOOD, GOOD	POOR	FAIR	FAIR
BALANCED (180°)	GOOD	FAIR, FAIR	VERY GOOD	FAIR	GOOD
DOUBLE-BALANCED	VERY GOOD	POOR, POOR	VERY GOOD	GOOD	VERY GOOD
IMAGE-REJECT	GOOD	GOOD, GOOD	GOOD	FAIR	GOOD
IMAGE-RECOVERY	EXCELLENT	GOOD, GOOD	VERY GOOD	FAIR	GOOD

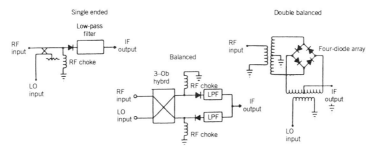

Figure 10.15 Mixer types and performance.

the output hybrid, and the image frequency shifted down to IF appears at the other port.

The image recovery mixer utilizes the fact that the mixer not only responds to a signal at the image frequency but also produces energy at the image frequency. With proper phasing, this energy can be shifted back to the signal frequency to improve the conversion loss of the mixer.

The harmonic mixer, not shown in Figure 10.16, operates with a LO at approximately half of the RF signal frequency. The LO frequency is doubled by the mixer diodes by the process described in Chapter 9 for frequency multipliers. The doubled LO signal then mixes with the RF signal to provide the IF signal. Harmonic mixers are used at high microwave frequencies in the EHF band because they simplify the local oscillator.

10.5 LOW-NOISE TRANSISTORS

Bipolar transistors, FETs, and HEMTs were discussed in Chapter 8. Low-noise transistors are made in the same way as the medium power transistors discussed in that chapter.

The major factor affecting noise figure is the transit time through the transistor. Thus, the noise figure can be reduced by decreasing the spacing between the internal elements to the lowest that can be fabricated. Low-noise transistors operate at low output power levels, and consist of only a few interdigital fingers.

Low-noise transistors must be specially biased and matched to achieve minimum noise performance. Normally the bias is about 20% of the current that gives maximum gain. They are also deliberately mismatched at their input to reduce the noise figure. Mismatching adjusts the phase between the input noise and the noise generated in the transistor and results in a partial cancellation of the total noise amplified by the transistor. Figure 10.16 illustrates this effect, showing contours of constant gain and constant noise figure on a Smith chart.

If the match were located as shown by the maximum-gain point on the Smith chart, the gain would be 14.7 dB. If the input match departed from this point, the gain would be reduced to 13.7 dB, 12.7 dB, and so on, to 6.7 dB, as shown by the constant-gain circles. However, to achieve the minimum noise figure, the input of the transistor should be matched as shown by the 2.5-dB noise figure point. If the input match departed from the optimal noise figure value, the noise figure would deteriorate to 3 dB. Achieving the minimum noise figure results in a gain reduction from 14.7 dB to 10.7 dB.

Associated gain is the gain that the transistor will have when it has been biased and matched for its lowest noise figure. This gain is less than the gain that could be obtained if the transistor were matched and biased for its highest gain. The most important requirement for the first amplifying stage in a low-noise receiver is low noise figure, so often a transistor amplifier is

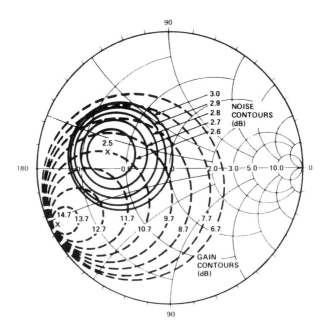

Figure 10.16 Input matching of low-noise transistors.

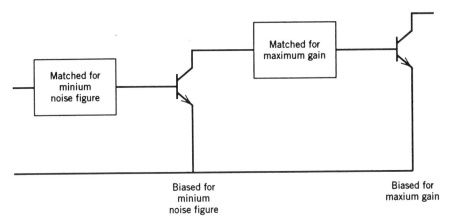

Figure 10.17 Two-stage transistor amplifier.

made of two stages, as shown in Figure 10.17. The first stage is matched and biased for minimum noise figure. Since the noise figure of the second stage is less important, because the signal has been amplified by the first stage, the second stage is matched and biased for maximum linear gain. The formula for calculating the total noise figure of two amplifiers together, from the noise figure of the individual amplifiers and the gain of the first stage is shown below. Note that to make this calculation, noise figure and gain must be expressed as a ratio rather than in dB format, which is the meaning of the asterisks in the formula.

$$NF_{total}^* = NF_1^* + \frac{NF_2^* - 1}{Gain_1^*}$$

where

$$NF_1^* = \text{noise figure of first amplifier}$$
$$NF_2^* = \text{noise figure of second amplifier}$$
$$Gain_1^* = \text{gain of first amplifier}$$

Example 10.8 In a two-stage amplifier, amplifier 1 has a noise figure of 3 dB and a gain of 20 dB. The second amplifier has a noise figure of 6 dB. Find the total noise figure.

Solution First, express NF and gain as ratios

$$NF_1^* = 2$$
$$NF_2^* = 4$$
$$Gain_1^* = 100$$

$$NF^*_{total} = NF^*_1 + \frac{NF^*_2 - 1}{Gain_1}$$

$$NF^*_{total} = 2 + \frac{4 - 1}{100}$$

$$= 2.03$$

$$= 3.1 \, dB$$

Note that the noise figure of the combination is hardly affected by the second amplifier.

Example 10.9   For the system previously discussed, NF_1 = 1 dB, NF_2 = 6 dB, and G_1 = 6 dB. What is the total noise figure?

Solution NF^*_1 = 1.26, NF^*_2 = 4.0, Gain = 4, so

$$NF^*_{total} = NF^*_1 + \frac{NF^*_2 - 1}{Gain_1}$$

$$= 1.26 + \frac{4 - 1}{4}$$

$$= 2$$

$$= 3 \, dB$$

Because the gain of amplifier 1 is so low, the effect of amplifier 2 is significant.

10.6   PARAMETRIC AMPLIFIERS

The parametric amplifier, commonly called a *paramp,* has a lower noise figure than any low-noise transistor. The noise figure of the paramp can be further lowered by cooling the amplifier to cryogenic temperatures.

A block diagram of a parametric amplifier is shown in Figure 10.18. The input microwave signal is applied through a circulator to a combiner, which contains a variable-capacitance diode commonly called a varactor. The microwave signal to be amplified is combined in the combiner with a pump signal, which is at approximately twice the frequency of the microwave signal. For example, if the microwave signal is at 10 GHz, the pump frequency must be at 20 GHz. The microwave signal is amplified by the parametric process that will be described shortly, and the amplified microwave signal then passes back into the circulator and into the output line.

The name *parametric amplifier* arises because a "parameter" of a tuned circuit is being varied to provide the amplification. The operating principles of the parametric amplifier are shown in Figure 10.19. To understand its operation, consider an *LC* circuit oscillating at its natural frequency. If the capacitor plates are physically pulled apart at the instant of time when the voltage between them is at a positive maximum, then work must be done on

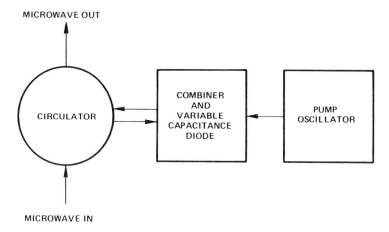

Figure 10.18 Parametric amplifier.

the capacitor, since a force must be applied to separate the plates. Just as the voltage between the plates passes through zero, the plates are now returned to their original separation. This involves no work, because there is no voltage on the plates at the time they are pushed back together. As the voltage then passes through a negative maximum, the plates are again pulled apart and the voltage increases once again. This process is repeated regularly as in Figure 10.19a–c. Figure 10.19a shows the voltage in the original resonant circuit, b the plate spacing as they are pulled apart at the peak of each cycle

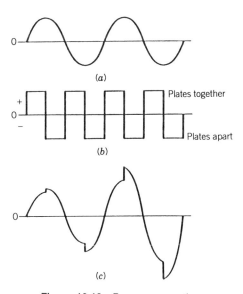

Figure 10.19 Paramp operation.

and pushed together at the zero of each cycle, and c the buildup of the signal voltage as the capacitance is successively changed.

In a microwave parametric amplifier, the capacitance is not varied mechanically but electronically by changing the capacitance of the varactor diode with the pump microwave signal.

This simple explanation requires that the pump frequency be exactly twice the signal frequency and in the correct phase relationship. This is not a basic requirement, although it is difficult to illustrate it graphically. Parametric amplifiers use an idler circuit to avoid pumping at exactly twice the frequency and at just the right phase. The pump frequency is made approximately twice the signal frequency, and in the parametric mixing process an idler frequency, which is the difference between the pump and signal frequencies, is generated. Since the pump is at approximately twice the signal frequency, the idler frequency is close to the signal frequency. The idler circuit establishes the right phase relationship for parametric amplification to occur.

ANNOTATED BIBLIOGRAPHY

1. S. J. Erst, *Receiving Systems Design,* Artech House, Dedham, MA, 1984.
2. S. A. Maas, *Microwave Mixers,* Artech House, Dedham, MA, 1986.

Reference 1 covers the design of low-noise receivers. Reference 2 covers mixers.

EXERCISES

10.1. Complete the following table. (Use Figure 10.7)

Noise Power Density (dBm/MHz)	Noise Temperature (K)	Noise Figure (dB)
−130		
	1500	
		4
	100	
−117		

10.2. Arrange the following low-noise receivers in order of decreasing noise: paramp, FET, bipolar transistor, HEMT, mixer.

10.3. Complete the following table for a low-noise amplifier at room temperature.

B (MHz)	S_{in} (dBm)	N_{in} (dBm)	$(S/N)_{in}$ (dB)	G (dB)	NF (dB)	S_{out} (dBm)	N_{out} (dBm)	$(S/N)_{out}$ (dB)
3	−80			30	4			
20	−84			10	2			
10	−90			25				11
1	−90					−30		20
4	−80				3	−40		

10.4. a. A source at room temperature ($T = 290°K$) delivers a signal of −90 dBm with a bandwidth of 2 MHz. What is the SNR of the system at this point?
   b. An amplifier is inserted to boost the signal to −20 dBm. What is the gain of the amplifier?
   c. The amplifier has a noise figure of 6 dB. What is the SNR at the amplifier output?
   d. What is the noise power in dBm at the amplifier output?

10.5. The SNR at a receiver input is 30 dB. If the receiver has a 5-dB noise figure, what is the SNR at the output?

10.6. How much noise power (in dBm) is generated by a noise source at $75°K$ over a 20-MHz bandwidth?

10.7. Name three parts of a mixer.

10.8. A mixer has a signal frequency of 10 GHz and an IF frequency of 10 MHz. What is the local oscillator frequency?

10.9. With the mixer characteristics of the previous problem, what is the image frequency?

10.10. What mixer type has the lowest spurious signals?

10.11. Name the four parts of a parametric amplifier.

11

MICROWAVE INTEGRATED CIRCUITS

Most microwave equipment is built up into subsystems or supercomponents that are combinations of many device types. This integration of many microwave devices to form a subsystem is a *microwave integrated circuit* (MIC).

11.1   TYPES OF MICROWAVE INTEGRATED CIRCUITS

The three types of microwave circuits are shown in Table 11.1. Discrete circuits consist of packaged diodes and transistors mounted in coaxial or waveguide transmission line assemblies. An example of a discrete microwave circuit is a *PIN* diode mounted in a coaxial transmission line to make an electronically controlled attenuator. Both the transmission line assembly and the diode are discrete parts, and the diodes can usually be removed from the assembly and replaced if required.

A hybrid MIC consists of diodes, transistors, and other circuit elements that are fabricated separately and a microstrip transmission line circuit consisting of microstrip circuit elements such as matching stubs, couplers, hybrids, and filters. The diodes and transistors are fabricated of silicon or gallium arsenide. The other circuit elements, such as dielectric resonators, ferrite circulators, and chip capacitors, are fabricated of the most appropriate material. All are then mounted into the microstrip circuit and connected with bond wires.

The monolithic microwave integrated circuit (MMIC) consists of diodes, transistors, microstrip transmission lines, microstrip circuits, and other circuit elements, such as lumped capacitors and resistors. All are fabricated simultaneously, including their interconnections, in a semiconductor chip.

Table 11.1   Microwave circuits

Discrete circuit	Packaged diodes/transistors mounted in coax and waveguide assemblies
Hybrid MIC	Diodes/transistors and microstrip fabricated separately and then assembled
MMIC	Diodes/transistors and microstrip fabricated simultaneously

The advantages of the hybrid microwave integrated circuits include:

1. Each component of the MIC can be designed for optimal performance. Each transistor can be made of the best material. Other devices (dielectric resonators and circulators) can be made of the most appropriate material. The lowest loss microstrip components can be made by choosing the optimal microstrip substrate.

2. It has high power capability since the high-power generating elements can be optimally heat-sinked.

3. Standard diodes and transistors can be used and made to perform different functions by using different circuit designs. Special-purpose devices for each function are not required.

4. Trimming adjustments are possible to optimize the performance of the circuit.

5. The hybrid microwave integrated circuit is the most economical approach when small quantities, up to several hundred, of the circuits are required. This quantity level is typical of most microwave equipment.

The disadvantages of hybrid microwave integrated circuits are

1. Wire bonds cause reliability problems. Each circuit element that is not part of the microstrip assembly must be attached to the microstrip by a wire bond, and if a single wire bond breaks the complete MIC is useless.

2. The number of devices that can be included is limited by the economics of mounting the devices onto the circuit and attaching them with wire bonds. The circuit is usually limited to a few dozen compartments.

The advantages of the MMIC are

1. The circuit can be designed with minimal mismatches and minimal signal delay because the parts can be placed extremely close together. They are connected during the fabrication process, and wire bonds with variable mismatch characteristics need not be used.

2. There are no wire bond reliability problems, since wire bonds are not used.

3. (Up to thousands) of devices can be fabricated at one time into a single MMIC.

4. It is the least expensive approach when large quantities are to be fabricated.

The disadvantages of MMIC are

1. Performance is compromised, since the optimal materials cannot be used for each circuit element.

2. Power capability is lower because good heat transfer materials cannot be used.

3. Trimming adjustments are difficult or impossible.

4. Unfavorable device-to-chip area ratio in the semiconductor material. The dielectric constant of GaAs used for MMIC is approximately the same as the dielectric constant of the alumina used for the hybrid MIC microstrip circuits. Consequently, the size of the circuits are approximately the same. With hybrid MIC, the expensive GaAs is used only for tiny transistor and diode chips. With MMIC, the expensive GaAs must be used for the entire IC substrate.

5. Tooling is prohibitively expensive for small quantities of MMIC.

In spite of the disadvantages of MMIC, its application in microwave equipment is growing, because of improved performance and reliability. MMIC offers the possibility of low-cost fabrication of microwave circuits in large quantities.

11.2 HYBRID MICROWAVE INTEGRATED CIRCUITS

Most MICs are hybrid types. Most microwave equipment is made in small quantities and requires optimal performance. The hybrid MIC can use standard diodes and transistors and the most optimal material for each circuit component, and each individual circuit can be "tweaked" for optimal performance.

11.3 MICROSTRIP MATERIALS AND DESIGN

Figure 11.1 shows a microstrip transmission line that consists of a dielectric substrate with a ground plane on one side and the microstrip line on the other. The electric field of the microwave is shown; most of the microwave field is contained in the dielectric substrate, but part of the field extends into the air region above the substrate. The advantage of microstrip is that complicated combinations of transmission lines that form directional couplers, hybrids, filters, and other devices can be fabricated as the transmission line itself is being fabricated by the photoetching process.

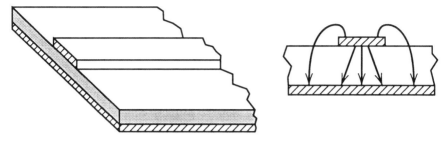

Figure 11.1 Microstrip transmission line.

Possible substrate materials for microstrip transmission lines are compared in Table 11.2. The first two materials are plastic—Teflon-fiberglass and epsilam 10, a ceramic-loaded plastic material that has the dielectric constant of alumina and the fabrication advantages of plastic material. The next three materials are ceramic, namely alumina, which is the most commonly used MIC material; beryllia, which is like alumina but has a very high thermal conductivity and can be used to heat-sink high power components; and ferrite, which allows the fabrication of microstrip circulators right on the substrate. To form a circulator, a small magnet must be added to the substrate to obtain the circulator action. In regions away from the magnet, where the ferrite is not magnetized, it acts as an insulator like any of the other substrate materials. Silicon and GaAs have the advantage that they can be used both for microstrip circuits and for active devices such as diodes and transistors.

The second column compares the dielectric constant of the various materials. The larger the dielectric constant, the smaller the parts will be at a given frequency. By proper loading of Teflon-fiberglass material with ceramic, any dielectric constant between 2 and 10 can be obtained, and different values are commercially available.

The most popular materials for microstrip transmission lines, and those used for hybrid MICs, are alumina ceramic and Teflon-fiberglass. Beryllia is sometimes used for high power applications, and the ferrite material is used

Table 11.2 Comparison of Substrate Materials for Microstrip Transmission Lines

Substrate Material	Dielectric Constant	Loss Tangent ($\times 10^{-4}$)	Thermal Conductivity (W/in:°C)	Maximum Temperature during Fabrication (°C)
Teflon-fiberglass	2.5	10	0.007	200
Epsilam 10	10	15	0.01	150
Alumina	10	1	0.1	500
Beryllia	6	2	1.0	500
Ferrite	15	2	0.1	500
Silicon	12	30	0.4	400
Gallium arsenide	12	16	0.1	400

when circulators must be fabricated along with the other microstrip circuit elements. Teflon-fiberglass is easy to machine and the circuit elements are larger, which may be an advantage or a disadvantage. Teflon-fiberglass substrates also have lower total loss, because, although their dielectric loss is higher, their smoother surface finish greatly reduces the microstrip conductor loss.

The advantages of the alumina substrate are its high temperature capability, which allows the parts that must be added to the microstrip circuit to be soldered to the circuit, its low thermal expansion, and its high dielectric constant, which leads to small size.

Microstrip characteristics are shown in Figure 11.2. Formulas were given in Chapter 6 for calculating microstrip line width as a function of substrate material and thickness for any desired impedance. A summary of some of these calculations is shown for alumina and Teflon-fiberglass substrates. For other materials or different values of impedance, the detailed formulas in Chapter 6 must be used. This summary table gives an idea of the size of the microstrip transmission lines in MICs.

The graph shows the guide wavelength as a function of frequency for the free-space case as a reference, and then for Teflon-fiberglass and alumina microstrip. The guide wavelength is reduced by a factor of 1.5 times in a 50-Ω Teflon-fiberglass microstrip line, and by a factor of 2.5 in a 50-Ω alumina transmission line, compared with the free-space value. A typical MIC microtrip line is .025 in wide on a .025 in thick alumina substrate. The characteristic impedance from Figure 11.2 is 50 Ω. Most microstrip circuits are based on quarter-wavelength dimensions; one quarter of a guide wavelength is 3 mm, at 10 GHz.

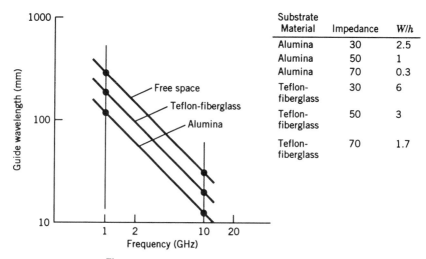

Substrate Material	Impedance	W/h
Alumina	30	2.5
Alumina	50	1
Alumina	70	0.3
Teflon-fiberglass	30	6
Teflon-fiberglass	50	3
Teflon-fiberglass	70	1.7

Figure 11.2 Microstrip line characteristics.

11.4 MICROSTRIP CIRCUIT ELEMENTS

The hybrid MIC consists of two parts: the microstrip transmission line circuit and the devices that are fabricated separately and added to the circuit.

Many microwave components are made of sections of microstrip lines and can be etched onto the substrate as the microstrip transmission line itself is being made. The components that can be fabricated as part of the microstrip transmission line are

- Matching stubs and transformers
- Directional couplers
- Combiners
- Resonators
- Filters
- Inductors and capacitors
- Thin-film resistors

Since they can be fabricated in the same fabrication process as the microstrip line, these components can be used either for hybrid MICs or for MMICs.

Figure 11.3 shows examples of microstrip circuit elements that are made of sections of microstrip line. These circuit elements have dimensions that are a fraction of a wavelength. The critical wavelength dimension is shown

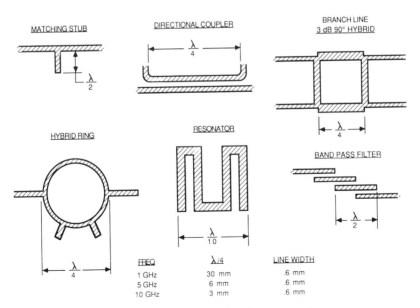

Figure 11.3 Microstrip circuit components.

on each element in Figure 11.3 to give an idea of the size of these elements in hybrid MICs. The table in Figure 11.3 shows the length and line width of a quarter-wavelength section of microstrip line on a 0.025-in. (0.6-mm) thick alumina substrate. (These dimensions were calculated from Figure 11.2.) At 10 GHz, a quarter wavelength is 3 mm, about the size of three paper clip wires.

Inductors and capacitors that can be fabricated with the microstrip line are illustrated in Figure 11.4, with the maximum capacitance or inductance that can be obtained. To understand the significance of these values, their reactances are shown at different frequencies in the following table, and can be compared to the 50-Ω characteristic impedance of microstrip transmission lines.

Frequency	Reactance (ohms)	
(GHz)	1 nH	1 pF
1	6	160
5	30	32
10	60	16

Inductors that can be fabricated during the photoetching process with the microstrip transmission line are shown on the left side of Figure 11.4. The loop and high impedance section of transmission line have inductances less than a nanohenry. If a narrow transmission line is formed into a spiral, high values of inductance can be obtained. The inductance of a spiral is

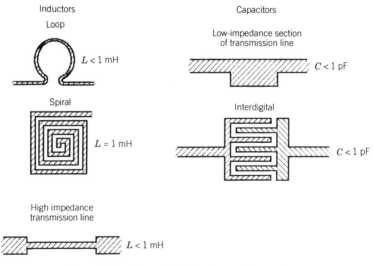

Figure 11.4 Microstrip inductors and capacitors.

proportional to the number of turns in the spiral squared. For the square spiral in Figure 11.4 the inductance is 20 nH.

Capacitors that can be fabricated with the microstrip line are shown on the right side of Figure 11.4. The low impedance section of transmission line can also be considered as a parallel plate capacitor, formed by the widened section of transmission line as the top plate and the microstrip ground plane as the bottom plate. Such capacitors have capacitances less than a picofarad. Another type of capacitor, which allows dc isolation between two circuits, is the interdigital capacitor. Here the capacitance is obtained by the fringing fields between one set of interdigital lines and the other. Again, the capacitance of the interdigital capacitor is less than 1 pF. If more than 1 pF is required, a MOS chip capacitor must be fabricated separately and added to the hybrid MIC.

Thin-film resistors, which can be fabricated as the microstrip lines are being fabricated, are shown in Figure 11.5. Thin-film resistors are used for microstrip line terminations and as the biasing resistors for transistors and diodes. Their fabrication also illustrates the fabrication process of the microstrip line. Figure 11.5 shows a cross section through a microstrip line, showing the alumina substrate and the various metallizings that form the resistors and the microstrip line. The thickness of the various metallizations are indicated, but are not drawn to scale. The substrate is 600 microns or 0.025 in. thick. After the substrate has been cleaned, a thin layer of tantalum nitride is sputtered onto the ceramic. This material forms the resistor, and its thickness is 0.06 microns. Sputtering thicknesses are usually specified in ang-

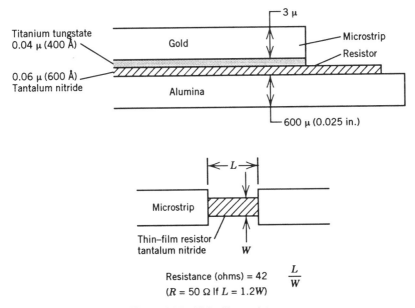

Resistance (ohms) = $42\ \dfrac{L}{W}$

($R = 50\ \Omega$ If $L = 1.2W$)

Figure 11.5 Thin-film resistors.

stroms, and 0.06 microns is equal to 600 Å. A layer of titanium tungstate is then sputtered on top of the tantalum nitride, with a thickness of 0.04 micron, or 400 Å. The titanium tungstate layer is used because the gold, which forms the microstrip conductor itself, adheres well to it. A layer of gold is then deposited with a thickness of 3 microns which is many skin depths. The gold forms the microstrip conductor.

After the substrate has been metallized, the microstrip transmission line patterns, including the microstrip transmission line itself, the various circuit elements, such as directional couplers, hybrids, and filters, and the inductors and capacitors shown in Figure 11.4 are photoetched. In the photoetching process, the metallized substrate is covered with a photoresist, a photomask containing the circuit elements is laid over the substrate, and the resist is exposed through the mask to ultraviolet light. The resist is then chemically removed in all areas except where the ultraviolet light has exposed it, and thus the conductor pattern is protected by the photoresist material. The gold and the titanium tungstate layer are then chemically removed in all areas except those covered by the photoresist, and this leaves the circuit pattern on top of the tantalum nitride layer. The photoetching process is then repeated to remove the tantalum nitride from all regions except where the thin-film resistors are to appear. The resulting resistors will appear as shown in the middle of Figure 11.5 connected to sections of microstrip line. The resistance of thin-film resistors that have a 600-Å-thick tantalum nitride layer is 42 L/W and is 50 Ω for a length-to-width ratio of 1.2.

11.5 COMPONENTS ADDED AFTER MICROSTRIP FABRICATION

The MIC components that are fabricated separately and added to the microstrip circuits are

- Bond wires
- Chip resistors
- Chip capacitors
- Dielectric resonators
- Circulators
- Diodes and transistors

These components are illustrated in Figures 11.6 through 11.9.

Bond wires are small, typically 0.0007 in. gold wires, which are used to connect the components to the microstrip line. Bond wires are attached by thermocompression bonding, which electrically heats the wire and applies pressure to bond the wire to the gold microstrip conductors. The bond wires act as small inductances, and so add inductance to the circuit, which usually is not wanted and must be compensated for in the microstrip circuit design. The inductance of a bond wire is typically 0.01 nH/mil. (1 mil = 0.001 in.)

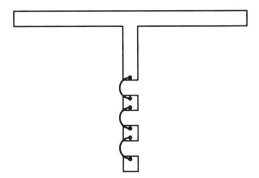

Figure 11.6 Bond wires.

For a 10 mil long bond wire (whose length is approximately three thicknesses of a piece of paper) its impedance at 10 GHz is 6 Ω, or approximately one tenth the impedance of a 50-Ω line. Often two bond wires in parallel are used to connect parts. The wires reduce the inductance to approximately one-half that of a single bond wire. Two wires also provide redundancy, so that if one bond wire breaks the circuit will still perform, although with a different value of connecting bond wire inductance. The trimming of MICs is accomplished by making or breaking the bond wire connections between components. Figure 11.6 shows a microstrip transmission line with a shunt capacitive tuning stub consisting of an array of individual capacitors connected by bond wires. The total capacitance connected in the circuit is more than is expected to be required. The circuit performance can be optimized by successively breaking the bond wires to reduce the amount of capacitance until the required value is obtained. The bond wires, since they add small amounts of inductance, can also be used for trimming.

If higher power resistors are required than can be obtained with the thin-film process, or if larger values of capacitance are required than can be obtained by the capacitors fabricated with the microstrip transmission line, then chip capacitors and resistors must be fabricated separately and added to the circuit. Examples of chip capacitors are shown in Figure 11.7, where the capacitors are connected to the microstrip transmission line by bond wires or bonding ribbons. The chip capacitors are formed by oxide layers between metal plates, and are called metal-oxide-semiconductor (MOS) capacitors. Their capacitance value can range from a fraction of a picofarad to a few hundred picofarads. Chip resistors are attached in the same fashion as chip capacitors.

Microstrip resonators are easily fabricated with the microstrip circuit, but have poor Q values and consequently do not provide sharp resonances. When a high Q is required, a dielectric resonator must be used. Dielectric resonators have high Q values of several thousand, almost as good as the Q of metal cavities, but are one sixth the size of a metal cavity resonator. Dielectric resonators appropriate for use with hybrid microwave circuits as shown in

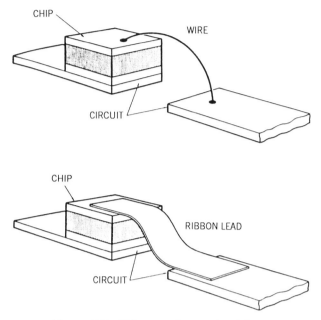

Figure 11.7 Chip capacitors and resistors.

Figure 11.8. The resonator is a cylinder made of barium tetratitanate. The resonator is mounted close to the microstrip line, and coupling is obtained between the microwave magnetic fields in the resonator and around the microstrip conductor. The dielectric constant of barium tetratitanate is about 36; therefore, the dielectric resonator is one sixth the size of a metal cavity resonator. The dielectric resonator is still relatively large, compared with the microstrip line, as shown in Table 11.3. Note that at 10 GHz, the diameter of the dielectric resonator is 150 miles and its height is 75 mils, whereas the microstrip transmission line is only 25 mils wide. At lower frequencies, the

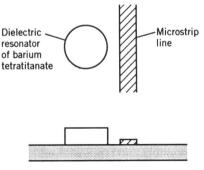

Figure 11.8 Dielectric resonators.

Table 11.3 Comparison of drelectric resonator diameter and height with frequency

Frequency (GHz)	Resonator Diameter (mils)	Resonator Height (mils)
2.5	600	300
5	300	150
10	150	75
15	100	50

size of the dielectric resonator is even larger and is much greater than the size of the microstrip line.

Y-junction circulators can be fabricated as part of the microstrip fabrication process if a ferrite substrate is used. Alternatively, circulators can be fabricated separately in a miniaturized format with the required ferrite material and magnet, and then added to the hybrid MIC. They can be used to form isolators by adding a resistor on one arm of the circulator.

Microwave diodes and transistors are the major circuit elements added to the microstrip circuit to form the hybrid MIC. The diodes and transistors used in MICs include varactors, PINs, Schottkys, bipolar transistors, FETs, IMPATTS, and Gunns.

The diodes and transistors may be mounted in the hybrid MIC as chips. The chips are soldered or epoxied to the substrate and are then connected to the microstrip transmission lines by wire bonds, as shown in Figure 11.9a.

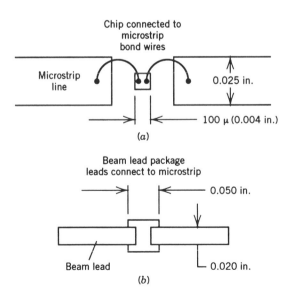

Figure 11.9 Connection of semiconductor devices to microstrip in a hybrid MIC.

This figure is drawn in relative proportion. The chips are typically 100 microns square, which is 0.004 in., whereas the microstrip transmission lines are 0.025 in. wide, or six times larger.

For ease in fabrication of the hybrid MIC, the diodes and transistors are often fabricated by their manufacturer in beam lead format. The diode or transistor comes in a package approximately 50 to 100 mils square with 20-mil-wide beam leads extending from the package, which connect easily to the 25-mil-wide microstrip line. A sketch of a beam lead package is shown in Figure 11.9*b*. Note that inside the package, bond wires must be used to connect the chip to the beam leads. The only difference between the two packaging schemes in Figure 11.9 is that in one case the manufacturer of the diode or transistor connects the beam lead to the chips inside the package, whereas in the other case the manufacturer of the hybrid MIC connects chips to the microstrip. In either case, bond wires must be used.

11.6 MOUNTING AND PACKAGING

The three levels of hybrid MIC packaging are illustrated in Figure 11.10. The first level is the packaging of the device itself. As just discussed, the transistor or beam lead diode chip is often mounted in its own beam package.

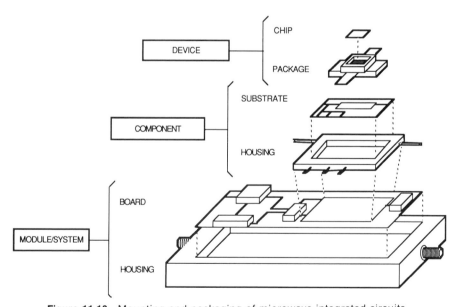

Figure 11.10 Mounting and packaging of microwave integrated circuits.

The next level involves mounting devices into the microstrip circuits. Diodes and transistors are included in this step, as well as the other devices used in the hybrid MIC, including chip capacitors, dielectric resonators, and circulators. Mounting is done with a thermosetting epoxy adhesive or by soldering the components to the substrate with a eutectic solder. Connections are made from the devices to the microstrip circuity by using thermal compression bonding of bond wires or ribbons. The bonding is accomplished by heat and pressure that fuse the wire or ribbon to the microstrip line. Another way to connect beam lead ribbons is with a conductive epoxy adhesive.

In the third level of integration, several hybrid microwave circuit components are assembled into a module or a subsystem. Several manufacturers may be involved in the packaging of a module. For example, the device manufacturer may do the packaging of the device, another component manufacturer may integrate the devices with the microstrip circuit to form a hybrid MIC, and a system manufacturer may combine various hybrid MICs to form the complete system.

11.7 MONOLITHIC MICROWAVE INTEGRATED CIRCUITS

In MMICs, the transistors and diodes and the microstrip transmission components are fabricated simultaneously. Therefore, the substrate material must have good insulating properties to serve as the dielectric for the microstrip transmission line, and it must also be a semiconductor that can be properly doped to make the diodes and transistors.

The losses of microstrip transmission lines substrates were compared in Table 11.3. Both silicon and gallium arsenide are poorer materials for a microstrip transmission line than alumina because they are not insulators but semiconductors.

Because of its very high attenuation, silicon is suitable for a MMIC only at low microwave frequencies. GaAs is commonly used for MMIC. Although its losses are higher than those of alumina, they are tolerable, and it can serve as the insulator for the microstrip line and as the active semiconductor material for the transistors and diodes.

The fabrication of different circuit elements that comprise an MMIC in a GaAs substrate are illustrated in Figure 11.11. Beginning in region 1, the basic substrate material is shown. The substrate is formed of undoped GaAs approximately 250 microns (0.010 in.) thick. The GaAs is metallized on the back surface to form a ground plane.

Region 2 shows the substrate used for a microstrip transmission line. A microstrip transmission line is formed by a metallized line on the upper surface. The GaAs serves as the microstrip insulating material.

Region 3 shows the use of the GaAs substrate for a Schottky diode. In this case the GaAs must be a semiconductor, so N doping is added in a small

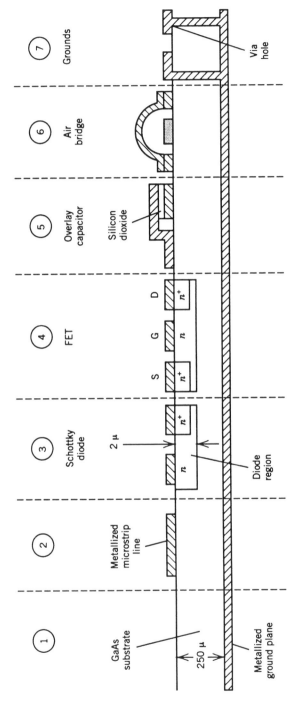

Figure 11.11 Monolithic microwave integrated circuits.

278

2 micron thick layer on the top surface of the substrate. A metal conductor is deposited on top of the N layer and an N-to-metal Schottky junction is formed. The other connection to the Schottky diode must be made through N^+ material, so a part of the substrate is further doped as N^+ and a metal contact is added.

Region 4 shows a FET. As with the Schottky diode, the semiconductor material is N-doped in a thin layer on the top surface of the substrate; N^+ doping for the source and drain connections are made, and then the metal connectors, which connect to the microstrip line, for the source and gate connections are made to the N^+ material. Finally, a Schottky barrier gate is made by depositing metal directly onto the N-doped semiconductor.

Region 5 shows the fabrication of an overlay capacitor. The capacitance that can be obtained by using microstrip lines is less than a picofarad. Thus, for higher capacitance values a silicon dioxide layer is deposited on top of a metallized conductor, and another metallized conductor, forming the upper plate of the capacitor, is deposited over the silicon dioxide layer.

Figure 11.12 shows a MMIC chip, illustrating the various elements just discussed. Note the use of air bridges, overlay capacitors, the inductance formed by a section of microstrip transmission line, and the thin-film resistor, used to provide the bias for the transistor. Notice also, along the front surface, an implanted resistor, which is obtained by N doping of the GaAs substrate itself. All of the circuit elements on the MMIC chip were formed by sequential doping and etching steps.

With MMIC techniques, complete microwave devices, such as amplifiers, oscillators, low-noise receivers, or complete microwave subsystems can be fabricated on a single chip.

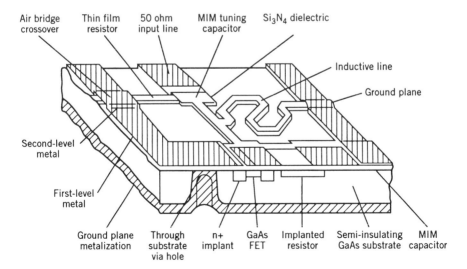

Figure 11.12 Typical MMIC.

ANNOTATED BIBLIOGRAPHY

1. P. H . Ladbrooke, *MMIC Design: GaAs FETs and MENTs,* Artech House, Dedham, MA, 1989.
2. J. Frey and K. Bhasin, *Microwave Integrated Circuits,* 2nd ed. Artech House, Dedham, MA, 1986.

References 1 and 2 cover the design and fabrication of microwave integrated circuits.

EXERCISES

11.1. What is the difference between a *hybrid* and a *monolithic* microwave integrated circuit?

11.2. What are five advantages of *hybrid* MICs compared to MMICs?

11.3. What are four advantages of MMICs compared to hybrid MICs?

11.4. What is the major advantage of each of the following substrate materials?

Teflon-fiberglass
Alumina
Ferrite
Gallium arsenide

11.5. What seven components can be fabricated as part of the microstrip line?

11.6. What six components are fabricated separately and added to the microstrip circuitry to form a hybrid MIC?

11.7. What is the advantage of a chip capacitor compared with a microstrip capacitor?

11.8. What is the advantage of a dielectric resonator compared with a microstrip resonator?

11.9. What is a bond wire?

11.10. How are tuning and tweaking accomplished in a hybrid MIC?

11.11. In what two ways are components mounted in a hybrid MIC?

11.12. In what two ways are components connected in a hybrid MIC?

11.13. Why is GaAs used in MMICs instead of alumina?

11.14. What is the purpose of each of the following components in an MMIC?
Air bridge
Overlay capacitor
Via hole

12

MICROWAVE TUBES

The advantages and disadvantages of microwave tubes and microwave semi-conductor devices are compared. Microwave tubes are then compared to each other with respect to their power generation capabilities, bandwidth, gain and efficiency. The operation of each tube is described. The microwave tubes discussed are gridded tubes, klystrons, helix traveling wave tubes, coupled-cavity traveling wave tubes, crossed-field amplifiers, and magnetrons.

12.1 ADVANTAGES AND DISADVANTAGES

Figure 12.1 compares the average power capability of tubes and microwave semiconductor devices as a function of frequency. The average power capability of microwave semiconductor devices is limited to about 100 W at 1 GHz, 10 W at 10 GHz, and 1 W at 100 GHz. In contrast, microwave tubes can provide over four orders of magnitude more power. If the power and frequency requirements of a given equipment lie below the solid-state curve, then solid-state devices should be used, since they have low weight, small size, and long life. If, however, the power and frequency requirements lie above the curve, as they do for most microwave systems, then a microwave tube must be used.

The much greater power generation capability of microwave tubes comes about because electrons travel approximately 1000 times faster in the vacuum inside the tube than they do in microwave semiconductor materials. In silicon or gallium arsenide, electrons travel at about 1/3000 the velocity of light. In a vacuum, electrons travel at approximately one third the velocity of light

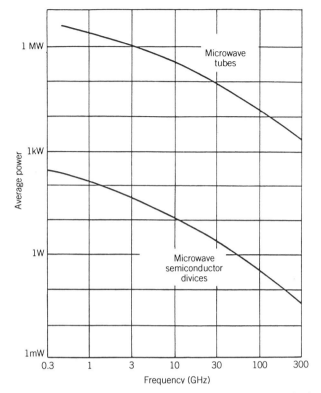

Figure 12.1 Comparison of microwave tube and semiconductor device power.

(depending on the exact voltage through which they have been accelerated). Therefore, at a given frequency, to obtain the same minimization of transit time effects, the internal dimensions of a vacuum tube can be 1000 times greater than the internal dimensions of a transistor, so several orders of magnitude more power can be obtained with a tube. The internal dimensions of a microwave tube are still very small. Realize that the internal dimensions of a microwave semiconductor device are approximately 1 micron or less, and although the tube is 1000 times larger its dimensions are still only millimeters. The very large microwave powers indicated in Figure 12.1 are generated inside microwave tubes that have only millimeter dimensions.

   The disadvantage of microwave tubes is their limited lifetime. The electrons in a microwave semiconductor device come from the semiconductor material itself, and there is therefore an unlimited supply. In contrast, the electrons in a vacuum tube must be emitted by a heated surface of barium metal, to 1000°C to permit sufficient electrons to be emitted. The barium continually evaporates from this hot surface as the tube is operated. When the barium has been depleted from the emitting surface, which is called the cathode, the tube can no longer perform, and this is the end of its life. Typical lifetimes of microwave tubes range from a few thousand to 100,000 hours. Note that one year of operating time is approximately 8500 hours, and many tubes have operating lifetimes of even less than a year.

The requirement for maintenance of a high vacuum of 10^{-11} atmospheres inside the tube is another disadvantage of microwave tubes. The loss of the vacuum by a small leak in the tube immediately renders it useless. Also, the requirement to obtain this vacuum complicates the fabrication of the tube. The tube must be made of special high-vacuum metal and ceramic materials. The metal and ceramics are brazed together at temperatures between 800 and 1000°C. The air must be pumped out of the finished tube while the tube is being heated to 600°C to drive out the gases that have been absorbed in the metal and ceramic parts. This complicated assembly process greatly increases tube cost. However, as shown in Figure 12.1, tubes are the only possible way of getting enough power to make most microwave systems practical.

Two possible arrangements for achieving high output power in a microwave system are shown in Figure 12.2. Figure 12.2*a* shows a low power microwave semiconductor oscillator generating a small microwave signal, which is amplified to the required high power level by an amplifier tube. Figure 12.2*b* shows a high power oscillator tube that generates the high required microwave power directly.

The low power oscillator–high power amplifier chain is used in most modern microwave systems because the microwave frequency must be carefully controlled, and it is difficult to control the frequency of a high power

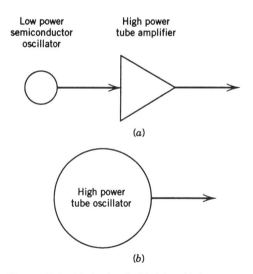

Low power
semiconductor
oscillator

High power
tube amplifier

(a)

High power
tube oscillator

(b)

Figure 12.2 Methods of obtaining high power.

oscillator. It is much easier to tune and stabilize the frequency of a low power microwave semiconductor source (using the various techniques discussed in Chapter 9), because the frequency-controlling elements operate at low power. Consequently, most microwave tubes are designed as amplifiers, taking the stabilized microwave signal generated by a low power microwave semiconductor oscillator and amplifying it up to the high power level required by the system.

In some systems microwave frequency stability is not particularly important so the high power oscillator represents a less expensive solution to the problem of generating the microwave power. Examples of such systems would be simple boat and commercial aircraft radars and microwave ovens. For these systems, the high power magnetron oscillator is used.

12.2 COMPARISON

Performance requirements for microwave tube amplifiers include

Peak power
Average power
Efficiency
Gain
Frequency
Bandwidth
Phase
Harmonic and spurious power
Intermodulation products.

These performance requirements were discussed for microwave semiconductor amplifiers in Chapter 8. The only difference in the requirements for tube and semiconductor amplifiers is the power level.

In many radar and other system applications, the transmitter is pulsed on for microseconds and remains off for thousands of microseconds before it is pulsed on again. (This allows the radar signal to be transmitted to a target, reflected by it, and then to return to be measured in the radar.) Microwave tubes are capable of providing several orders of magnitude more pulsed power than their already very high average power illustrated in Figure 12.1. Therefore both the peak (pulsed) and the average power capability of tubes must be considered.

The various types of microwave tubes are compared in Figures 12.3 and 12.4 and Table 12.1. Figure 12.3 compares the average power capability of microwave tubes. Figure 12.4 compares their peak power capabilities, and Table 12.1 compares their remaining characteristics, such as bandwidth,

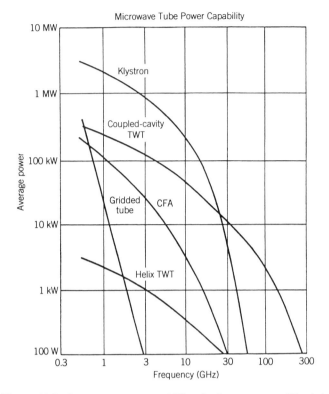

Figure 12.3 Average power capability of microwave amplifier tubes.

efficiency, and gain. Gridded microwave tubes, which are just like conventional low frequency tubes except that their internal elements are spaced close together to reduce transit time effects, provide both high average and high peak power, but only at the low frequency end of the microwave band. Gridded tubes are only useful below 1 GHz.

Klystrons provide the highest peak and average power, with over 1 MW of average power and up to 100 MW of peak power at the low end of the microwave band. Even at 10 GHz, klystrons are capable of providing almost 0.5 MW of average power. Coupled-cavity traveling wave tubes (TWTs), helix TWTs, and crossed-field amplifiers (CFAs) provide lower average and peak power, but have other advantages, such as wide bandwidth, high efficiency, and high gain.

Although klystrons provide the highest microwave peak and average output power, they are limited in bandwidth to only a few percent. In contrast, the helix TWT provides 100 times more bandwidth, but at lower efficiency and lower peak and average power levels. The coupled-cavity TWT has almost as good a power capability as the klystron (and even better power

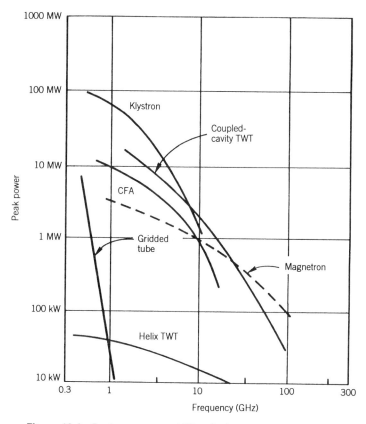

Figure 12.4 Peak power capability of microwave amplifier tubes.

Table 12.1 Comparison of microwave tubes

Type	Bandwidth (%)	Efficiency (%)	Gain (dB)	Relative Spurious Signal Level*	Relative Operating Voltage	Relative Complexity of Operation*
Gridded tube	1–10	20–50	6–15	2	Low	1
Klystron	1–5	30–70	40–60	1	High	2
Helix TWT	30–120	20–40	30–50	3	High	3
Coupled-cavity TWT	5–40	20–40	30–50	3	High	3
CFA	5–40	40–80	10–15	5	Low	4

* 1 = Best, 5 = poorest.

capability above 30 GHz), and it also has 10 times more bandwidth than a klystron, although at poorer efficiency. The crossed field amplifier has as good a bandwidth as the coupled-cavity TWTs and the best efficiency of any microwave tube—as high as 80%. However, the CFA has low gain, high noise, and only moderate average power.

Since some tubes are better in one characteristic, others in another, various types of microwave tubes are available to optimize the performance of a microwave system.

12.3 GRIDDED TUBES

Microwave gridded tubes operate according to the same principles as low frequency gridded tubes. The gridded tube consists of a cathode, grid, and plate. The weak microwave signal to be amplified is applied between the grid and the cathode. The signal applied to the grid controls the number of electrons drawn from the cathode. During the positive half of the microwave cycle more electrons are drawn; during the negative half less electrons are drawn. This modulated beam of electrons passes through the grid and goes to the plate. A small voltage on the grid controls a large amount of current; As this current passes through an external load it produces a large voltage, and the gridded tube therefore provides gain. The problems of lead reactance and transit time (which are not important for a low frequency tube) must be solved for a microwave gridded tube.

Gridded microwave tubes have small size, low weight, and low operating voltage; they are very efficient, easy to operate, and inexpensive. However, due to their limited frequency capability, they are useful only at the low end of the microwave frequency range.

A typical microwave gridded tube is shown in Figures 12.5 and 12.6. The tube consists of a microwave cavity assembly and the tube itself. Lead reactance is minimized by mounting the tube inside the microwave cavity. Transit time is minimized by using very small spacing between the tube electrodes. Figure 12.5 is a cross section of the tube and cavity assembly, and Figure 12.6 shows the tube in the foreground and the microwave cavity assembly into which it fits. The tube illustrated is designed to provide 200 W of CW power in the 850–870 MHz band, with a gain of 13 dB.

Gridded microwave tubes can only be used at the low end of the microwave frequency range because of their long transit time. Transit time limitation is illustrated in Figure 12.7, which shows the grid voltage, plate current, plate voltage, and electron trajectories from the cathode, through the grid, to the plate. All are shown as a function of time during one cycle. The dotted curves indicate plate current and plate voltage for the low frequency case; the solid curves show the effects at microwave frequencies.

Electron current is drawn from the cathode whenever the grid goes posi-

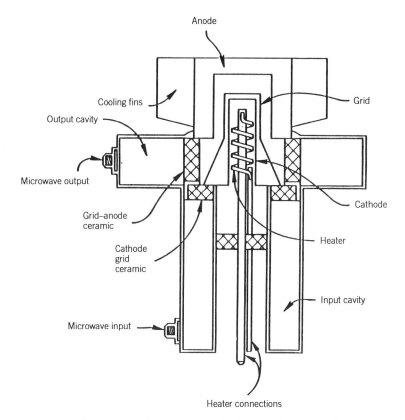

Figure 12.5 Cross section of gridded microwave tube.

tive. However, as shown in Figure 12.7*c*, electrons do not immediately reach the plate because of their finite transit time from the cathode through the grid to the plate. At low frequencies this transit time is an insignificant fraction of an RF cycle, so the plate current is a faithful reproduction of the grid voltage. Consequently, the plate voltage is also a faithful reproduction of the grid voltage. However, at microwave frequencies, the situation is much different. Due to the transit time effects the plate current and plate voltage are reduced in amplitude and have a distorted waveform.

Most of the electrons travel from the grid to the plate during the part of the microwave cycle when the grid is positive. As electrons pass from grid to plate, they generate a large microwave field in the output cavity. See Figure 12.8. The left side shows a bunch of electrons about to pass through the cavity. They have induced positive charges on the wall of the cavity near the grid. The center sketch shows the electrons midway across the cavity, and equal positive charges are induced on the cavity grid and plate walls.

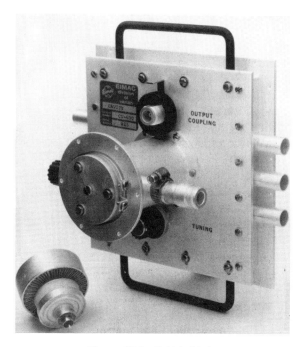

Figure 12.6 Gridded tube.

As the electrons travel across the cavity, electron current flows in the cavity walls to properly balance the positive charges. These induced wall currents generate the microwave field in the cavity.

The right side of Figure 12.8 shows the situation as the electrons are about to strike the plate side of the cavity. All the positive charges have been drawn to the cavity plate side. The output cavity is designed so that the microwave voltage across the cavity at the maximum of the microwave cycle is almost equal to the plate voltage.

As electrons pass through the cavity, they are slowed down by the microwave field, and they transfer their energy to the microwave field in the cavity. If the electrons could transfer all their energy to the microwave field, the efficiency of the gridded tube would be 100%. They cannot, however, because they would be slowed down so much that they could not get through the output cavity before the microwave field changed its direction. The typical efficiency of a microwave tube is therefore 30–50%.

Transit time effects can be reduced by moving the cathode, grid, and plate closer together, but even with a minimum grid-to-cathode spacing of a few tenths of a millimeter (a few thousandths of an inch), gridded microwave tubes are limited in performance to the low frequency end of the microwave band. All other microwave tube types solve the transit time problem by using velocity modulation. This technique actually uses transit time to generate a bunched beam of electrons rather than trying to minimize transit time effects.

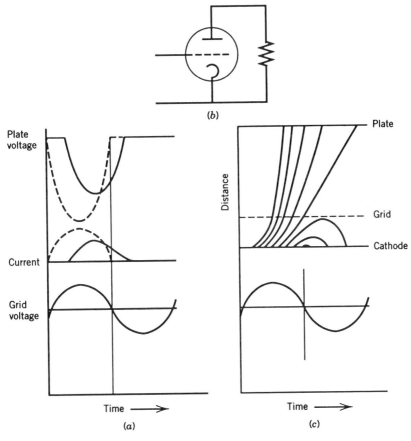

(b)

Plate voltage

Current

Grid voltage

Time ——→

(a)

Distance

Plate

Grid

Cathode

Time ——→

(c)

Figure 12.7 Transit time.

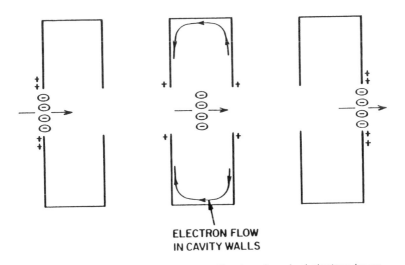

ELECTRON FLOW
IN CAVITY WALLS

Figure 12.8 Microwave power generation by a bunched electron beam.

12.4   KLYSTRONS

Klystrons have high peak power, high average power, good efficiency, high gain, and low spurious signals. They provide the highest peak and average power of any microwave amplifier tube up to about 30 GHz. The only disadvantage of klystrons is their narrow bandwidth of only a few percent.

Figure 12.9 is a schematic of a klystron amplifier. The major elements are

1. An electron gun to form and accelerate a beam of electrons
2. A focusing magnet to focus the beam of electrons through the cavities
3. Microwave cavities where the electron beam power is converted to microwave power
4. A collector to collect the electron beam after the microwave power has been generated
5. A microwave input where the microwave signal to be amplified is introduced into the klystron
6. A microwave output where the amplified microwave power is taken out

The microwave signal to be amplified is introduced into the input cavity, where the input signal modulates the electron beam. The bunched electron beam is successively amplified in the intermediate cavities. The bunched electron beam generates a large microwave signal in the output cavity. The

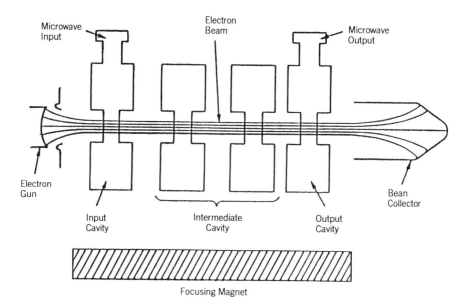

Figure 12.9   Klystron.

output cavity of a klystron amplifier is similar to the output cavity of the gridded tube. In both types of tubes, the modulated electron beam generates a microwave field as it passes through the output cavity. This microwave field decelerates the electrons, causing them to give up their energy to the microwave field. The microwave power is then taken out of the cavity into a transmission line. The major difference between the klystron and the gridded tube is the physical mechanism by which the bunched beam is formed before it enters the output cavity.

Figure 12.10 is a cross section of a klystron, and Figure 12.11 shows the tube mounted in its focusing magnet. This klystron provides 5 kW of average power at 10 GHz with 50% efficiency and 50 dB gain.

The basic velocity modulation process is illustrated in Figure 12.12 and 12.13. The process starts with the formation of a beam of electrons accelerated from a cathode by a high voltage. The accelerating voltage ranges from a few thousand volts up to several hundred thousand volts, depending on the power level of the microwave tube. The cathode, the accelerating anode, and the beam-forming electrode between the cathode and the anode that shape the electron beam comprise the electron gun.

In a velocity modulation tube the electron beam is drawn from the cathode with a constant voltage, so the number of electrons leaving the electron gun is constant with time, and all electrons have the same velocity. This is in contrast to gridded microwave tubes, where both the number of electrons and their velocity depend on the time instant during the microwave cycle that the electrons were drawn from the cathode.

The electron beam emerging from the electron gun is then shot through the input cavity of the klystron. A small microwave signal to be amplified is

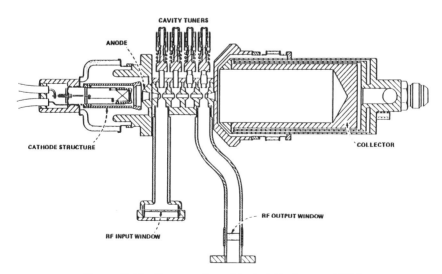

Figure 12.10   Cross section of typical klystron amplifier.

A. TUBE

B. TUBE IN FOCUSING MAGNET

Figure 12.11 Klystron and focusing magnet.

ELECTRON DISTANCE VS TIME

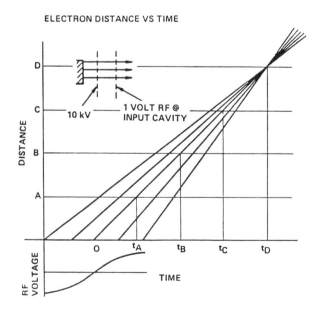

Figure 12.12 Velocity modulation.

applied to the input cavity, and the small electric field set up across the cavity by this input signal interacts with the electron beam. This interaction process is shown schematically in Figure 12.12 where trajectories of the electrons are shown as a function of time during one RF cycle. The electrons are initially accelerated by 10,000 V, and they reach approximately 20% of the velocity of light as a result. The microwave voltage across the input cavity is typically only 1 V, which is very small compared with 10,000 V, but it speeds up some electrons in the electron beam and slows down others, depending on the phase of the microwave signal in the input cavity as the individual electrons come through.

Electron 3 comes through the input cavity at the time in the RF cycle when the microwave field in the cavity is zero. Consequently, electron 3 is neither speeded up nor slowed down. It leaves the input cavity and travels down the axis of the klystron with its initial velocity, and at times T_A, T_B, T_C, T_D, and so on, reaches distances A,B,C,D, respectively.

In contrast, electron 1 entered the input cavity $\frac{1}{4}$ cycle earlier, when the microwave field was decelerating (that is, the voltage was negative), so electron 1 is slowed down and takes longer to reach a point down the tube. However, since it came through the input cavity sooner than electron 3, both electrons reach point D at the same time.

Similarly, electron 5 enters the input cavity $\frac{1}{4}$ cycle later than electron 3, when the field is accelerating, so electron 5 is speeded up. Electron 5 takes less time to reach point D because of its higher velocity. It too reaches point D at the same time as electrons 1 and 3. Likewise, electron 2 is slightly decelerated and electron 4 is slightly accelerated, and all of these electrons reach point D at the same time.

The velocity modulation process causes some electrons to be speeded up, others slowed down, depending on when during the RF cycle they come through the input cavity. As the electrons travel down the tube beyond the input cavity, the fast electrons catch up with the slow electrons and form a bunched electron beam.

The velocity modulation process is illustrated in another way in Figure 12.13, which is a three-dimensional plot showing the time during two microwave cycles (horizontally), distance along the tube into the plane of the paper, and the number of electrons (vertically).

Note that the same number of electrons arrive at the input cavity at all times during the microwave cycle, since the electrons are accelerated from the cathode by a constant voltage. As the electrons pass through the input cavity, some are slowed down and some are speeded up, depending on when they pass through the input cavity. At some distance A beyond the input cavity, the fast electrons are beginning to catch up with the slow electrons, and more reach point A at one time during the microwave cycle than at other times.

When the electron beam has traveled to a distance B beyond the input cavity, more of the fast electrons have caught up with the slow electrons

CURRENT MODULATION

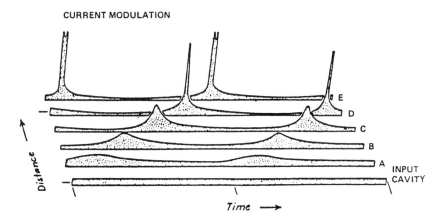

Figure 12.13 Additional explanation of velocity modulation.

and a more tightly bunched beam is formed. The bunching process continues to point C, and at point D almost all of the fast electrons have overtaken the slow electrons, so a tightly bunched beam is formed, with almost all of the electrons reaching point D at one instant of time during the microwave cycle.

One complete cycle later, a second bunch of electrons arrives at point D. At point E the faster electrons have passed the slow electrons, and the bunched beam is beginning to disperse.

Figure 12.13 suggests that the electron beam could be completely bunched even with a very weak microwave signal if the electron beam were allowed to travel a sufficient distance to give time for the slightly faster electrons to catch up with the slightly slower electrons. This would be true except for space-charge forces within the electron beam. As the fast electrons begin to catch the slow electrons, and the bunch begins to form, the mutual repulsion of the negatively charged electrons tends to push the bunch apart, so the velocity modulation process must be enhanced by adding intermediate cavities along the klystron.

The velocity modulation interaction process in a multicavity klystron is illustrated in Figure 12.14. This figure shows the microwave voltage, the electron velocity, and the microwave current as functions of the distance along the klystron, from the gun, through the four cavities, to the collector. The upper curve shows the voltage applied to the electrons relative to the cathode voltage. The middle curve shows the velocity of the electrons, and the lower curve shows the microwave beam current. The electrons are all accelerated from the cathode through the same velocity as they enter the first cavity. As the electrons pass through the input cavity, some see an increased voltage if they pass through the cavity at a time when the microwave signal is accelerating.

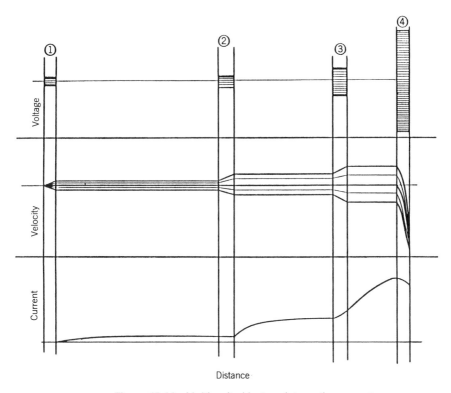

Figure 12.14 Multicavity klystron interaction.

Other electrons are decelerated if they pass through the input cavity when the phase of the microwave signal is decelerating. As a result of this velocity modulation, some electrons are speeded up as they pass through the input cavity, and others are slowed down, as shown by the middle curve.

As the partially bunched electron beam travels through the second cavity, the small RF current that exists because of the partial bunching induces a microwave field into this cavity. This microwave field applies additional velocity modulation to the electron beam. Note that the accelerating and decelerating voltages are larger in the second cavity than they were in the input cavity; consequently, the faster electrons are speeded up more, and the slower electrons are slowed down by a greater amount. In the input cavity the microwave field was supplied by the external microwave signal that was to be amplified. In the second cavity, however, the microwave field is generated by the partially bunched electron beam, and the resulting field interacts with the beam to further speed up the fast electrons and further slow down the slow electrons.

This greater velocity modulation allows the electrons to better override the space-charge forces, and the fast electrons come closer to catching the slow electrons, as the electrons travel to the third cavity.

The space-charge forces, however, still prevent the electrons from becoming completely bunched, and at the point where the maximum current is reached the third cavity is located. The velocity modulation process is enhanced in the same way in the third cavity, and by the time the electron beam reaches the output cavity most of the fast electrons have caught up with the slow electrons and the electron beam is completely bunched.

A large microwave field is generated as the tightly bunched beam passes through the output cavity. This microwave power is taken out of the cavity through the output window. The output cavity is designed so that the microwave voltage across the cavity, at the location of the electron beam, is approximately equal to the original accelerating voltage, shown in the upper curve of Figure 12.14. The microwave field in the output cavity is generated by the bunched beam of electrons, and the resulting microwave field is decelerating at the time in the microwave cycle when the electron bunch is passing through the cavity, so the electrons are slowed down, as shown in the middle curve.

All the electrons lose energy as they go through the output cavity, but some electrons lose more energy than others, depending on their phase as they enter the output cavity, and their phase depends on the time during the microwave cycle that they entered the input cavity. This reduction of the electron beam energy in the output cavity is the source of power that supplies the amplified microwave power.

In a typical klystron, the electron beam still has 70% to 50% of its initial energy as it leaves the output cavity, so typical klystron efficiencies range from 30% to 50%. Note that no matter how weak the input signal is, the velocity modulation process can be enhanced up to its optimal value at the output cavity by adding more intermediate cavities.

Figure 12.15 shows a cutaway photograph of a high peak power klystron. This klystron provides 1 MW of peak power at 3 GHz with 50 dB gain. Also shown is the detail of the klystron cavities. The cavities are cylindrical resonators, with noses projecting from each side in the center of the cavity to concentrate the electric field at the electron beam. The cavities contain tuning mechanisms so that the center frequency of the klystron can be mechanically adjusted over a 10% range. The instantaneous bandwidth of the klystron is only a few percent. The input microwave signal is introduced into the first cavity from a coaxial transmission line by the coupling loop.

The microwave power generated in the output cavity is coupled from the cavity into a waveguide and out of the tube through a ceramic window, which allows the microwave power to pass into the external waveguide.

After the electron beam is formed, it must be focused through the several klystron cavities from the gun to the collector. Without this magnetic focusing field, the beam would hit the klystron cavities due to mutual repulsion of the negatively charged electrons. Four methods of obtaining the required focusing field are shown in Figure 12.16. An electromagnet yoke is shown in A, and a photograph of such a yoke was shown in Figure 12.11. The yoke magnet is formed by two electromagnet coils in an iron yoke that carries the

Figure 12.15 Multicavity klystron.

magnetic field from the coils to the cavity region of the klystron. The gun and collector are inside the iron yoke and so are shielded from the main magnetic focusing field. The advantage of the electromagnet yoke is that the klystron cavities are accessible for tuning. Its disadvantage is its large size and weight and the 500–2500 W of power required to operate the coils.

The size and weight of the focusing structure can be greatly reduced by using a solenoid (B) instead of a yoke. The solenoid requires power to operate it. An opening must be cut into the solenoid to provide access for tuning the klystron cavities.

The power requirements of the yoke or solenoid can be eliminated by using a permanent magnet (C), either of a barrel or yoke design. Its disadvantages are the large fringing magnetic field, which can adversely affect surrounding electronic equipment, and the weight of its yoke.

The size and weight of the permanent magnet can be drastically reduced by using "periodic" permanent magnets for focusing (D). This design is used on some klystrons, and is widely used on traveling wave tubes. With periodic

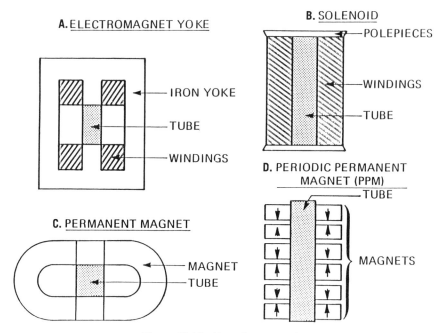

Figure 12.16   Focusing magnets.

permanent magnet focusing, an array of ring magnets, with alternating polarities, is used. The electron beam is focused by the first magnet, allowed to expand, focused again by the next magnet, allowed to expand, focused again by the third magnet, and so on. By alternating the magnet polarities, the external field, and hence the magnet size and weight, are greatly reduced.

The collector collects the electron beam after it leaves the output cavity where the maximum microwave power has been extracted. Because klystron efficiency is 30–50%, an appreciable amount of beam power is still left in the electron beam as it enters the collector. The collector can be made as large as necessary to obtain the required reduction in beam power density at the collector surface. As the electrons strike the collector surface, they transfer their power as heat, and this heat must be carried away from the collector by air or liquid cooling.

12.5   TRAVELING WAVE TUBES

The main disadvantage of a klystron is its small bandwidth. In contrast, the traveling wave tube (TWT) has wide bandwidth (30–120%), high gain, moderate peak power, and moderate average power. Its main disadvantage is low efficiency, the worst of all microwave amplifier tubes. However, in many applications, its wide bandwidth greatly outweighs its poor efficiency.

A schematic of a TWT is shown in Figure 12.17. The major elements include:

1. An electron gun to form and accelerate a beam of electrons
2. A focusing magnet to focus the beam of electrons through the interaction structure
3. A collector to collect the electron beam after the microwave power has been generated
4. An input window where the small microwave signal to be amplified is introduced to the interaction structure
5. An interaction structure, where the electron beam interacts with the microwave signal to be amplified
6. A microwave output window, where the microwave power is taken out of the tube
7. An internal attenuator, to absorb the power reflected back into the tube from mismatches in the output transmission line.

The electron gun, focusing magnet, collector, and input and output windows of a TWT are the same as those used in a klystron and have identical functions. The major difference between a klystron and a TWT is that the klystron cavities are replaced by the TWT interaction structure.

Figure 12.18 is an exploded photograph of a TWT. The tube provides 200 W of CW power, with 40 dB gain, and an overall efficiency of 25% from 5 to 10 GHz. It uses the same basic velocity modulation process as the klystron. The process starts with the formation of a beam of electrons by the electron gun. All electrons have the same velocity. The electron beam

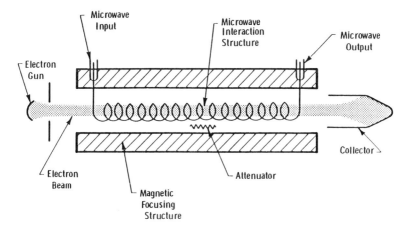

Figure 12.17 Traveling wave tube

Figure 12.18 Typical helix traveling wave tube.

emerging from the electron gun is then shot into the interaction structure. The microwave signal to be amplified is put into the input end of the interaction structure. The electrons in the electron beam and the microwave signal interact as they travel through the interaction structure. The key feature of the TWT is that the electrons travel at the same velocity as the microwave signal on the interaction structure.

If an electron enters the interaction structure during the positive phase of an RF cycle, it stays in the positive phase of the cycle as it travels with the microwave signal through the interaction structure. It is continuously accelerated. If the electron enters the interaction structure during the negative phase of a cycle, it stays in the negative part of the cycle and is decelerated continuously as it travels through the interaction structure. Therefore, depending on the phase at which they enter the interaction structure, some electrons are speeded up and others are slowed down. As the electrons progress along the length of the TWT, the fast electrons begin to catch the slower electrons, and a bunched beam is formed that excites an increasing microwave signal on the interaction structure. This interaction process is shown schematically in Figures 12.19 and 12.20.

Figure 12.19 shows the electric field in a TWT interaction structure at three instants of time, and the location of two representative electrons. At time 0, electron 1, which had entered the interaction structure earlier, is in a position where the electric field is accelerating. Electron 2 has not yet entered the interaction structure. At a time $\frac{1}{2}$ RF cycle later (middle figure) electron 1 has traveled further down the interaction structure. The microwave signal, which was fed into the input end of the interaction structure at the same velocity as the electron, has traveled the same distance, so electron 1 is still in an accelerating field. Electron 2 by this time has entered the interaction structure and is in the same position where electron 1 was at time 0. Note, however, that the electric field at this same position is now decelerating, so

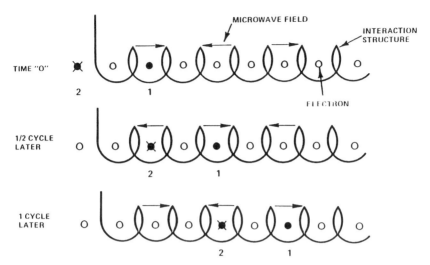

Figure 12.19 Velocity modulation in a TWT.

electron 2 is being slowed down. At a time one cycle later (bottom figure) electrons 1 and 2 have moved further down the interaction structure and the microwave signal has moved down the interaction structure at the same velocity, so the electric field is still accelerating electron 1 and decelerating electron 2.

The TWT interaction process is shown in another way in Figure 12.20. The interaction effects of the microwave signal, electron density, and electron velocity are shown as a function of distance along the TWT. The dark areas represent the decelerating phase, the light areas the accelerating phase. At the input end of the tube (left side) the microwave signal is small, the electron density in the electron beam is uniform, and the electron velocity is constant. As the electrons and the microwave signal progress through the interaction structure, some electrons are slowed down, some are speeded up, and bunches of current begin to form. These bunches excite an increasing microwave signal onto the interaction structure. Near the output end of the TWT the microwave power on the interaction structure grows to about 20% of the beam power. At this point the beam is tightly bunched, with most of the electrons grouped together in the decelerating phase of the RF cycle. As the energy has been extracted from the beam, almost all of the electrons have been slowed down so much that they no longer travel at the same velocity as the microwave signal, and traveling further through the interaction structure they would get into the accelerating field and absorb power from the microwave field rather than generate it. The microwave signal is therefore taken off of the interaction structure at the point where the maximum signal level occurs.

A. **MICROWAVE SIGNAL**

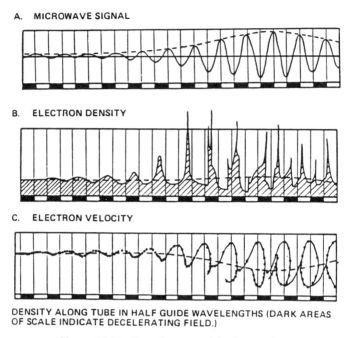

B. **ELECTRON DENSITY**

C. **ELECTRON VELOCITY**

DENSITY ALONG TUBE IN HALF GUIDE WAVELENGTHS (DARK AREAS OF SCALE INDICATE DECELERATING FIELD.)

Figure 12.20 Traveling wave tube interaction.

In part C, although most electrons are slowed down as they give up energy to the RF wave, some are actually speeded up, since they have gotten into the wrong phase. As the electrons leave the interaction structure to enter the collector, they have a large range of velocities. Some even have a greater velocity than when they were initially accelerated from the cathode.

The klystron achieves high gain per unit length, and high efficiency, at the expense of bandwidth by using resonant cavities, which provide a high microwave interaction field. The TWT, since it does not use resonant cavities, can provide extremely wide bandwidth, but the microwave interaction fields are much weaker. However, because the interaction is allowed to build up continuously, high gain and power can be achieved.

All of the klystron power is extracted from the output cavity. The beam can thus be tightly bunched and remain that way as it passes through the cavity. In contrast, the buildup of power in the TWT is over a considerable length of the interaction structure, and the electrons get out of phase. Therefore, TWT efficiency is limited to 10–30% compared with the klystron's 30–50%.

All klystrons use the same basic types of cavities. In contrast, two types of interaction structure are used for traveling wave tubes (see Figure 12.21). The helix is made of tungsten wire and is supported on ceramic rods inside

A. HELIX

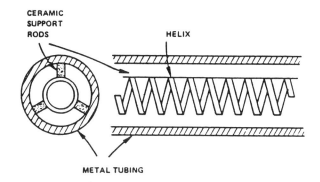

B. COUPLED CAVITY

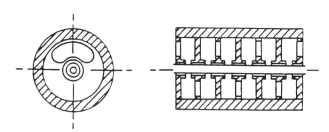

Figure 12.21   Traveling wave tube interaction structures.

a metal tubing, that forms the vacuum envelope of the tube. The helix can cover a wide bandwidth. For example, the entire frequency range of 2 to 8 GHz can be covered with a single TWT using a helix interaction structure. However, it has low average power capability because of its fragile nature.

The coupled-cavity interaction structure, on the other hand, solves the average power limitations of the helix, but at the expense of bandwidth. It consists of an array of klystron-like cavities, with each cavity coupled to successive cavities through a kidney-shaped opening in the cavity wall. Since the interaction structure is made entirely of metal, heat can easily be conducted from the region surrounding the electron beam out to the outer walls of the cavities that form the vacuum envelope. This structure's average power capability is practically as good as the klystron's, and its bandwidth can be from 5% to 40%, depending on the size of the coupling hole.

A cross section of a coupled-cavity TWT is shown in Figure 12.22, along with a piece of waveguide for size comparison. This tube is designed to provide 10 kW of peak power and 1 kW of average power over a 20% frequency range at 10 GHz. The electron gun can be seen on the left, and the

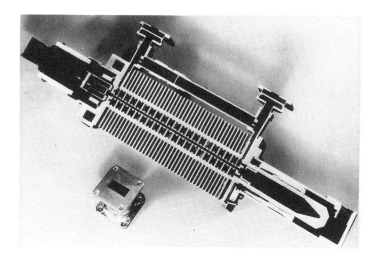

Figure 12.22 Coupled-cavity TWT.

collector at the right. The coupled-cavity microwave interaction structure is in the center of the tube between the input and output waveguides.

The electron gun and focusing magnets of a TWT serve the same function as those elements in a klystron. Most TWTs, including those in Figures 12.18 and 12.22, use periodic permanent magnetic focusing.

As in a klystron, the collector of a TWT collects the electron beam after the power has been extracted from the beam in the interaction structure. In the klystron, about 50% of the electron beam power is converted into microwave power, whereas in a TWT only about 15% is. The electron beam, therefore, still contains 85% of its original power when it enters the collector.

The overall efficiency of the TWT can be increased by depressing the collector—that is, operating the collector at approximately half of the interaction structure voltage. A block diagram of a TWT with a depressed collector, illustrating the relationships between the tube and its power supplies, is shown in Figure 12.23. The electron beam is accelerated from the cathode to the interaction structure by a 10,000-V interaction structure power supply. As electrons pass through the interaction structure, they are slowed down as they convert 15% of their power to microwave power.

Very few of the electrons strike the interaction structure, and they have 85% of their original power as they leave the structure and enter the collector. This power can be regained by slowing down the electrons, by operating the collector at a voltage midway between the cathode and interaction structure voltages. The electron beam is ultimately collected at half the original accelerating voltage. In this manner, the efficiency of the TWT can be approximately doubled, from 15% to 30%.

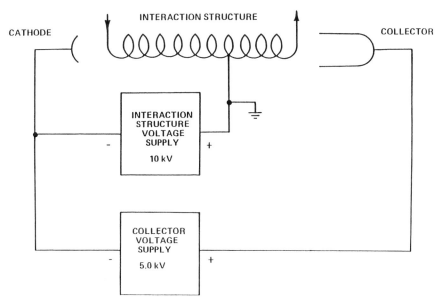

Figure 12.23 Depressed collector operation of TWT.

12.6 CROSSED-FIELD AMPLIFIERS

The low efficiency problem of the TWT is solved, while retaining its good bandwidth capability, by the crossed-field amplifier (CFA). The CFA has high efficiency (up to 80%—the highest of any microwave amplifier), high peak power, and wide bandwidth. However, it has low gain, high noise output, and low average power capability.

The CFA is shown in Figure 12.24. Its major elements are

1. An electron gun to form and accelerate a beam of electrons
2. An interaction structure, where the electron beam interacts with the microwave signal to be amplified
3. A microwave input where the weak signal to be amplified is introduced into the interaction structure
4. A microwave output where the microwave power is taken out of the tube
5. A collector to collect the electron beam after the microwave power has been generated
6. A sole electrode
7. An external magnet to provide a magnetic field in the interaction region.

The CFA has the same basic parts as the TWT, but a different type of velocity modulation interaction is used to overcome the TWT's low efficiency. The

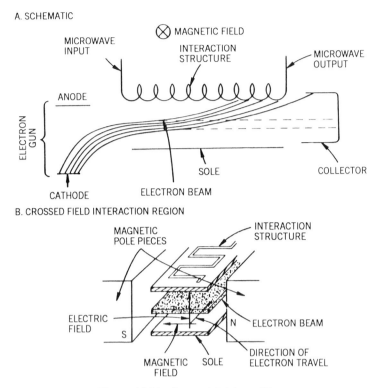

A. SCHEMATIC

Figure 12.24 Crossed-field amplifier.

velocity modulation process in the CFA occurs in a region of crossed electric and magnetic fields; the interaction region is shown in Figure 12.24B. This interaction structure is shown at right angles to the schematic in A. The electron beam in a CFA is a strip beam that travels in the region between the interaction structure and a negative electrode called the sole. The electron beam is shown moving into the plane of the paper. The sole provides an electric field at right angles to the direction of electron beam travel. This field forces the electrons toward the interaction structure. The magnetic field in a CFA provides a magnetic field at right angles to the direction of electron travel and at right angles to the direction of the electric field. The combined action of the moving electron beam and the magnetic field provides a force that counteracts the electric force developed by the sole and pushes the electrons away from the interaction structure. The voltage on the sole and the strength of the magnetic field are adjusted so that the two forces exactly counteract each other. When no microwave input is applied to the tube, the electrons move in a straight path along the tube axis (dashed line representing the electron beam in A). The electric field between the sole and the interaction structure, the magnetic field, and the electron velocity, are all "crossed"; that is, they are at right angles to each other. This is the key feature of the interaction process, which gives the CFA its name and its high efficiency.

The magnetic field in a klystron or TWT is used just to focus the electron beam through the cavities or the interaction structure. These tubes would still work without magnetic focusing, for example, if the beam were electrostatically focused through the tube. In the CFA, however, the magnetic field not only confines the beam but it is an essential part of the velocity modulation process. The crossed-field interaction process is illustrated in Figure 12.25. Part A shows the microwave field in the interaction region and the bunching forces on the electrons. If no microwave field were present, the electric force from the electric field between the sole and interaction structure would be exactly balanced by the magnetic force of the magnetic field and the moving electron beam. Therefore, the electron would move in a straight line along the tube axis (dashed line).

When a microwave field is applied to the interaction structure, the electrons at position A are speeded up, those at position B are slowed down. Just as in a TWT the velocity of the microwave signal along the tube axis is made equal to the velocity of the electrons. An electron in position A will therefore see the same field as it moves along the tube. As electron A is speeded up, the magnetic force, which depends on the strength of the magnetic field and

A. MICROWAVE FIELD AND FORCES ON ELECTRONS

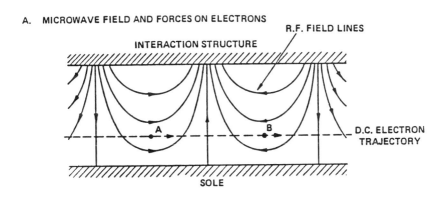

B. ELECTRON BEAM BUNCHING

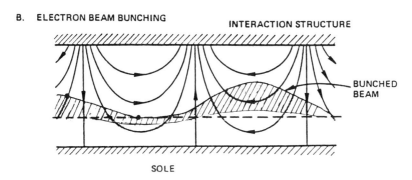

Figure 12.25 Crossed-field interaction.

the magnitude of electron velocity, increases and becomes greater than the electric force created by the negative sole. So the electron, instead of moving in the axial direction, moves away from the circuit. In the same way, the electron at position *B* slows down, and the electric force from the sole becomes greater than the magnetic force, and the electron moves toward the interaction structure.

This differential movement, depending on the position of the electron during the phase of the microwave field, causes a velocity modulation bunching to occur, and the resulting bunched beam is shown in Figure 12.25B. Because of the crossed electric and magnetic fields, the electrons do not bunch axially as the beam moves along the interaction structure, but they bunch closer to, or further from, the interaction structure. As in a TWT, as the bunches form they interact with the interaction structure to generate an increasing microwave field, which further bunches the beam. Unlike a TWT, however, the electron velocity along the tube axis remains the same. The electrons do not slow down as power is transferred from the electron beam to the microwave field. The power for the microwave field comes from the potential energy of the electrons. Because the electrons don't slow down as they travel along the interaction structure, they don't lose synchronism. The interaction process can be continued until almost all the power is transferred from the beam to the interaction structure. Consequently, the CFA has extremely high efficiency.

This basic crossed-field interaction process, which provides the high efficiency, causes the CFA to have low gain, high noise, and only moderate average power capability. The beam moving in the crossed electric and magnetic field region is basically unstable; that is, any small noise modulation will enhance itself. Consequently, if the CFA's interaction region is made very long, the noise that is always present in the beam due to nonuniform emission of electrons from the cathode will be amplified and become greater than the desired signal.

The interaction region in a klystron or TWT can be as long as necessary to get the required gain. But the length of a crossed-field interaction region must be limited so that noise will not build up. Consequently, CFAs can have no more than 10–15 dB gain, whereas klystrons and TWTs can have 50–60 dB. A CFA must have a chain of other CFAs or a TWT to drive it, whereas a klystron or TWT, because of their high gain, can be driven from a low-level solid-state source. Even at the low gain levels of 10–15 dB, the CFA still has a noisier output than a high-gain klystron or TWT at the same output power level.

For highest efficiency, the electron beam in a CFA must move very close to the interaction structure, and in most CFAs the beam actually strikes the interaction structure. (See Figure 12.24.) Most of the power has been extracted from the beam, but the fraction that remains is converted to heat, which must be absorbed by the interaction structure. In contrast, in a klystron or TWT only a small fraction of the beam strikes the interaction structure.

Consequently, the average power of a CFA is much less than that of a klystron or coupled-cavity TWT, which collect the unused electron beam in a separate collection region rather than on their interaction structures.

A coupled-cavity interaction structure is used for a CFA to provide peak power levels up to several megawatts with 20% bandwidth. The electron gun of a CFA is different from that of a klystron or a TWT. As shown in Figure 12.24, as electrons are emitted from the cathode they are initially accelerated to the anode. Note that the cathode is approximately at the same voltage as the sole, and the anode is at the same voltage as the interaction structure. However, as electrons begin to move away from the cathode, they are acted on by the magnetic field and their trajectories are bent, so the electron beam enters the interaction region and moves at right angles to the cathode.

Some CFAs are actually made as in Figure 12.24, with the electron beam injected into the interaction structure. This figure, however, was mainly used to compare the CFA to the klystron and TWT. Most CFAs use an emitting-sole type of electron gun, as shown in Figure 12.26. The sole is made to be

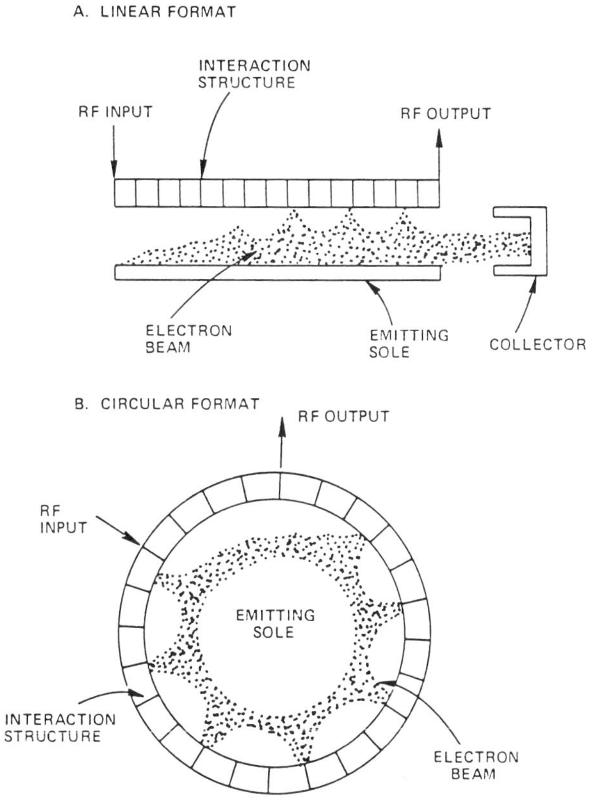

A. LINEAR FORMAT

B. CIRCULAR FORMAT

Figure 12.26 Emitting sole crossed field amplifer.

the cathode. Electrons are emitted from it and, because of the action of the magnetic field, are bent to move down the tube axis. The electron beam current is small at the input end and increases toward the output end. The emitting sole can provide much higher current than the gun of a klystron or TWT. Consequently, for a given power level the CFA can operate at a much lower voltage than other tubes. For example, a klystron providing 1 MW of peak power requires a beam voltage of 100 kV, whereas a CFA providing 1 MW of peak power operates at a beam voltage of 40 kV.

For practical construction reasons, the emitting-sole CFA is usually made in a circular format (Figure 12.26B). The emitting sole becomes a cylinder, which is easier to heat to obtain electron emission. The interaction structure in a circular format is more compact, and it is much easier to apply the crossed magnetic field. In Figure 12.26 the magnetic field is applied into the plane of the paper. The collector in the circular format CFA is eliminated, because most of the electrons strike the interaction structure. A typical CFA using an emitting sole in circular format, with a coupled-cavity type of interaction structure, is shown in Figure 12.27. The tube itself is shown in Figure 12.28. This tube provides 1 MW of peak output power from 5.4 to 5.9 GHz, with 13 dB gain and 60% efficiency.

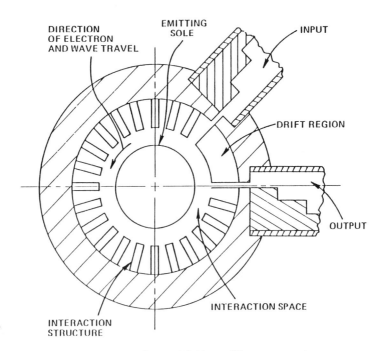

Figure 12.27   Crossed-field amplifier cross section.

Figure 12.28 Typical crossed-field amplifer.

12.7 MAGNETRONS

The gridded tubes, klystrons, helix TWTs, coupled-cavity TWTs, and CFAs are used to amplify microwave signals generated by semiconductor oscillators and amplifiers at the watt level up to the high power levels required for the transmitters in microwave systems. The magnetron, in contrast, is a high power tube oscillator. It has high peak power, moderate average power, low cost, and the high efficiency—over 80% in some tubes. However, its noise output is high, and it is difficult to maintain a stable output frequency with it. Magnetron operation can best be understood by thinking of it as a CFA with its output connected directly to its input.

Figure 12.29 is a cross section of a magnetron. The major elements are

1. The cylindrical cathode where electrons are emitted (like the emitting sole in a CFA)
2. The anode, where the microwave signal is propagated. (like the interaction structure in a CFA)
3. The output, where the microwave signal generated in the magnetron is taken out into an external transmission line
4. The magnet, which provides the magnetic field needed for the crossed-field interaction

The magnetron is similar to the emitting-sole CFA. The magnetron has no input because it is an oscillator, not an amplifier. The interaction process is the same in the magnetron as in the CFA. The microwave signal travels along the anode, which is a coupled-cavity type of interaction structure, with the vanes forming the cavities. Electrons are emitted from the cathode, but instead of being accelerated to the anode they are bent by the combined effect of the electric and magnetic fields to move around the cathode.

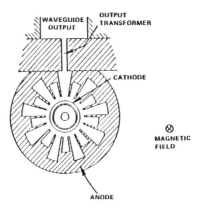

Figure 12.29   Magnetron.

As the electrons bunch, under the influence of the microwave field on the anode, they move toward the anode, exchanging potential energy to build up the microwave field, just as in a CFA, and finally they strike the anode. The output of the magnetron is fed directly into its input. In fact, the interaction region is continuous. Oscillation therefore occurs, and power is coupled from the anode circuit to an output transmission line.

A magnetron can be fixed-tuned or mechanically tuned. A typical fixed-tuned magnetron is shown in Figure 12.30. This magnetron provides 600 W of CW output power at 2450 MHz and is designed for use in microwave ovens. Several million of these magnetrons are made each year, and the large-quantity cost is less than $20.

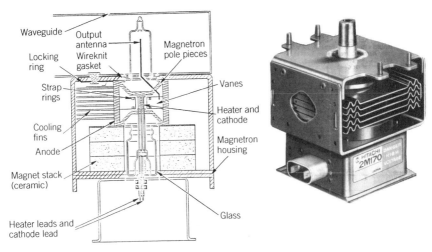

Figure 12.30   Magnetron for microwave ovens.

12.8   HIGH POWER MICROWAVE TUBES

The microwave tubes described in the previous sections are used in microwave communications, radar, electronic warfare, and navigation systems, and provide adequate transmitter power for all these applications.

Even higher power tubes, providing over 100 MW of peak power, are currently being developed for special applications, which include

1. Microwave beam weapons to destroy the electronic guidance and control systems of enemy missiles
2. Power sources for accelerating nuclear particles for nuclear research
3. Trigger power sources to start the thermonuclear power generation processes
4. Power sources to radiate nuclear waste to reduce its half-life

Further details of some of these scientific uses of microwaves are given in Chapter 19.

These high power microwave tubes include the gyrotron, free-electron laser, relativistic magnetron and others. They use variations of the same velocity modulation techniques that are used in conventional microwave tubes. They all operate at such high beam voltages that their electron velocity approaches the velocity of light, and relativistic effects must be considered in the tube design. In fact, some of these tubes would not work at all except for the effects of relativity. The operation of these high power tubes is not discussed in this book, but is described in reference 2.

ANNOTATED BIBLIOGRAPHY

1. A. S. Gilmour, *Microwave Tubes*, Artech House, Dedham, MA, 1986.
2. V. L. Granatstein and I. Alexeff, *High-Power Microwave Sources*, Artech House, Dedham, MA, 1987.

Reference 1 covers the microwave tubes discussed in this chapter. Reference 2 covers "high-power" microwave tubes at peak power levels above 100 MW.

EXERCISES

12.1. What advantage do microwave tubes have over microwave semiconductor devices?

12.2. What is the major disadvantage of microwave tubes?

12.3. What is the advantage of using a low power semiconductor oscillator driving a high power tube amplifier, instead of using a high power oscillator?

12.4. Name the five types of microwave amplifier tubes.

12.5. What average power can be obtained from the following microwave tubes at 1 GHz? (*Hint:* Use Figure 12.3.)

Gridded tube

_____

Klystron

_____

Helix TWT

_____

Coupled cavity TWT

_____

CFA

_____

12.6. What average power can be obtained from the following microwave tubes at 10 GHz?

Klystron

_____

Helix TWT

_____

Coupled-cavity TWT

_____

CFA

_____

12.7. What is the major advantage of a gridded tube?

12.8. What is the major advantage of a klystron?

12.9. What is the major advantage of a helix TWT?

12.10. What is the major advantage of a coupled-cavity TWT?

12.11. What is the major advantage of a crossed-field amplifier?

12.12. What is the major disadvantage of a gridded tube?

12.13. What is the major disadvantage of a klystron?

12.14. What is the major disadvantage of TWTs?

12.15. What are two disadvantages of a CFA?

12.16. What limits the frequency capability of gridded tubes?

12.17. How do klystrons, TWTs, and CFAs avoid the transit time problem?

12.18. What is the function of each of the following klystron parts?

Electron gun
Focusing magnet
Beam collector
Input cavity
Intermediate cavity
Output cavity

12.19. Name the four types of focusing magnets used with a klystron or a TWT.

12.20. What is the function of each of the following TWT parts?

Electron gun
Focusing magnet
Beam collector
Microwave interaction structure

12.21. What is the purpose of the windows in microwave tubes?

12.22. What type of magnetic focusing is most commonly used in a TWT?

12.23. What is the advantage of a helix interaction structure over a coupled-cavity interaction structure?

12.24. What is the advantage of a coupled-cavity interaction structure over a helix interaction structure?

12.25. What is the purpose of a depressed collector in a TWT?

12.26. What is the function of each of the following CFA parts?

Electron gun
Interaction structure
Sole electrode
Magnet

12.27. What is the difference in function of the magnetic field in a TWT and in a CFA?

12.28. What is an emitting-sole CFA?

12.29. What is a magnetron?

12.30. Name three advantages of the magnetron.

13

MICROWAVE ANTENNAS

The function of a microwave antenna is twofold. At the transmitter end of a microwave system, the antenna broadcasts the microwave signal in the direction of the receiver. At the receiver end, the antenna picks up as much power as possible from the transmitter. Antenna requirements are discussed first. Then the different antenna elements (dipoles, slots, horns, spirals, helix) are considered, and how they can be combined into arrays to provide increased gain or greater receiving area.

Since the most common microwave antenna is the parabolic dish, its characteristics and formulas for calculating its design are discussed.

In a radar system, the microwave beam from the antenna must be moved to determine the angular location of the target: the bigger the antenna, the smaller the beam, and the better the radar's angular resolution. However, the bigger the antenna, the more difficult it is to move the antenna to move the beam. Also, in military radar applications, the antenna beam must be moved in fractions of a second to cope with various target threats. The only possible way to meet these radar requirements is to move the antenna beam without moving the antenna, which is accomplished with a phased array antenna, discussed at the end of the chapter.

13.1 REQUIREMENTS

Antenna requirements include gain, receiving area, beamwidth, polarization, bandwidth, and sidelobes.

The gain of an antenna is a measure of how much the antenna concentrates its transmitted microwave power in a given direction. It is defined as the power level along the axis of the antenna, compared with the power level that would exist at the same point if the antenna had not been used, but if the power had been uniformly radiated in all directions. (An antenna that radiates power uniformly in all directions is called an *isotropic* antenna) The gain of microwave antennas varies from 0 dB (isotropic antennas) to 60 dB (parabolic dishes). Gain is proportional to antenna area divided by the square of the wavelength of the transmitted frequency. One of the major advantages of microwaves is that their short wavelength allows large gains to be obtained with antennas of reasonable size.

Although microwave antennas have high gains and concentrate the transmitted microwave power in a direction toward the receiver, the transmitted microwave power spreads over a large area at large distances from the antenna. For example, for a search radar antenna 6 ft in diameter and operating at 3 GHz, the transmitted power is spread over an area of approximately six square miles at a distance of 50 miles from the antenna.

The major function of the antenna used at the receiver end of a system is to collect as much of this transmitted power as possible. This is accomplished by making the receiving area or *aperture* of the antenna as large as possible. Gain and receiving area are directly related.

The properties of microwave antennas are reciprocal. A given antenna can be used as a transmitting antenna (in which case its gain is the important characteristic) or as a receiving antenna (in which case its receiving area is the important characteristic). All other characteristics are the same. Figure 13.1 shows an antenna pattern, which is the power transmitted by the antenna as a function of the direction of transmission. Along the horizontal axis to the right is a high concentration of power, while along this axis to the left there is no power. Most of the power is concentrated in the direction of the main beam. Notice, however, that there are also "side" lobes where power is transmitted in other than the desired direction. As stated above, the transmitting pattern of an antenna is identical to the receiving pattern of the antenna.

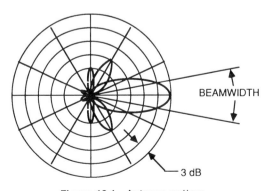

Figure 13.1 Antenna pattern.

The beamwidth of an antenna is a measure of the angular width of the beam. Beamwidth is defined as the angle where the transmitted power has dropped by 3 dB from the maximum power in the direction at which the antenna is pointing.

An antenna may have different antenna patterns and different beamwidths in its azimuth and elevation planes. In Figure 13.2a the antenna shape is wider than it is high, and in Figure 13.2b this antenna generates a fan-shaped beam, which is very narrow in its azimuth or horizontal dimension and large in its elevation or vertical dimension. Such an antenna would therefore have a different antenna pattern in its elevation plane than in its azimuthal plane, as illustrated in Figure 13.2c.

Antenna polarization specifies the orientation of the electric field of the microwave signal propagated from the antenna. Polarization can be vertical—perpendicular to the earth's surface—or horizontal. It can also alternate between horizontal and vertical, in which case the antenna beam is said to be circularly polarized. Microwave signals transmitted with different directions of polarization are independent of each other and thus do not interfere, since the electric and magnetic fields of one signal are perpendicular to the corresponding fields of the other signal. Some parabolic dish antennas are even made with two feeds so that one set of signals can be propagated with vertical polarization and a second independent set with horizontal polarization. In circular polarization, the electric and magnetic fields rotate in a left-hand direction or a right-hand direction about the beam axis as the wave travels through space. A circularly polarized beam rotating in one direction is independent of a circularly polarized beam from the same antenna rotating in the opposite direction.

The bandwidth of an antenna is the measure of the frequency range over which the antenna properties, such as gain, area, beamwidth, and polariza-

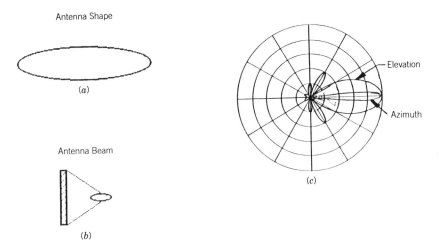

Figure 13.2 Antenna patterns in azimuth and elevation planes.

tion, remain satisfactory for the application. For example, if a certain gain is required to make a system function, then the bandwidth is determined by the frequency range over which the gain can be obtained. For communication and radar systems, a bandwidth of 10% is usually adequate. For electronic warfare receivers, bandwidths of several octaves are necessary to that the antenna can receive radar signals from any radar.

13.2 TYPES

The various types of microwave antennas are

- Dipole
- Slot
- Horn
- Spiral
- Helix
- Arrays
- Parabolic dish
- Phased Arrays.

The first five antennas are individual antenna elements that may be used separately but are usually combined to form arrays, parabolic antennas, or electronically steered phased arrays.

A half-wave dipole antenna is shown in Figure 13.3. It is the common antenna building block for low frequency antennas, including AM and FM radio transmitting and receiving antennas, television transmitting and receiving antennas, and mobile radio antennas. The half-wave dipole consists of a two-wire feedline, which opens at the antenna. Each feedline extends at right angles for one quarter wavelength. Thus, the total length of the antenna is a half wavelength.

The radiation pattern of a dipole antenna is shown in Figure 13.3b. The dipole antenna radiates no power in the direction of the dipole wires, and radiates its maximum power in a direction at right angles to the wires. Consequently, the antenna pattern resembles a doughnut. A cross section through the doughnut shows the antenna pattern to be a figure-8. A dipole has this radiation pattern for all lengths, up to about $\frac{5}{8}$ wavelength. It is not necessary to make the dipole a half-wavelength long to achieve the pattern. The antenna impedance, however, varies with its length; if the dipole antenna is small, in terms of a wavelength, the impedance is very low and it is difficult to match the antenna to the feedline. When the antenna is a half wavelength long, its impedance is 75 Ω and is easy to match.

The table shows the length of half-wave dipoles at different frequencies below the microwave band and up through the microwave band. At 1 MHz,

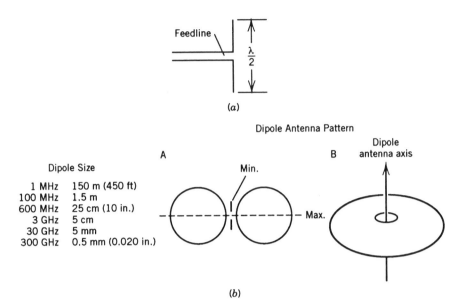

Figure 13.3 Half-wave dipole.

a half-wave dipole is 150 meters long. A half-wave dipole, with half of the half wavelength extending above ground is used for an AM radio transmitter. At 100 MHz, the half-wave dipole is 1.5 m long, which is about 60 in. Antennas of this length are used for FM radio receivers. At 600 MHz, the low end of the microwave band, a half wavelength is 25 cm (10 in.), and antennas of this size are used as the receiving antennas for UHF TV sets. At 3 GHz, a half wavelength is 5 cm, at 30 GHz a half wavelength is 5 mm, and at 300 GHz, the upper end of the microwave band, a half wavelength is $\frac{1}{2}$ mm.

Half-wave dipole antennas are used up through the low frequency end of the microwave band. However, according to the table of Figure 13.3, at the high end of the microwave band the sizes of the half-wave dipoles become impractically small. At high microwave frequencies, a half-wavelength slot is used in place of the dipole. A slot antenna is shown in Figure 13.4. It has the same radiation pattern as the dipole antenna, except that the electric field pattern of the slot antenna corresponds to the magnetic field pattern of the dipole antenna, and vice versa.

The horn antenna is illustrated in Figure 13.5. It is formed as an extension of a waveguide; one or both dimensions of the waveguide are tapered so that the microwave power is transitioned from the waveguide into the radiating horn with no mismatch. The gain of the horn antenna depends on the ratio of the horn opening to the square of the wavelength, so as much gain as desired can be obtained by making the horn large enough. Normally, horn antennas are limited to about 20 dB of gain, because if the horn opening is

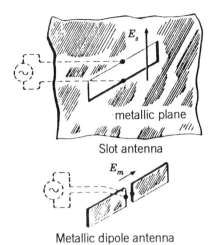

Slot antenna

Metallic dipole antenna

Figure 13.4 Equivalence between dipole antenna and slot antenna.

made larger to increase the gain the length of the horn becomes excessive. For higher gain requirements, a better solution is to have the horn feed a parabolic reflecting dish, which will be discussed later. The horn antenna in Figure 13.5 covers the 8–12 GHz frequency range and typically tapers from the 0.9 in. × 0.4 in. cross section of the waveguide to a 2 in. × 3 in. opening

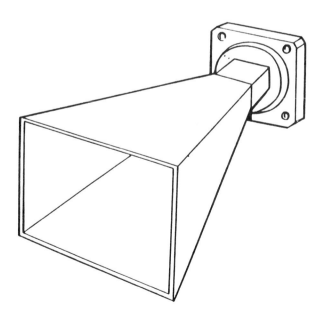

Figure 13.5 Horn antenna.

in a 5-in. taper length. Such a horn has 14 dB gain at 8 GHz, increasing to 18 dB at 12 GHz.

Dipole, slot and horn antennas are narrowband antennas covering only a 2 : 1 frequency range. For broader band requirements, a spiral antenna (Figure 13.6) is often used. The spiral is fed from the center, and the microwave power travels around the spiral until the circumference of the spiral is equal to one wavelength, and then the microwave power is radiated. By making the antenna in a spiral pattern there will be a one-wavelength circumference in different turns of the spiral at different frequencies. Consequently, the spiral can cover a large frequency range. The radiation pattern from the spiral is circularly polarized. The spiral antenna in Figure 13.6 covers from 2 to 40 GHz. The beamwidth of the spiral antenna is broad and the antenna has about 0 dB gain. The antenna would be expected to have at least 3-dB gain, since the power is radiated in the forward direction only, but the microwave signal undergoes a 3-dB loss in traversing the long length around the spiral, and these two factors give the antenna approximately 0 dB gain.

Another type of broadband antenna is the helical antenna (Figure 13.7). The effective radiating area of the helix antenna is made up of the several turns of the helix, so the helix antenna provides greater gain for a given cross-sectional area than does a horn or a parabolic reflecting dish. This feature makes helical antennas particularly useful at the low end of the microwave band, because they are smaller in diameter than other types for a given gain. The gain of the helix antenna varies from 10 to 14 dB across a 50% bandwidth. The reflector at the feed end of the helix antenna reflects the power so that the antenna radiates only in one direction.

13.3 ARRAYS

Many antennas are formed of arrays of the various antenna elements. Array antennas have higher gain, larger receiving areas, and greater bandwidth than do the individual elements. Several examples of array antennas are shown in Figure 13.8 and 13.9. The upper sketches of Figure 13.8 show an array of three AM radio towers. An AM radio transmitter tower is half of a half-wave dipole antenna. The tower is therefore a quarter wavelength high. The other half of the half wavelength would extend below ground. Actually it doesn't, but is formed by an image of the tower itself reflected from the ground plane. At AM broadcast frequencies around 1 MHz, the earth is a good conductor and appears like a metal-reflecting surface.

The antenna pattern of a single AM transmitter tower is half of the dough-nut-shaped pattern shown in Figure 13.3, so the tower broadcasts equally in all directions. This type of AM broadcast antenna may waste power by broadcasting into areas where there are no listeners, or it may interfere with other distant stations operated on the same frequency. Consequently, most AM radio transmitters use an array of towers, such as shown in the top of

Figure 13.6   Spiral antenna.

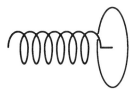

Figure 13.7   Helix antenna.

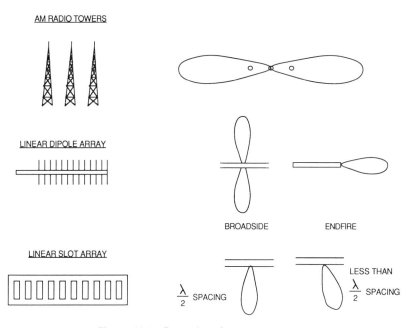

AM RADIO TOWERS

LINEAR DIPOLE ARRAY

LINEAR SLOT ARRAY

BROADSIDE   ENDFIRE

$\frac{\lambda}{2}$ SPACING

LESS THAN
$\frac{\lambda}{2}$ SPACING

Figure 13.8   Examples of array antennas.

Figure 13.8. Adjusting the phase between the towers allows any desired radiation pattern to be obtained. In the figure the transmitter power is divided equally between the three towers, and the phase between the signals is 160°, which produce the radiation pattern shown. Changing the phase permits other patterns to be obtained. Some AM radio transmitters actually change the phase from daytime to nighttime to avoid interference at night when greater transmission distances can be achieved.

Figure 13.8 (center) shows a linear array of dipoles attached to a common feedline. The gain of the linear array and its receiving area is increased directly as the number of dipoles. In the figure the gain and receiving area are six times (or 8 dB) greater than the gain and area of a single dipole. By adjusting the phase between the dipoles to be 90°, the broadside radiation pattern can be obtained. By adjusting the phase to be 180°, the endfire pattern can be obtained.

A linear array of slots can be used in place of the dipoles (see bottom figure). This type of an array antenna would be appropriate at the high frequency end of the microwave band. The slots could actually be cut in the wall of a waveguide, and the waveguide then used as the transmission line feed for the slots. If the slots were a half wavelength long and spaced a half wavelength apart along the waveguide, then the transmission would be at right angles or broadside to the waveguide, as shown. If the frequency were lowered so that the spacing between the slots were less than a half wavelength, then the beam pattern will be at an angle to the array of slots. Thus, changing the frequency allows the direction of beam pointing to be varied. This feature is used in a frequency-scanned array, discussed later.

Two examples of log-periodic arrays are shown in Figure 13.9. They consist of different lengths of dipoles along a common feedline. The log-periodic antenna has a wide bandwidth capability, because a dipole approximately a half wavelength long exists for each range of frequencies to be covered by the antenna. The log-periodic antenna array, Figure 13.9a, is familiar because of its use as a VHF TV receiving antenna, from 50 to 200 MHz, a 4 : 1 bandwidth. A log-periodic microwave antenna, covering

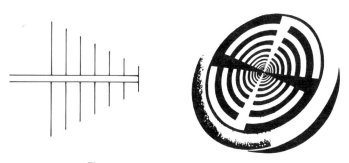

Figure 13.9 Log-periodic antennas.

from 1 to 20 GHz, is shown in Figure 13.9*b*. The dipoles are formed by the vertical stripline and the curved sections. Note that there is a dipole for each band of frequencies in the 1–20 GHz range.

13.4 PARABOLIC DISH

Perhaps the most common microwave antenna, certainly the most well known, is a parabolic dish. The principles of parabolic antennas are illustrated in Figure 13.10. The parabolic antenna consists of an antenna feed, which radiates its power into a parabolic-shaped reflecting surface. The parabolic reflector concentrates the microwave power back along the antenna axis. The parabolic shape has the unique property that all rays from the feed to any point on the parabolic surface travel an equal path length from the feed to a plane at right angles to the direction of propagation. This feature gives the parabolic reflecting surface its focusing characteristics and allows it to concentrate the microwave power from the feed horn along the antenna axis. The gain, receiving area, and beamwidth are determined by the parabolic reflector. Antenna polarization and bandwidth are determined by the feed.

Figure 13.11 shows four examples of microwave parabolic antennas. A microwave relay antenna is shown in the upper left. The parabolic surface is fed by a waveguide horn. The polarization with the orientation shown is horizontal because the field is horizontally directed as it comes out of the feed horn. A radar antenna is shown in the upper right. It has different horizontal and vertical dimensions to give a fan-shaped beam as was explained in Figure 13.2. The parabolic antenna in the lower left is for a satellite earth station. It is 30 ft in diameter, and its surface is so large that a simple horn or dipole would not provide a wide enough beam to spread the power over the entire dish surface. Consequently, this antenna uses a Cassegrain

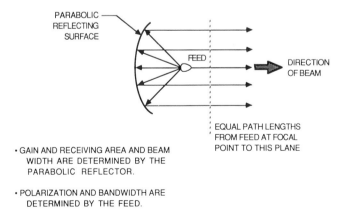

Figure 13.10 Parabolic dish antenna.

Figure 13.11 Typical parabolic dish antennas. (Photos courtesy of Andrew Corporation.)

feed, which consists of a horn that radiates the microwave power to a convex subreflector, which then distributes it over the entire parabolic surface. The antenna shown in the lower right has a log-periodic feed, which provides a wide bandwidth from 1 to 12 GHz, but only 3-dB gain. To increase its gain (to approximately 13 dB), the power is reflected from the parabolic surface.

The design of parabolic antennas is illustrated in Figure 13.12. The feed horn must be designed so that the dish illumination is tapered to control the sidelobes. The horn is designed so that the maximum power is radiated into the center of the dish, and the illumination decreases toward the edges of the dish. If uniform illumination were used, there would be no way to stop the feed pattern at the edge of the dish, so a certain amount of power from the feed would slop over the edge of the antenna and form sidelobes. Typically, the illumination pattern is tapered so that the microwave field intensity is 10 dB down from its maximum value at the edge of the parabolic dish. Tapering reduces antenna gain but it is needed to control the sidelobes.

With 10-dB tapering, the gain of a parabolic dish antenna is reduced to about 60% of the ideal value with uniform illumination. The following formulas apply:

$$\text{Gain} \approx 0.6 \left(\frac{4\pi A}{\lambda^2} \right) \approx 6 \left(\frac{D}{\lambda} \right)^2$$

where D = dish diameter
A = antenna area
λ = wavelength

$$\text{Aperture (Effective Area)} \approx 0.6A \approx 0.6\left(\frac{\pi D^2}{4}\right)$$

$$\text{Half-Power Beamwidth} \approx 60°\left(\frac{\lambda}{D}\right)$$

Example 13.1 A parabolic dish antenna has a diameter of 1 m. It operates at 10 GHz, where the free-space wavelength is 0.03 m. Find the antenna's gain, beamwidth, and aperture.

Solution

$$\text{Gain} = 6\left(\frac{D}{\lambda}\right)^2 = 6\left(\frac{1}{0.03}\right)^2$$
$$= 6670 = 38 \text{ dB}$$
$$\text{Beamwidth} = 60°\left(\frac{\lambda}{D}\right) = 60°(0.03)$$
$$= 1.8°$$
$$\text{Aperture} = 0.6\left(\frac{\pi D^2}{4}\right) = 0.6\left(\frac{\pi}{4}\right)(1)^2$$
$$= 0.5 \text{ m}^2$$

Dish illumination must be tapered to control side lobes

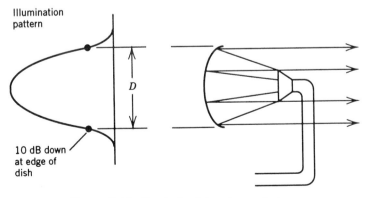

Illumination pattern

D

10 dB down at edge of dish

Figure 13.12 Parabolic dish antenna design.

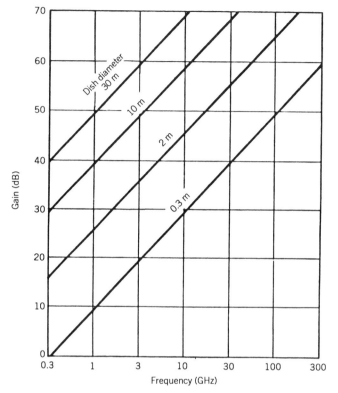

Figure 13.13   Gain of parabolic dish antennas.

Example 13.2   A parabolic dish antenna has a diameter of 3 m, and operates at 4 GHz, where the wavelength is 0.075 m. Calculate the antenna gain, beamwidth, and aperture.

Solution

$$\text{Gain} = 6\left(\frac{D}{\lambda}\right)^2 = 6\left(\frac{3}{0.075}\right)^2$$
$$= 9600 = 40\,\text{dB}$$
$$\text{Beamwidth} = 60°\left(\frac{\lambda}{D}\right) = 60°\left(\frac{0.075}{3}\right)$$
$$= 1.5°$$
$$\text{Aperture} = 0.6\left(\frac{\pi}{4}D^2\right) = 0.6\left(\frac{9\pi}{4}\right)$$
$$= 4.2\,\text{m}^2$$

This antenna is typical of a TVRO earth station antenna.

At any frequency, antenna gain can be made arbitrarily large and beamwidth arbitrarily small by increasing the antenna diameter. Note that the gain and the beamwidth depend on the *ratio* of diameter to wavelength, not the diameter alone. Another way of stating this is that the gain and beamwidth depend on the number of wavelengths across the antenna diameter. Note also that the aperture of the receiving antenna depends only on the antenna diameter.

Figure 13.13 plots the gain of a parabolic dish antenna as a function of frequency with the beam diameter as a parameter. Note that with a 30-m diameter (about 100 ft) a gain of about 60 dB is obtained at 4 GHz. This 100-ft diameter antenna is used for international satellite communication earth stations. However, a 60-dB gain could also be obtained with a 0.3-m antenna (about 1 ft in diameter) by operating the antenna at 300 GHz, which is the advantage of using millimeter frequencies to obtain high angular resolution with small antennas.

Figure 13.13 suggests that an antenna gain can be made arbitrarily large simply by increasing the antenna size. This is true in principle, but it becomes difficult to hold the necessary mechanical tolerances in an antenna with a diameter more than a few hundred wavelengths. Consequently, the practical gain from parabolic dish antennas is about 60 dB.

13.5 PHASED ARRAY

In all radar systems, and in many navigation and communication systems, the antenna beam must be moved in only fractions of a millisecond. It is obviously not possible to physically move parabolic antennas so rapidly, especially when its diameter is several hundred wavelengths. Hence, a phased array antenna, which moves the beam in space without moving the antenna, is used. Figure 13.8 showed that the direction that the beam pointed in array antennas depended on the phase relationships between the individual transmitting elements. Therefore, if these phase relationships can be electronically varied the beam can be moved without physically moving the elements.

Figure 13.14 shows a linear array of antennas, each fed by a phase shifter. The radiating elements can be dipoles, slots, or horns. The direction in which the beam formed by this array of radiating elements points is determined by the relative phase of the power leaving each antenna element. The angle of the beam from the antenna axis is determined by the operating wavelength of the microwave signal, the spacing between the antenna elements, which is usually half a wavelength, and the phase shift between the signals in the individual elements.

Another way of visualizing how a phased array antenna works is shown in Figures 13.15 and 13.16. Figure 13.15*a* is the same as Figure 13.10, which illustrates parabolic antenna operation. The parabolic shape of the reflector

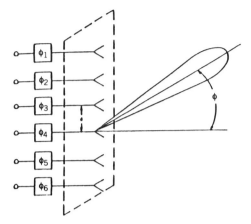

Figure 13.14 Phased array antenna.

directs the beam along the axis of the parabola. To move the beam to another direction (Figure 13.15b), the entire reflecting surface must be moved.

Figure 13.16 shows how the parabolic dish could be simulated with a flat reflecting surface by using phase shifters along the surface to adjust the effective path length of each ray from the feed horn out to a plane perpendicular to the antenna axis. By the proper adjustment of phase, all rays proceeding perpendicular to the antenna axis could be made to have the same path length, and the parabolic surface could then be replaced by a flat reflecting surface covered with phase shifters. The phase shift at each point on the flat

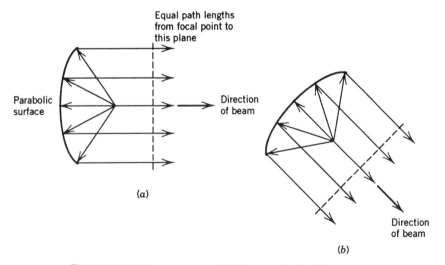

Figure 13.15 Beam movement with a parabolic dish antenna.

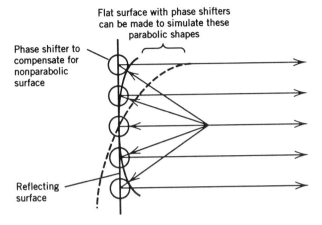

Figure 13.16   Beam movement with a phased array antenna.

surface could then be changed to simulate a parabolic surface whose axis was pointing in a different direction. Therefore, by changing the phase of the phase shifters, the direction of the beam can be changed without moving the reflecting surface.

Two phased array antennas using these principles are shown in Figure 13.17. Figure 13.17a is a reflective array antenna, which uses a horn feed to illuminate the antenna array. The radiation pattern of the horn is adjusted so

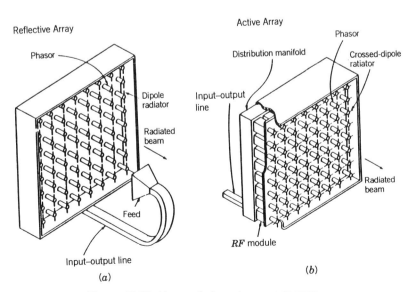

Figure 13.17   Types of phased array antennas.

that the total power is divided uniformly among the 64 array elements. At each element, part of the power from the feed horn is received by the small dipole antenna, travels through a phase shifter, is reflected from the reflecting surface, travels again through the phase shifter, and is radiated from the dipole. The phase of the reflected signal from each array element is adjusted electronically by the phase shifter, and the total beam from all antenna elements can be made to point in any direction simply by controlling the phase of the radiating elements.

The gain and beamwidth of a phased array antenna depend on antenna diameter in exactly the same way as for a parabolic antenna. Gain, beamwidth, and effective receiving area are therefore calculated from the formulas for the dish antenna.

The number of radiating elements used in the phased array antenna should be minimized, but they must be spaced no more than half a wavelength apart. Otherwise the antenna pattern will have holes in it.

The reflective array in Figure 13.17a can also be modified to be a "lens" array. In this case, the surface is fed by the feed horn, but instead of reflecting the power after it is received by the individual dipoles at each element and passed through the phase shifter, the power is transmitted, with the correct phase relationships, out the other side of the array.

An active array is shown in Figure 13.17b. A low-level microwave signal is divided among the array elements. This signal can be supplied to each element by radiating the low-level power from a feed horn, as for the lens array. The power can also be distributed by a distribution manifold. Each array element contains the phase shifter to provide the beam steering and an amplifier to amplify the power in each element up to the required transmitter power. In this way, the total radiated power is not supplied by a single high power transmitter but by the sum of many low power elements. Some long-range search radars have as many as 10,000 transmitting elements. The transmitter power of each element is about 100 W, which can be obtained at frequencies below 1 GHz from a microwave bipolar transistor, and the total transmitted power from the array is 10,000 elements times 100 W per element, which is equal to 1 Megawatt.

The active phased array offers the advantage of "graceful degradation," which means that if one active transmitter element of the array fails, the array continues to perform with just slightly degraded performance. This is in contrast to the reflective or the lens array where the entire microwave system stops working if the transmitter fails.

Since antenna characteristics are reciprocal, the phased array antenna can also be used as a receiving antenna. With a particular phase adjustment in each element of the array, the array will accept power only from a given direction. Power coming in from other directions will be received by the array elements, but since these individual powers are not in phase they cancel out.

The key component of phased array antennas is obviously the phase shifters. *Pin* diode or ferrite phase shifters are used. These devices were described in Chapter 7, and can shift the phase is discrete steps in fractions of a microsecond. Two other designs of phased array antennas reduce the number of phase shifters required. In the frequency-scanned phased array in Figure 13.18*a* the microwave signal is transmitted down a helical transmission line. At each turn of the transmission line part of the power is taken out and fed to one of the phased array antenna elements. The phase relationship between each antenna element varies with the frequency. Hence, if the transmitter frequency changes, the angle of the transmitted beam changes. This type of phased array antenna permits one-dimensional scanning of the antenna beam. Some antennas use phase shifters to move the beam in azimuth, and frequency scanning to move the beam in elevation. This approach greatly reduces the number of phase shifters required. For example, if an antenna requires 100 elements in its vertical dimension for elevation scanning and 100 elements in its horizontal dimension for azimuth scanning, the total number of phase shifters in the antenna is 10,000. If the antenna uses phase-shift scanning in azimuth and frequency scanning in elevation, then the number of phase shifters needed is only 100.

In the lens type of phased array antenna in Figure 13.18*b*, a dielectric lens is used. Half of the lens contains inputs, and half of the antenna contains output elements. The input power is switched between the various input ports, depending on the direction in which the output power is to point. As the power radiates through the lens, the phase at any point on the output surface is automatically adjusted by the lens. Hence, the beam coming from the sum of the output radiating elements adds up in the correct phase to point the beam in the complementary direction to the input feed. At each output element, an amplifier can be provided so that low-level beam switching on the input side of the lens can control high output powers on the output side.

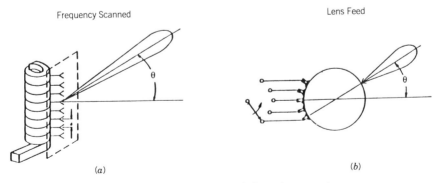

Frequency Scanned

Lens Feed

(a)

(b)

Figure 13.18 Alternate types of phased array antennas.

ANNOTATED BIBLIOGRAPHY

1. E. A. Wolff, *Antenna Analysis*, 2d ed. Artech House, Dedham, MA, 1988.
2. L. V. Blake, *Antennas*, Artech House, Dedham, MA, 1984.
3. K. C. Gupta, and A. Benella, *Microstrip Antenna Design*, Artech House, Dedham, MA, 1988.

References 1 and 2 cover all types of antennas, including wire antennas, antenna arrays, and parabolic dishes. Reference 3 covers microstrip antennas.

EXERCISES

13.1 Define the following antenna requirements:
1. Gain
2. Receiving area
3. Beamwidth
4. Polarization
5. Bandwidth
6. Sidelobes
7. Antenna pattern

13.2. What is an isotropic antenna?

13.3. What is the gain of an isotropic antenna?

13.4. Complete the table.

Frequency (GHz)	Antenna Area (m^2)	Gain (dB)
1	1	
10	1	
1		20
10		30

13.5. Name the five types of antenna elements.

13.6. How long is a half-wave dipole at the following frequencies?

Frequency (GHz)	Length of Half-Wave Dipole (m)
2	
5	
12	

13.7. How long is a half-wave slot at the following frequencies?

Frequency (GHz)	Length of Half-Wave Slot (m)
6	
10	
20	
100	

13.8. An antenna is formed of an array of 20 dipoles. Each dipole has a gain of 1 dB. What is the total gain of the array?

13.9. An antenna is formed of four helix antennas, each of which has a gain of 14 dB. What is the total gain of this antenna array?

13.10. Complete the following table.

Antenna Diameter (m)	Frequency (GHz)	Gain (dB)	Effective Area (m²)	Beamwidth (deg)
1	4			
10	4			
30	4			
1	10			
3	10			
10	10			
.5	30			
1	30			

13.11. What is the gain at 6 GHz of a parabolic antenna with a diameter of 2 m?

13.12. A parabolic antenna has a gain of 20 dB at 4 GHz. What is the gain of this antenna at 10 GHz?

13.13. A parabolic antenna has a gain of 30 dB. How much must its diameter be increased to raise its gain to 40 dB?

13.14. What is the purpose of a phased array radar?

13.15. How far apart must the antenna elements of a phased array radar be placed?

13.16. A square phased array radar is to have a gain of 30 dB.
   a. What must be the width dimensions in wavelengths?
   b. If the antenna elements are spaced a half wavelength apart, how many elements are required?

Define the following types of phased array antennas.

13.17. Reflective array

13.18. Lens array

13.19. Active array

MICROWAVE SYSTEMS

14

INTRODUCTION TO MICROWAVE SYSTEMS

This chapter covers the nature of the electronic signals used in microwave communication, radar, electronic warfare, and navigation systems. Electronic signals as a function of time and as a function of frequency are analyzed.

The nature of electronic communication system signals, including audio and telephone signals, video (TV) signals, and data, is discussed. Multiplexing many signals so that they can be transmitted through the same communication system without interference is discussed, and signal-to-noise requirements for communication systems are analyzed. Pulse code modulation is studied.

Amplitude, frequency, and phase modulation are described, including modulation sidebands, signal-to-noise improvement provided by the modulation, and transmission bandwidths. The different modulation systems are compared with regard to bandwidth, signal-to-noise improvement, and equipment complexity. Carrier modulation with digital baseband signals is also reviewed.

14.1 SPECTRUM ANALYSIS OF ELECTRONIC SIGNALS

Figure 14.1 illustrates spectrum analysis of complex electronic signals. Figure 14.1a is an electronic signal plotted as a function of time for a period of 1 second. What is its frequency? Clearly the frequency is difficult to define because the signal is not a simple single-frequency waveform. This complex signal is actually a mixture of the three signals shown in Figure 14.1b–d, with frequencies of approximately 1 Hz, 5 Hz, and 16 Hz. The complex

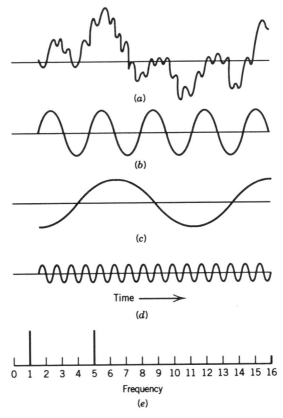

Figure 14.1   Spectrum analysis of complex signal.

signal *a* was obtained by adding the amplitudes of three single-frequency signals *b–d* at each point in time.

Figure 14.1 illustrates that any electronic signal that repeats its waveform in time can be analyzed as a sum of single-frequency signals. This procedure is called *spectrum analysis*. The spectrum of signal *a* is given in Figure 14.1*e*, where the amplitude of each of the three single-frequency signals that make up the complex signal are shown as a function of frequency. The first signal, at 1 Hz, and the second signal, at 5 Hz, have approximately the same amplitude. The higher frequency signal at 16 Hz has a much smaller amplitude.

Spectrum analysis is widely used to analyze microwave systems because any microwave system or device can be analyzed at a single frequency, or over a range of frequencies, by measuring its performance at one frequency at a time. For example, the gain of an amplifier can be determined over a range of frequencies by measuring its gain one frequency at a time. This is called *network analysis*.

To analyze how a microwave system or device will respond to a complex waveform, the complicated waveform is broken down into individual fre-

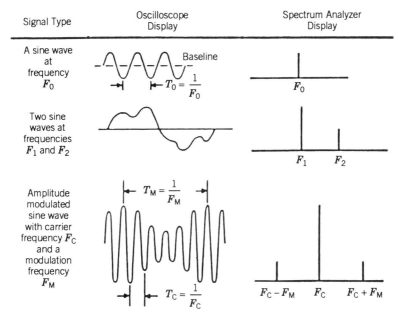

Signal Type	Oscilloscope Display	Spectrum Analyzer Display

Figure 14.2 Comparison of time and frequency representation of electronic signal.

quencies, then the effect on each frequency by the system or device, and hence the overall performance, can be determined.

Any electronic signal can be analyzed as a function of time or frequency. Both analyses of some typical electronic signals are shown in Figure 14.2. The first signal analyzed is a single frequency. On an oscilloscope, which displays a waveform as a function of time, the signal appears as a single-frequency sine wave at a frequency F_0. On a spectrum analyzer, which displays the single-frequency signals composing the signal waveform, the signal appears as a single-frequency line.

The second signal is composed of a large signal at F_1 and a smaller signal at F_2. These two signals form a complex waveform, as viewed on the oscilloscope, but the spectrum analysis display shows that the complex waveform is simply composed of two single-frequency signals.

The third signal is a frequency F_0 amplitude modulated by a frequency F_M. When displayed on a spectrum analyzer, this complex waveform, is composed of just the carrier and upper and lower sidebands.

14.2 COMMUNICATION SYSTEM SIGNALS

Figure 14.3 illustrates three telecommunication systems. The telephone (top) converts the speaker's voice to an electronic signal that is transmitted through the telecom system to the telephone at the other end. That phone takes the received electronic signal and reconverts it to an audio signal that the listener can hear.

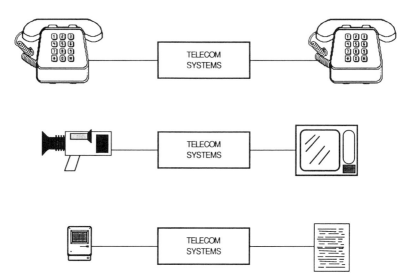

Figure 14.3 Telecommunication systems.

The television camera (center) converts a visual image to an electronic signal that is transmitted by the telecom system to the TV monitor at the receiver. The monitor reconverts this electronic signal to a visual display.

The computer terminal in the data communication system (bottom) converts letters and numbers, which represent information, to electronic signals that are transmitted by the telecom system. At the receiving end, the signals are reconverted in a printer to the original data. The telecom transmission system doesn't know, or care, where the signal comes from. The signal from the telephone, TV camera, or computer is simply an electronic voltage that varies with time. The frequency spectrum of the signal determines the frequency performance characteristics of the telecommunication transmission system, so the frequency characteristics of these signals need to be analyzed.

Figure 14.4 shows the characteristics of the audio signal from a microphone or telephone. In Figure 14.4a the electronic signal of a human voice from a microphone is shown as a function of time. The major peaks are approximately 1 millisecond apart, which corresponds to a frequency of 1 kHz. The frequency spectrum of the voice signal is shown in Figure 14.4b. The frequencies in the human voice extend from 30 Hz at the low end to 6 kHz at the high end, with the largest signals around 1 kHz. These characteristics of the audio signal of a human voice are determined by the human vocal chords. A person cannot speak at frequencies below 30 Hz or above 6 kHz.

The frequency components of the audio signal from musical instruments extend over a much wider frequency range. Musical instruments can generate frequencies below 30 Hz and well above 6 kHz. However, the average human ear cannot hear audio signals below 30 Hz or above 15 kHz; therefore the frequency range of music is taken to be 30 Hz–15 kHz, and equipment such as stereos and hifis must operate in this range.

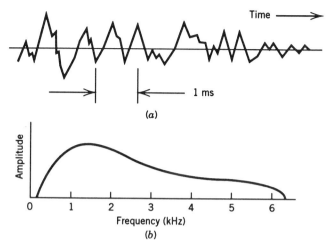

Figure 14.4 Audio signals.

A telephone signal ranges from 300 Hz to 3.4 kHz. The telephone set deliberately filters out frequencies from the human voice below 300 Hz and above 3.4 kHz, which is why a person's voice over a telephone can be recognized but the telephone signal does not sound exactly like the person. The telephone set does this to reduce the signal bandwidth so that more telephone calls can be fitted into a given communication system. The bandwidth of a telephone call is conventionally taken to be from 0 to 4 kHz. The actual audio signal is 300 Hz to 3.4 kHz, and this signal is fitted into the 0–4 kHz telephone channel. The narrower frequency range of the actual signal compared with the telephone channel gives guard frequency bands so that the signal from one channel does not overlap into another channel when frequency-division multiplexing is used.

A radio station usually broadcasts a single voice or music signal. However, for economic reasons a microwave relay or a satellite communication system cannot be built to transmit just a single telephone call. To make these systems economically feasible, hundreds or thousands of telephone calls must be combined into the signal to be transmitted. However, each telephone call is in the same frequency range; hence, if all of them are combined, it would sound as if thousands of people were talking at once. The technique used to transmit many independent signals from many telephones, or from several TV cameras, or from many data terminals, is called *multiplexing*. One common form of multiplexing is frequency-division multiplexing (FDM), shown in Figure 14.5.

In the figure 1000 telephone calls are combined. Each call has frequency components in a 0–4 kHz bandwidth. If they were combined directly, they would interfere with each other. With FDM each telephone call is shifted from its normal 0–4 kHz band up to a higher frequency band. For example, channel 1, which operates from 0 to 4 kHz is transmitted in its original band,

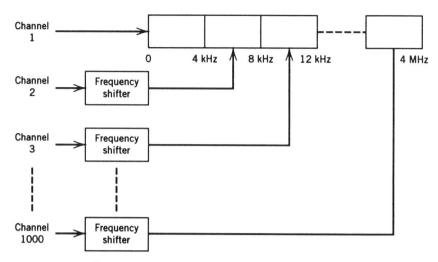

Figure 14.5 Frequency-division multiplexing (FDM).

but channel 2, which occupies the same frequency range, is shifted into the 4–8 kHz band, and channel 3 is shifted into the 8–12 kHz band. In like manner the other channels are shifted up in frequency into adjacent 4-kHz bands. Channel 1000 is shifted up in frequency to 4 MHz. The upshifted signals do not sound right; in fact, none of the upshifted channels above channel 4 can be heard because they are beyond the audio range of the human ear, but the signal information in them is preserved. The resulting multiplexed signal varies with time, with frequency components from 0 to 4 MHz.

This FDM signal is transmitted, received, and demultiplexed. At the receiving end, each of the 1000 individual telephone calls is put back into its 0–4 kHz band onto a separate telephone line. How multiplexing is accomplished is discussed later. A typical telephone multiplexer costs $500 to put each channel into the multiplexed signal and $500 to get the channel back out again into its original 0–4 kHz band. Therefore the cost of a 1000-channel multiplexer is $1 million.

Figure 14.5 shows channel 1 occupying the original 0 to 4-kHz band, channel 2 shifted up by 4-kHz and so on. This figure was used to illustrate the technique. Telephone channels actually shifted up in the same way, but are shifted to different frequencies. The number of channels combined and the frequencies to which they are shifted must be standardized so that the multiplexed signal can be transmitted from one telephone company to another. This FDM standardization is called a *hierarchy,* and the hierarchy used in the United States is shown in Figure 14.6. Twelve telephone channels are combined to form a basic group covering a frequency range of 48 kHz. These 12 channels are fitted into a 48-kHz range from 60 to 108 kHz. Then

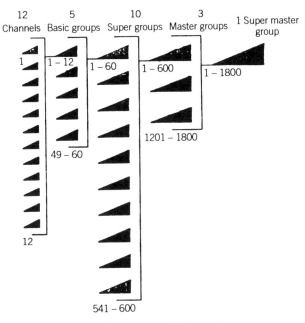

Figure 14.6 FDM hierarchy.

as shown in Table 14.1 five basic groups are combined to form a super group, which consists of 60 telephone channels. Ten supergroups are combined into a 600-channel master group. Three master groups are combined to form a super master group, which contains 1800 telephone calls, and the super master group is fitted into the frequency range from 564 kHz to 8.284 MHz. The multiplexed signal is shifted to these megahertz frequencies because it is easier to transmit higher frequency signals through coaxial cable without interference between adjacent cables, which would occur if the frequencies were in the low kilohertz range.

The electronic signal of television is shown in Figure 14.7. The television camera converts a visual image into an electronic signal by projecting the

Table 14.1 FDM hierarchy

Name	Total Number of Channels	Frequency Range (kHz)
Channel	1	0–4
Basic group	12	64–108
Super group	60	312–552
Master group	600	654–3084
Super master group	1800	564–8284

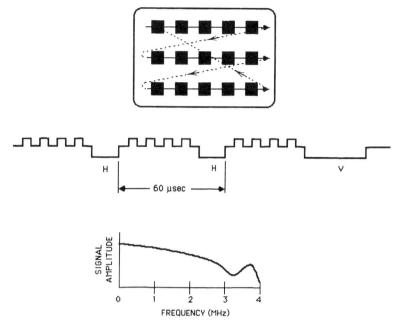

Figure 14.7 Television (video).

visual image onto the face of the camera tube. An electron beam scans the back of the camera tube face one line at a time from top to bottom. When the scanning beam reaches the bottom of the tube face, it returns to the top and repeats the scanning process. A simple three-line scanning pattern is shown in Figure 14.7. If a given spot on the image is bright, many electrons are emitted from the camera tube face as the electron beam hits that spot during its scan. At this instant of time, the TV camera tube produces a large output signal.

Four bright areas between the dark areas are shown in the figure, and the resulting electrical signal as the electron beam scans across the face of the camera tube are also shown. Note the four higher amplitude regions of the signal during each scan corresponding to the four brighter spots, and the five lower amplitude regions corresponding to the dark spots.

A broadcast television camera tube actually has 525 scan lines. The scan must be accomplished so fast that the human eye cannot see it. Therefore, the 525 lines must be scanned in $\frac{1}{30}$ of a second, and the beam must return to the top of the camera tube and repeat the scanning process. Therefore, 60 microseconds are allowed for each line to be scanned across the picture tube. In the middle of Figure 14.7, 10 microseconds of this time, shown in the regions marked H, are used for the scanning beam to return to scan the next line. (The beam returns from the bottom to the top of the screen to repeat the complete scanning process in the interval marked V.) Therefore each line of the picture is scanned in 50 microseconds. Broadcast television has a resolution capability of 400 light and dark spots across one line. This must be scanned in 50 microseconds, so the highest frequency component, if the finest detail in the picture is used, is 4 MHz. In a normal broadcast television display, part of the screen contains large objects where the signal does not change from bright to dark over most of its scan. These regions give lower frequency components below 4 MHz, so the TV signal contains, as shown at the bottom of Figure 14.7, frequency components extending from 0 to 4 MHz. (The concentration of signals at 3.5 MHz is the color information.)

The video signal for broadcast television requires 4 MHz of bandwidth. (Recall that 1000 telephone calls can be multiplexed together into this same bandwidth—easy to remember by noting that "one picture is worth a thousand words.") The 525 lines could be scanned at a slower rate than $\frac{1}{30}$ second, but then the picture would not seem to be continuously moving. However, a slower scan speed reduces the bandwidth requirements of the transmission system, and is used in videoconfererncing, where a series of views, rather than a continuously moving picture, are presented. An extreme example of slower scanning is the TV pictures sent back from the deep-space planetary probes. One complete picture requires 5 minute to transmit, since the signals are sent one line at a time over a 5-minute interval. This must be done to reduce the bandwidth requirements in order to reduce noise. At the opposite extreme, more scan lines are used to obtain high-definition television (HDTV). The bandwidth requirements are then greater than 4 MHz.

At the receiving end, the process is reversed and the electrical signal is reconverted to a visual image in the TV monitor. The intensity of the electron beam as it scans the picture tube is high when the signal is strong and low when the signal is weak: the stronger the electron beam, the brighter the spot on the picture tube. The picture is therefore reproduced as a series of bright and dark spots on the picture tube face.

Whether standard broadcast television or high-resolution television is considered, the video signals require a much larger bandwidth for their transmission, since they contain far higher frequency components than does a single telephone call. Just the opposite is true with data transmission, where a great deal of data can be transmitted in far less bandwidth than is required for a single telephone call.

Figure 14.8 shows the digital representation of the letters U and F, the number 3, and the carriage return (CR) control signal. These digital representations are taken from the ASCII code, the basic digital code used for data transmission. The letter U is represented by 1010101. In Figure 14.8a, U can be represented as an electronic signal where the positive voltage represents a 1 and the negative voltage a 0. The digital electronic signal representing the digits for F, 3, and the carriage return are also shown. The ASCII code has seven digits, so the number of characters is 2^7, or 128 characters. These

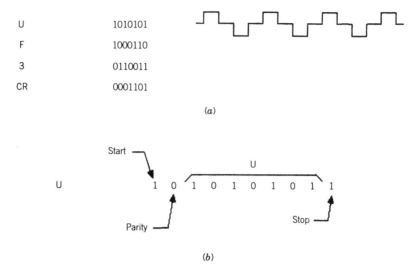

U	1010101
F	1000110
3	0110011
CR	0001101

(a)

(b)

Figure 14.8 Data signals.

characters include capital letters, small letters, numbers, symbols, and control signals, and represent the keys on a standard computer keyboard.

An error in any of the seven digits would give a completely different meaning to the symbol. For example, if the last digit of the letter U, which is a 1, was 0, the digital code would represent an entirely different letter. To eliminate this type of error, most data transmission systems add a parity digit, as shown in Figure 14.8b, so that the total number of 1's in the code is always even. The letter U already has an even number of 1's, so the parity digit is 0. The letter F has an odd number of 1's, so the parity digit is 1. The number of digits is counted when the data signal is received; if the number of 1's is not even, then the receiver knows there has been a transmission error. The receiver does not know which digit is wrong, but it knows that the transmission of that symbol is incorrect and requires it to be retransmitted. The parity digit increases the number of digits from 7 to 8.

Finally, a start and a stop digit are required in most data transmission systems to distinguish between one symbol and the next. The total number of digits then required for transmission of a single symbol (a letter or number) is 10.

For the transmission of 1 word per minute, using the standard ASCII code with parity and a stop and start digit, with each word consisting of six letters, and each letter consisting of 10 bits, 1 word/min requires 1 b/s, which is equivalent to 1 Hz. Therefore, a 120-word/min message, about the rate of human speech, can be fit into a 120 Hz bandwidth. Note that a single telephone call fits into a 4-kHz bandwidth, and a single television signal fits into a 4-MHz bandwidth. The bandwidth advantage for the transmission of data is clearly evident.

Audio and video signals are in analog form. The signal is a voltage that varies with time, and its amplitude is proportional to the amplitude of the audio or video information. These analog signals are often coded into digital signals before transmission, to make the signal immune to noise interference.

14.3 SIGNAL-TO-NOISE REQUIREMENTS

The problem of noise in electronic communication systems is illustrated in Figure 14.9. A telephone system is shown transmitting a single-frequency signal. The signal is clear and undistorted as it leaves the telephone. However, in the transmission system, noise is added so that the signal plus the noise, when it is received, is a badly distorted replica of the original signal.

The requirements for an acceptable signal to noise ratio are shown in Figure 14.10. For telephone service, the required SNR is 50 dB. This is the standard for common carrier telephone service. Any departure can be detected as noise on the audio signal. This means that when a telephone signal is transmitted from one location to another, and in the process passes through several different transmission systems, such as telephone wires, coaxial cable, microwave relay, and communication satellites, the signal to noise ratio after the signal is received, amplified, and arrives at the receiving telephone hand set, must be 50 dB.

Signal to noise requirements for television transmission are less severe. If the signal to noise ratio is 40 dB, there is no noticeable degradation of the TV picture, if the signal to noise ratio is 35 dB, there is some snow, if the signal to noise ratio is 30 dB, there is objectionable interference with the picture, and at a 25 dB signal to noise ratio the picture is all snow.

The 20-dB signal-to-noise ratio illustrated shows the voltage waveform of the signal and the noise as functions of time. The signal power is 100 times

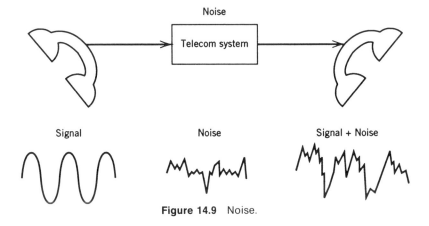

Figure 14.9 Noise.

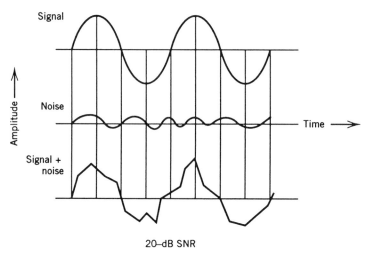

Figure 14.10 Signal-to-noise requirements.

greater than the noise power, so the signal voltage is 10 times greater than the noise voltage.

The signal-to-noise ratio can obviously be improved by more signal and less noise. Another way to reduce the signal/noise ratio is to make the signal immune to noise interference by pulse coding the analog signal into a digital signal.

14.4 PULSE CODE MODULATION

The advantage of pulse coding analog signals to eliminate noise interference is illustrated in Figure 14.11. Figure 14.11*a* shows the analog signal (say an electronic signal from a telephone set or television camera) to be transmitted as a voltage varying with time. If considerable noise is added during transmission, the signal is distorted as shown in Figure 14.11*b* and is entirely useless. However, pulse code modulation in the same high noise environment eliminates the distortion, as shown in Figure 14.11*c,d*. To do this, the analog signal is not transmitted directly. Instead, before transmission the analog signal is measured (See Figure 14.11*c* for a measurement at one point in time). Rather than transmitting the 5-V signal, the 5 V is coded into a digital signal, represented as 1001110, which is transmitted as a series of positive and negative pulses. When the digital signals, which represent the value of the analog signal at a given instant of time, are transmitted and received (Figure 14.11*d*) they are badly distorted. Note that some of the 1's are large and some are small, but there is no question which digits are 1's and which are 0's. Although a great deal of noise has been added to the digital signal, there is no question that this sequence of digits is 1001110, which represents

Transmitted

Received

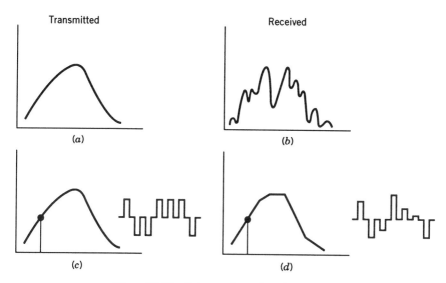

Figure 14.11 Pulse coding of analog signals.

the value 5. Therefore the value of the original analog signal is known at the receiving end of the system at this instant in time. At several instants of time, the value of the analog signal to be transmitted must be measured, converted into a digital signal that represents the value, and transmitted. At the receiving end, the original analog signal must be reconstructed from the digital code. Note that the reconstructed signal is nearly a perfect replica of the original signal.

Figure 14.12 shows the analog signal to be transmitted as a function of time. The signal varies over the amplitude range from 0 to 7, and a three-digit representation of the numbers from 0 to 7 is shown. At various sampling times, the signal is measured, represented by its digital code, and the digits

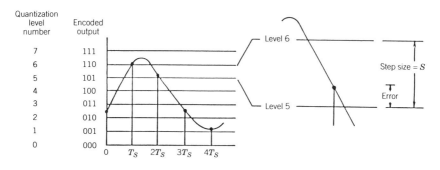

Output = 010110101011001

Figure 14.12 Pulse code modulation (PCM).

are transmitted. Therefore the transmitted output is 010110101011001, the value of the analog signal at the five time instants when it is measured. The 1's and 0's are transmitted and noise adds to them, so upon reception the amplitudes of the 1 and 0 pulses are distorted but there is no question as to which is a 1 and which is a 0. The reconstructed received analog signal is almost an exact replica of the transmitted signal. The reconstructed signal is not perfect due to "quantization" error. The quantization error is illustrated in the expanded view of the analog signal measured at time $4t_s$. At this sampling time, the signal is between 5 and 6, but closer to 5. Since the digital code only has the numbers 5 and 6, the analog signal must be quantized and given the value 5. Thus, there is an error. In a PCM system, no error is added by the signal transmission, but the error is inserted when the signal is coded, and this error is called the *quantization error*. When the digital signal is reconstructed into analog form at the receiver, the analog signal is given the value of 5, which is not the true value that the original signal had. The error can be reduced to an arbitrary small value by making the step size smaller, which requires more and more digits. If eight digits were used, 256 quantizing values are produced instead of the eight values in the example; the signal-to-quantization-noise ratio would then be 50 dB. Consequently, the error would meet the signal-to-noise requirements for telephone transmission. Therefore, an eight-digit representation of the analog signal is used in most PCM systems.

The signal must be sampled at twice the highest frequency present in the analog signal, to obtain both the amplitude and phase information in the signal. Sampling at a lower rate does not provide a faithful time representation of the signal. Sampling at a higher rate requires more data to be transmitted than is necessary.

The PCM process consists of sampling, quantizing, and coding. Figure 14.13a shows the analog signal to be transmitted. It is sampled at a sampling rate of twice the highest frequency component in the signal. The next step is to quantize the samples, that is, to give them a value that can be represented by the digital code. The third step is to code the signals, that is, to represent them by a series of 1's and 0's so that they can be transmitted as a digital code. With an eight-digit code, the quantization signal-to-noise ratio is 50 dB, which means that the reconstructed signal will have an effective noise 50 dB less than the signal itself. The required bandwidth is 16 times the highest frequency in the signal. The 16-fold increase in bandwidth comes about because the signal must be sampled twice at the highest frequency, and eight digits must be used to represent the signal at each sampling.

It might appear that if PCM is used, the received signal to noise ratio is no longer important. However, if the signal to noise ratio is bad enough, so much noise is added that a 1 could be mistaken for a 0, and vice versa. The likelihood of this happening is a function of the signal to noise ratio as shown in Figure 14.14. The bit error rate (BER) is the probability that an error will occur. If the received signal to noise ratio is 12 dB, the BER is 10^{-8}, which

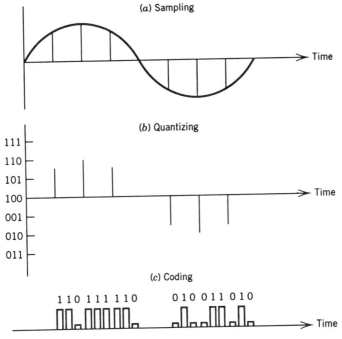

(a) Sampling

Time

(b) Quantizing

111
110
101
100
001
010
011

Time

(c) Coding

1 1 0 1 1 1 1 1 0   0 1 0 0 1 1 0 1 0

Time

Figure 14.13 Additional explanation of PCM.

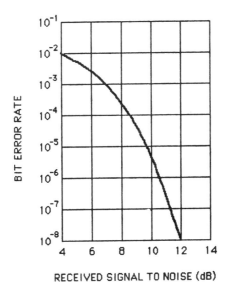

Figure 14.14 Bit error rate (BER).

is 1 part in 100 million. This may look like an extremely good bit error rate, but millions of bits are normally being transmitted per second in a PCM system. In general, if the signal to noise ratio is 20 dB, then the bit errors are negligible.

Hundreds of independent signals, each in the same frequency range, can be combined for transmission by using frequency division multiplexing (FDM). The signals can be multiplexed in a much less expensive way by PCM coding the analog signals and combining their digital representations. This process is time division multiplexing or (TDM), and is illustrated in Figure 14.15. The figure shows the multiplexing of four independent telephone signals. The first step in TDM is to sample the signals and combine the samples (*a*). A sample from channel 1 is taken, then a sample from channel 2, one from channel 3, and one from channel 4. These samples appear consecutively in time. Then a second sample each from channels 1 to 4 are taken. Then a third sample is taken from each channel.

A series of pulse amplitude modulated signals comes from the samples. The first one in the series is from the first channel, the second is from the second channel, and so on. The series then repeats itself with the second sample from the first channel. These signals are sent into a quantizer and coder (Figure 14.15*b*), and each sample is coded into a digital representation. The first sample from channel 1 is represented by 0010111, and the first sample from channel 2 by 1011011 (Figure 14.15*a*). This quantizing and coding process is the same as that for PCM, but now it is done with successive samples of the four channels. The digital bit stream is then transmitted. As shown in Figure 14.15*c*, the process is reversed at the receiver. The digital code is used to reconstruct the sampled signals and then they are separated out, so the samples of channel 1 appear in the channel 1 output line, the samples of channel 2 appear in the channel 2 output line, and so on.

The PCM and the TDM processes look very complicated, but they are all digital functions, involving only the switching of 1 and 0 signals, and thus can be accomplished with digital integrated circuits. Standard digital circuits contain 16,000 or more gates on a single chip; so once designed, the PCM and the TDM digital chips can be economically made. Multiplexing using PCM–TDM costs approximately $50 per channel, compared with $500 per channel for frequency division multiplexing. This significant advantage is in addition to the noise immunity advantage.

Any number of channels can be combined, but it must be done in a standard way so that all telephone systems are compatible. This standard digital hierarchy is shown in Figure 14.16. The multiplexing always starts with 24 voice frequency telephone channels, each of which has a 0–4 kHz band. The sampling rate must be 8 kHz and eight digits must be used to represent each sample. The multiplexing equipment that combines the 24 channels is called a *D channel bank,* and the resulting channel is called a *T1 line.* The data rate is 1.544 Megabits per second. If the number of channels is multiplied by the sampling rate and by the number of bits per sample, the result is slightly

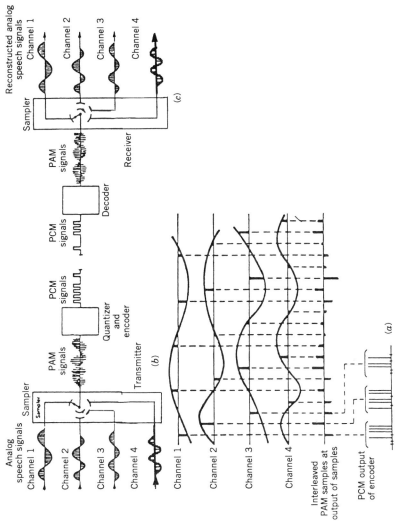

Figure 14.15 Time-division multiplexing (TDM).

357

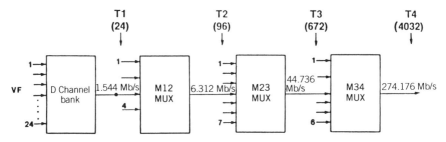

Figure 14.16   TDM hierarchy.

less than 1.544 Megabits. The additional bits are added as timing pulses to separate the signals from the individual channels.

The next step is to combine four T1 lines to obtain 96 multiplexed channels, which is called a *T2 line*. the bit rate of this data stream is 6.312 Mb/s. Again additional timing pulses are added to the T2 signal. Finally, seven T2 lines are combined to form a *T3 line*, which is 672 telephone channels. Then six T3 lines are combined to form a *T4 line*, which is 4032 telephone channels. The bit rate of the T4 line is 274 Mb/s. Different hierarchies are used in Japan and Europe.

The advantages of pulse code modulation over analog signals are noise immunity, less expensive multiplexing, less expensive signal switching, and compatibility of the PCM analog signals with data signals. Its only disadvantages is increased bandwidth. As shown in Figure 14.17*b*, the analog signal must be sampled at twice the analog frequency, and each time the sampling is done a series of eight digits must be transmitted. Thus, the low frequency analog signal is replaced with a high frequency digital signal, which is 16 times higher in frequency (two times for the sampling rate and eight times for the eight digits). This bandwidth disadvantage, however, is not as bad as it seems, as will be shown later.

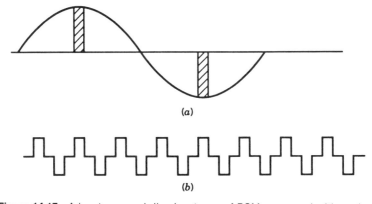

Figure 14.17   Advantages and disadvantages of PCM compared with analog.

14.5   BASEBAND SIGNALS

The electronic signal that needs to be transmitted is called the *baseband* signal. Various possible baseband signals are

Analog	Digital
Audio	Pulse-coded audio, video, and telephone
Telephone	TDM telephone
FDM telephone	Digital data
Video	
Modem tones from data	

The baseband signal can be analog or digital. Examples of analog signals are audio, telephone, FDM telephone, or video. Even computer data can be in analog format. Data are represented by digital pulses of 1's and 0's, but many data systems transmit the data by first using a modem to convert the digital signal into an audio tone in the 0–4 kHz range, which is compatible with a telephone voice transmission line.

Examples of digital baseband signals are pulse-coded audio, video, and telephone signals, TDM telephone or TV signals, and digital data.

Older telecommunication equipment was designed to handle analog baseband signals. However, the many advantages of a digital baseband, such as noise immunity and reduced multiplexing and switching costs, are resulting in greater and greater use of digital basebands in microwave communication systems.

14.6   TRANSMISSION SYSTEMS

Types of transmission systems are shown in Table 14.2. In a cable system the baseband (an analog or digital electrical signal) is simply transmitted down the cable from the transmitter to the receiver. As the signal is transmitted it gets attenuated by the cable, so amplifiers must be placed at intervals along the cable to maintain a satisfactory signal to noise ratio. Since the cables are handling the baseband directly, cable systems are sometimes called *baseband* systems. Cable systems can transmit large amounts of information. If the amount of information exceeds the capacity of a single cable, parallel cables can be added. The disadvantages of cable systems are that the cable must be laid from the transmitter to the receiver. A right-of-way must be obtained, and amplifiers must be placed at various points along the transmission path. The cable must be laid under ground or suspended on poles. This is difficult to accomplish over rugged terrain and across oceans.

Broadcast systems need only transmitter and receiver locations, so terrain is not a problem. Broadcast systems also provide multipoint distribution, in

Table 14.2 Transmission systems

Type	Advantages	Disadvantages
CABLE SYSTEMS		
Telephone wires	High capacity	Cable must be laid
Coaxial cable		Must have right of way
Fiber optics		Difficult to lay over terrain, in congested downtown areas, and across oceans
BROADCAST SYSTEMS		
Radio, TV	Need only transmitter and receiver sites	Limited capacity
Microwave relay		
Troposcatter	Solves terrain and congested downtown installation problems	
Satellite		
	Multipoint distribution	

which one transmitter sends its signal to many receivers. For example, cable TV programs are distributed by satellite from one transmitter location to thousands of receivers in the United States. Broadcast systems, however, have limited capacity. After all the assigned frequency bandwidth has been used, no more information can be transmitted.

Because of the advantages and disadvantages of cable and broadcast systems, each has their place for telecommunication transmission. Where large amounts of information need to be transmitted, for example, from one large metropolitan area to another, cable transmission is most appropriate. Where less information needs to be transmitted, to eliminate the problem of laying cable, and also to provide multipoint distribution, broadcast tranmission is used.

The principles of broadcast transmission systems are shown in Figure 14.18. The broadcasted microwave signal serves as the carrier. The baseband signal must be amplitude-, frequency-, or phase-modulated onto the microwave carrier. Amplitude and frequency modulation of the carrier by the baseband are shown in Figure 14.18a. The carrier frequency and the modulating baseband signal are shown. The baseband signal is to be sent from the transmitter to the receiver. It cannot be broadcast directly, but is used to modulate the broadcast carrier. The carrier is transmitted, and the baseband information is then removed from the carrier at the receiver. With amplitude modulation, the carrier frequency remains constant, but the amplitude is varied by the modulation signal. With frequency modulation the amplitude of the carrier remains constant, but its frequency is changed. When the modulating signal is large and positive, the carrier frequency is decreased. When the modulating signal is large and negative, the carrier frequency is increased.

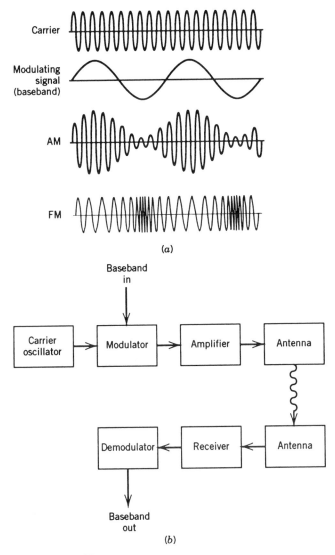

Figure 14.18 Broadcast systems.

Broadcast transmission begins by generating the broadcast carrier in an oscillator. The carrier is then amplitude- or frequency-modulated. The modulated carrier is amplified and transmitted from the transmitting antenna to the receiving antenna. The weak received signal is amplified and appears after amplification with its AM or FM modulation. In the demodulator the baseband is removed from the carrier.

14.7 MODULATION

Figure 14.19 shows a carrier amplitude-modulated by a single frequency. Part of the unmodulated carrier is shown at the left. The carrier frequency remains constant, but its amplitude is modulated by the baseband signal. The amplitude modulated carrier is no longer a single-frequency signal. A spectrum analysis of its waveform is shown in Figure 14.19b. When the carrier is modulated by just a single frequency, the spectrum analysis shows that the AM signal consists of three frequencies: the original carrier (f_c) and two sidebands, one above the carrier frequency by the amount of the modulation frequency ($f_c + f_m$), and one below the carrier frequency by the amount of the modulation frequency ($f_c - f_m$). For example, if f_c = 4 GHz and f_m = 4 MHz, then the spectrum analysis of the AM wave shows the carrier at 4 GHz, the upper sideband at 4.004 GHz, and the lower sideband at 3.996 GHz.

As shown in Figure 14.19c, the modulating signal is usually not just a single frequency but a complex signal. The modulating signal can be analyzed into a sum of individual modulating frequencies. The resulting AM signal is a set of upper sidebands and a set of lower sidebands, with a pair of sidebands for each frequency present in the modulating signal. The spectrum of the AM wave is then a set of frequencies above the carrier and a set of frequencies below the carrier, with each set consisting of all frequencies in the modulating wave.

Various types of amplitude modulation are illustrated in Figure 14.20, which shows the amplitude of the carrier and the sidebands as a function of frequency around the carrier. Figure 14.20a shows a double-sideband transmitted carrier (DSBTC). This is the amplitude modulation just discussed. Figure 14.20b shows a double-sideband suppressed carrier (DSBSC). The carrier contains no information and need not be transmitted, which conserves transmitter power. Figure 14.20c shows the extremely important case of single sideband (SSB). Not only does the carrier contain no information, but the lower sideband is an image of the upper sideband and both contain the same information. Therefore only one set of sidebands needs to be transmitted to get all the information from the transmitter to the receiver. If only one set of sidebands is transmitted, the bandwidth is reduced. Single sideband modulation is used as the frequency-shifting process in the frequency division multiplexing discussed previously.

Figure 14.20d shows vestigal sideband, where all of the upper sideband and part of the carrier and a small part of the lower sideband are transmitted. Vestigal sideband is used primarily in broadcast TV transmission so that the carrier need not be generated in the TV receiver and expensive filters need not be used.

Frequency modulation is shown in Figure 14.21. The waveforms show the carrier, the modulating signal, and the FM modulated carrier. The spectrum analysis of the FM waveform consists of an infinite set of upper and lower

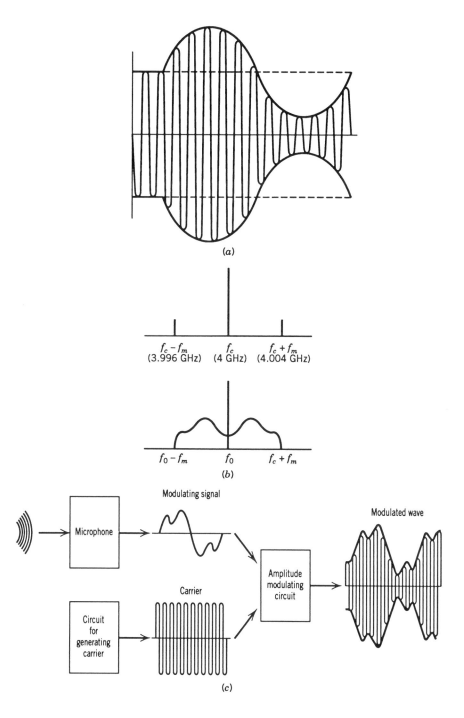

Figure 14.19 Amplitude modulation.

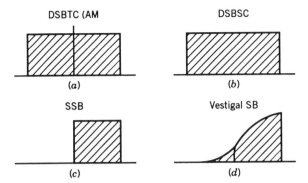

Figure 14.20 Types of amplitude modulation.

sidebands whose amplitude are symmetrically distributed about the carrier and separated from each other by the modulation frequency.

With AM only two sidebands were produced: one at frequency f_m above the carrier, the other at frequency f_m below the carrier. With FM an infinite set of sidebands occur, each spaced f_m apart above and below the carrier. The sidebands decrease in amplitude away from the carrier, and for practical system design purposes only the sidebands whose amplitudes are within 20 dB of the original unmodulated carrier need be used, as shown in Figure 14.21*b*.

The characteristics of the FM signal depend on the modulation frequency

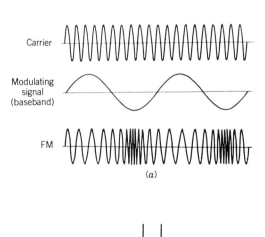

Figure 14.21 Frequency modulation.

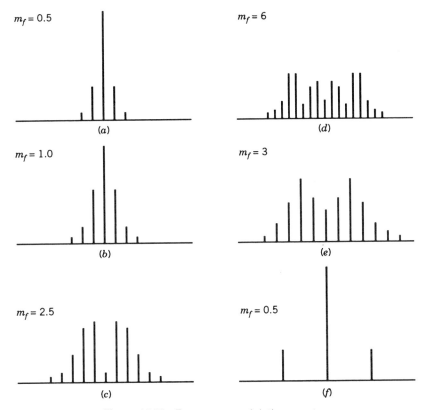

$m_f = 0.5$

(a)

$m_f = 1.0$

(b)

$m_f = 2.5$

(c)

$m_f = 6$

(d)

$m_f = 3$

(e)

$m_f = 0.5$

(f)

Figure 14.22 Frequency modulation spectra.

f_m and the carrier deviation δ. The carrier deviation is the amount that the carrier frequency is varied above or below its unmodulated value by the modulating process. For example, if f_c = 4 GHz and δ = 10 MHz, then the microwave frequency deviates or varies from 3.99 GHz to 4.01 GHz, and the rate at which the microwave frequency is varied is the modulation frequency f_m = 4 MHz. Most characteristics of the FM signal are determined by the modulation index $m_f = \delta/f_m$. Here m_f = 10 MHz/4MHz = 2.5. The bandwidth BW of a FM signal includes all upper and lower sidebands whose amplitudes are within 20 dB of the original unmodulated carrier. BW = $2 f_m (1 + m_f)$, so in this example BW = 28 MHz.

Some typical FM spectra are shown in Figure 14.22. Figure 14.22a–c are spectra with a constant modulation frequency but an increasing deviation. Note that all the lines are spaced by the modulation frequency. As the deviation increases, the number of sideband lines whose amplitude is within 20 dB of the original unmodulated carrier increases, so the required bandwidth increases. When the modulation index is 2.5, the carrier is extremely small and the sidebands are larger than the carrier. The spectra in d–f are for the case where the deviation is kept constant but the modulation frequency is increased. The bandwidth remains approximately the same, but the spacing between the sideband lines increases as the modulation frequency increases. Figures 14.22*a* and *f* have the same modulation index but different spectral lines because of their different modulating frequencies.

Up to now it seems that FM requires increased bandwidth, so AM is more desirable. However, in FM the signal to noise ratio is enhanced by the demodulation process. FM creates many modulation sidebands. During demodulation, all of the sidebands are combined coherently in phase to reconstruct the modulating signal. However, the noise combines randomly and tends to cancel during the demodulation process. Thus, the signal to noise ratio is improved by demodulation. The greater the carrier deviation, the greater the number of sideband lines that are formed, and so the signal-to-noise is better. However, the greater the carrier deviation, the greater is the bandwidth required for the signal to be transmitted. Consequently, there is a signal-to-noise improvement versus bandwidth trade-off for frequency modulation as shown in Table 14.3.

Table 14.4 shows the improvement that can be obtained in signal to noise ratio for various modulation techniques. The signal to noise ratio after demodulation is shown for a broadcast carrier to noise ratio before demodulation of 20 dB. For telephone systems, a signal to noise ratio after demodulation of 50 dB is required and a 40 dB signal to noise ratio is required for television systems. With amplitude modulation or single side band modulation (which is a form of amplitude modulation), no improvement in signal to noise ratio is obtained by the demodulation process. With frequency modulation with $m_f = 1$, a 4-dB improvement is obtained; with $m_f = 5$, there is a 20-dB improvement. Consequently, if the carrier to noise ratio is 20 dB, the signal to noise ratio after demodulation is increased to 40 dB by using FM with $m_f = 5$. Also shown is the case of PCM with an eight-digit code. If the signal to noise ratio is adequate, so that the bit error rate is low, no noise is added by the transmission process, and the signal to noise ratio remains at 50 dB, which is quantization signal to noise ratio. Consequently, if the carrier to noise ratio is 20 dB, the signal to noise ratio after reconstruction of the pulse code modulated signal is 50 dB.

Table 14.4 also compares the bandwidth required for these four modulation cases.

Table 14.3   Signal to Noise versus bandwidth trade-off for FM

Modulation Index	SNR Improvement	Bandwidth (relative to Modulation bandwidth)
1	4 dB	4
3	12 dB	8
5	20 dB	12

Table 14.4 Comparison of modulation techniques (carrier to noise ratio = 20 dB)

Modulation	Signal to Noise Ratio	Bandwidth
SSB	20 dB	1
FM ($m_f = 1$)	24 dB	4
FM ($m_f = 5$)	40 dB	12
PCM	50 dB	16

14.8 CARRIER MODULATION WITH DIGITAL BASEBAND SIGNALS

Special carrier modulation techniques are used to modulate digital baseband signals onto a carrier, since the baseband signal has only one of two values, either a 1 or a 0. Some of the most common digital modulation techniques are illustrated in Figure 14.23. Amplitude modulation is shown in Figure 14.23*a*. The amplitude has a value of 1 unit for a 1 digit and is 0 for a 0 digit. The four-digit sequence 1001 requires 4 units of time for its transmission.

A multilevel amplitude modulation technique is shown in Figure 14.23*b*. Two digits are sent in each unit of time. The signal level has four possible values. A 0 signal level corresponds to two digits, 00. The next higher signal

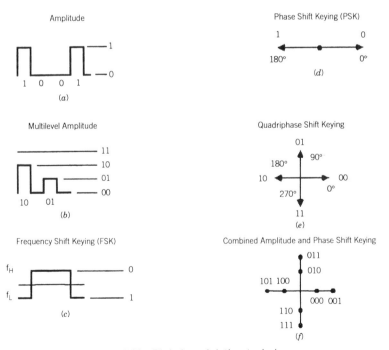

Figure 14.23 Digital modulation techniques.

level corresponds to 01, the next to 10, and the highest to 11. The sequence 1001 can be transmitted in only 2 units of time with a two-level amplitude signal. An even greater number of amplitude levels could be used, which allows even more digits to be transmitted in a unit of time. However, the bit error rate would increase and the transmission system would require a higher signal to noise ratio.

Figure 14.23c shows frequency shift keying. The carrier is shifted from one frequency to another. The higher frequency corresponds to a 0, the lower to a 1. To transmit the series of digits 1001, four time intervals are required.

Phase shift keying is shown in Figure 14.23d. The amplitude and frequency of the carrier remain constant, but its phase is shifted. A 0° phase shift represents a 0, and a 180° phase shift represents a 1.

One of the most popular digital modulation techniques is quadraphase shift keying (QPSK). The amplitude and frequency of the carrier remain constant, but its phase is shifted to one of four values: 0°, 90°, 180°, and 270°. As shown in Figure 14.23e, each phase represents two digits: 0° represents 00, 90° represents 01, 180° represents 10, 270° represents 11.

Multilevel phase shift keying can be used to represent digital information. In some digital microwave communication systems, as many as 16 phase states are used, but the greater the number of phase states, the greater is the bit error rate, and the greater the signal to noise ratio required by the system. Of course, the greater the number of digits that can be represented by a single amplitude, frequency, or phase state, the lower the system bandwidth requirements. A trade-off must therefore be made between bandwidth and signal to noise requirements.

It is also possible to combine amplitude levels with phase shifting. An example is shown in Figure 14.23f. Two amplitude levels and four phase positions are used. The lower amplitude level with 0° phase shift represents 000, the higher amplitude level with 0° phase shift represents 001, and so forth.

ANNOTATED BIBLIOGRAPHY

1. A. M. Noll, *Introduction to Telecommunication Electronics*, Artech House, Dedham, MA, 1988.
2. R. L. Freeman, *Telecommunication Transmission Handbook*, 2d ed., Wiley, New York, 1985.
3. A. M. Noll, *Television Technology: Fundamentals and Future Prospects*, Artech House, Dedham, MA, 1988.
4. A. M. Noll, *Introduction to Telephones and Telephone Systems*, Artech House, Dedham, MA, 1986.

Reference 1 simply explains telecommunication principles. Reference 2 provides a comprehensive technical explanation of all communication systems. Reference 3 simply explains television signals. Reference 4 simply explains telephone signals.

EXERCISES

14.1. Define the communication systems terms in the following list:
1. Baseband
2. Carrier
3. Modulation
4. Demodulation
5. Frequency modulation (FM)
6. Amplitude modulation (AM)
7. Sidebands
8. Bandwidth
9. Single Sideband (SSB)
10. Multiplexing
11. Frequency-division multiplexing (FDM)
12. Pulse code modulation (PCM)
13. Time-division multiplexing (TDM)
14. Carrier-to-noise ratio (C/N)
15. Signal-to-noise ratio (S/N)

14.2. The frequency range of a single telephone signal is from _____ Hz to _____ Hz.

14.3. A single telephone signal is fitted into a frequency band from _____ Hz to _____ Hz.

14.4. How many telephone signals are frequency-division-multiplexed together to form an FDM basic group?

14.5. What is the frequency range of an FDM basic group?

14.6. The frequency range of a single TV signal is from _____ Hz to _____ MHz.

14.7. How many digital bits are required for a single letter in an asynchronous ASCII code with parity?

14.8. What signal-to-noise ratio is required for satisfactory telephone service?

14.9. What signal-to-noise ratio is required for satisfactory television reception?

14.10. Coding of an analog signal into a series of digital pulses is called _____.

14.11. The combining of many telephone signals by first coding each into a series of digital pulses and then interleaving the pulses series of each signal is called _____.

14.12. What must be the PCM sampling rate on a 4 kHz analog signal?

14.13. What is the quantization signal-to-noise ratio if an eight-digit PCM code is used?

14.14. What is the digital bit rate for a standard PCM telephone channel?

14.15. A special TDM system combines six telephone channels using a six-digit PCM code. What is the bit rate of this TDM?

14.16. A high-speed TDM system combines four broadcast TV channels using a six-digit PCM code. What is the bit rate of this TDM?

14.17. What is the bit error rate (BER) if the received signal-to-noise ratio is 10 dB? (Use Figure 14.14)

14.18. In the TDM hierarchy:
  a. How many telephone channels are combined into the smallest TDM group?
  b. What is the bit rate for this TDM group?
  c. What is this signal formating called?

14.19. In the TDM hierarchy:
  a. How many telephone channels are transmitted in a T2 line?
  b. What is the T2 bit rate?

14.20. Name four advantages of PCM compared with analog.

14.21. What is the disadvantage of PCM compared with analog?

14.22. The changing of digital signals from a computer into frequency tones for transmission down a conventional voice grade telephone circuit is done in a _____.

14.23. What is the advantage of cable communication systems compared with a broadcast system?

14.24. Name two advantages of broadcast systems compared with cable systems.

14.25. What is the advantage of FM compared with AM?

14.26. What is the advantage of AM compared with FM?

14.27. Why is the requirement for a wide bandwidth a disadvantage in a communication system?

14.28. Draw the spectrum of a 4-GHz carrier amplitude modulated by a 4-MHz baseband signal.

14.29. Draw the spectrum of a 4-GHz carrier frequency modulated by a 4-MHz baseband signal if the carrier deviation is 4 MHz.

14.30. Draw the spectrum of a 4-GHz carrier frequency modulated by a 4-MHz baseband signal if the carrier deviation is 20 MHz.

14.31. What bandwidth is required for the AM signal of Exercise 14.28?

14.32. What are the modulation index and bandwidth for the FM signal of Exercise 14.29?

14.33. What are the modulation index and bandwidth for the FM signal of Exercise 14.30?

14.34. What is the approximate S/N improvement factor for an FM signal with a modulation index of 5?

14.35. Name four methods for modulating digital signals onto a microwave carrier.

15

MICROWAVE RELAY

The chapter explains the purpose and operation of microwave relay. The various types of microwave relay, including baseband radios, heterodyne repeaters, and digital microwave relay, and the antennas and antenna accessories required are discussed.

Path loss calculations to determine the received carrier power and the carrier to noise ratio are carried out to determine how much transmitter power, antenna gain, and receiver noise figure are needed to make the system work. Multipath loss is discussed. Diversity systems to improve microwave relay reliability are also described.

Troposcatter systems are described.

The chapter concludes with a discussion of wireless local area networks.

15.1 INTRODUCTION

A microwave relay is illustrated in Figure 15.1. It is often called terrestrial microwave relay because it is located on the surface of the earth. The figure exaggerates why microwave relay is necessary. Microwaves travel in approximately a straight line, but since the earth is curved the transmission distance for microwave relay along its surface is only 10 to 50 mi. The actual transmission distance depends on the height of the antenna above the earth's surface. If the antennas are mounted on towers on a building on a hill, distances up to 50 mi can be attained. If, the microwave antennas are only a few feet off the ground, the transmission distance is only 10 mi.

To transmit over greater distances, microwave must be relayed from point to point. Hence, they are also called point-to-point microwave relay systems.

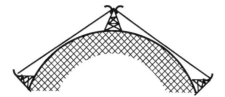

Figure 15.1 Terrestrial microwave relay.

Sections of the microwave frequency band have been assigned for micro-wave relay. The four types of microwave relay are common carrier, industrial, government, and studio transmitter link (STL).

The common carrier bands are assigned to communication companies such as AT&T, MCI, or Sprint, who are authorized to provide communication service to the public. The major common carrier bands are 3.7–4.2 GHz, 5.925–6.425 GHz, and 10.7–11.7 GHz. These bands are completely utilized in most areas of the country.

Industrial bands, including the major industrial band from 6.575 to 6.875 GHz, are used for private commercial purposes by railroads, pipeline companies, and others.

Government bands, including the major band from 7.125 to 8.4 GHz, are used for government and military communication purposes.

Studio Transmitter Link (STL) bands are used by television and radio broadcast stations to send programs from the studio to the transmitter and from remote locations (where news events are being televised) to the studio.

Channelization of the microwave frequency bands is shown in Figure 15.2. The total microwave common carrier band, from 5.9 to 6.4 GHz, is 500 MHz wide and is divided into 20-MHz-wide channels. Into one of these channels, 1000 frequency-division-multiplexed telephone channels, or one TV program video, or 20 MHz of digital data can be fitted.

The baseband width of 1000 FDM telephone channels, or one TV program, is 4 MHz. Why must 20 MHz of microwave frequency range be used for this 4-MHz baseband? The reason is that when the baseband is frequency

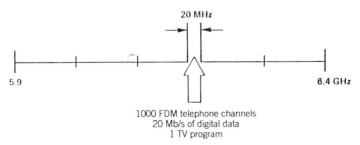

Figure 15.2 Channelization of bandwidth.

modulated onto the microwave carrier, sidebands are produced and the microwave relay channel must be wide enough for all the required FM sidebands.

In a typical microwave relay system, several 20-MHz channels are used, each for a different set of 1000 telephone calls or one TV program. In most metropolitan areas, the full 25 channels in the 5.925–6.425 GHz band are used. Once a specific channel has been used for transmission in a particular direction to a receiver, it cannot be used to transmit other information, because of interference. The same channel can be used in another direction, provided that the microwave relay antennas are carefully designed to not have sidelobes, which propagate some of their signal in directions other than along their axis. This limitation on information carrying capacity is the major disadvantage of microwave relay as contrasted to cable transmission systems. If fiber optic or coaxial cable capacity has been completely utilized, another cable can be run along the same right-of-way to increase capacity, and the number of cables can be increased as required to handle hundreds of thousands of telephone channels or many television programs.

Actually, a given frequency can be used twice for microwave relay by using dual polarization. Microwave signals in the same frequency range and transmitted in the same direction can be used for two sets of information if one set is transmitted with vertical polarization and the other with horizontal polarization. The microwave fields in the two polarizations are independent and do not interfere.

15.2 BLOCK DIAGRAMS

Figure 15.3 is a baseband microwave radio block diagram. The system begins with a baseband input at the transmitter and ends with a baseband output at the receiver. Beginning at the top left, the baseband input is sent to a modulation amplifier, and the amplified baseband signal frequency modulates a 2-GHz oscillator. The 2-GHz signal is amplified by a power amplifier, and a sample of the signal, with its frequency modulation, is taken out through a directional coupler and applied to an automatic frequency control (AFC) circuit. The AFC circuit compares the microwave signal to a harmonic of a quartz crystal oscillator and develops an error signal that is used to control the frequency of the 2-GHz oscillator to ensure that it stays on frequency. The microwave signal is then sent through an isolator to a frequency multiplier, which multiplies the microwave frequency with its frequency modulation sidebands up to the 6-GHz common carrier band. At this point, the frequency modulated microwave signal extends over a 20-MHz channel. It then passes through an isolator and channel filter to a circulator. The filter and circulator form a multiplexer, allowing only signals in the frequency range of the filter to pass from the transmitter into the transmitter feedline. The signal from the transmitter then passes through the T/R combining

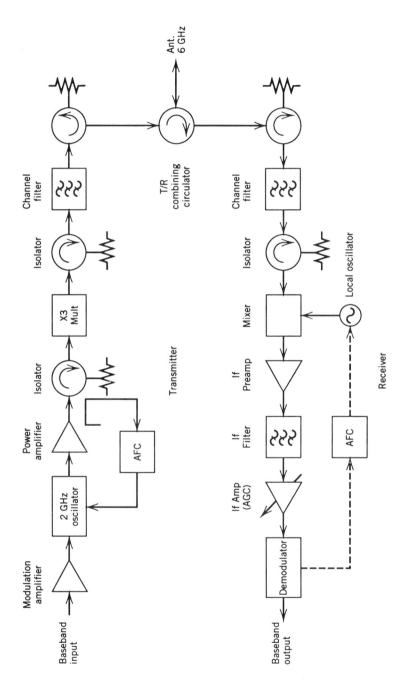

Figure 15.3 Baseband microwave radio.

circulator, which allows the transmitted power to travel up the antenna feedline to the antenna and allows the received signal from the same antenna to travel down the feedline to the receiver.

The signal comes from a distant location into the antenna and down the feedline to the receiver. The channel filter and circulator form a multiplexer, which allows only microwave frequencies in the particular 20-MHz filter bandwidth to pass into the receiver. The signal passes through an isolator and into a mixer, where it is mixed with a local oscillator to change the microwave frequency with its modulation sidebands to an IF frequency so the signal can be more easily amplified. The IF frequency is amplified in an IF amplifier, passed through an IF filter, and goes to an automatic gain control (AGC) amplifier. The gain of this amplifier is automatically adjusted so that as the level of the microwave signal changes due to changes in propagation conditions, the level of the IF signal can be maintained constant. The IF signal then goes to a limiter, which limits its amplitude variation, and to a demodulator, from which the baseband output is obtained. A sample of the demodulator output is sent back to an AFC circuit to control the LO frequency. The baseband output of the receiver is identical to the baseband input at the transmitter, and the signal has thus been sent from a distant transmitter to the receiver.

The baseband microwave radio just discussed shows one 20-MHz channel out of the total available 500-MHz bandwidth that extends from 5.9 to 6.4 GHz. Identical microwave radios operating in other 20-MHz channels within the total bandwidth can be combined to be transmitted or received from the same antenna by RF multiplexing, as shown in Figure 15.4. Different baseband signals are sent into each transmitter (TX). One transmitter operates in a 20-MHz microwave channel centered around frequency f_1, the second in a 20-MHz channel centered around frequency f_2, and the third in a 20-MHz channel centered around f_3. The filter and circulator in each channel form multiplexers that work as follows: The signal from the lower transmitter at a frequency f_3 passes through the filter and into arm 1 of its circulator and out of arm 2 into the transmission line. This signal then enters the middle circulator and tries to enter the middle transmitter. However, the signal is reflected at the filter, which can only pass frequency f_2. The signal from the lower transmitter then reenters the circulator and passes out its upper arm into the transmission line. The signal from the lower transmitter then tries to enter the upper transmitter. However, it is reflected at the upper filter back into the circulator and into the main feedline. In a similar fashion, the signal from the middle transmitter passes out through its filter and through the circulator into the feedline. It travels up the feedline to the upper transmitter which it tries to enter, but is reflected by the upper filter back into the feedline. The signal from the upper transmitter passes through its filter into its circulator and into the feedline. Consequently, the signals from each transmitter, located at different 20-MHz frequency channels within the total common carrier band, are combined into the output feedline and transmitted to the antenna.

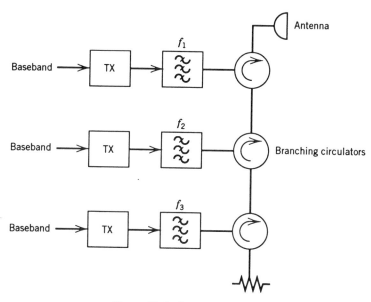

Figure 15.4 RF multiplexing.

Obviously, the same RF multiplexing technique can be applied to the receiver.

Microwave signals must often be sent over longer distances than can be obtained with a single microwave relay hop. In this case, the signal must be sent from one repeater to the next, and so on. In the *baseband repeater* in Figure 15.5 a signal is received in the east antenna and, by the circulator, is

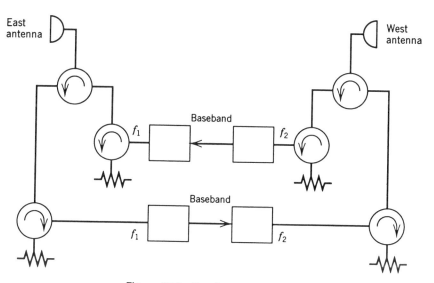

Figure 15.5 Baseband repeater.

routed into the lower receiver. It is then amplified and converted to a baseband signal. The baseband signal is then sent to a transmitter that operates at frequency f_2. The baseband information is modulated onto the transmitter carrier and transmitted through the circulators out the west antenna. A signal at the same frequency, but carrying different information, enters the west antenna and goes to the receiver in the upper branch, where it is converted to baseband, sent to the transmitter, and retransmitted from the east antenna at frequency f_1.

If the repeater station is operated at a remote site, there is really no need to convert the microwave signal to baseband, because there is no use for the baseband frequency at the remote site. In this situation a heterodyne repeater is used. In a *heterodyne repeater* the signal is not converted to the baseband but to the 70-MHz IF frequency, and is sent from the receiver to the repeating transmitter.

An *RF repeater* is sometimes used. The microwave signal with its 20-MHz passband is sent through a channel filter and amplified directly at the microwave frequency. This appears to be the simplest repeater method, but it is not used as often as the heterodyne repeater. The heterodyne repeater allows the transmission frequency at each repeater station to be changed. The mixer–IF amplifier combination is also less expensive than high gain microwave amplifiers.

Figure 15.6 is a microwave relay system that combines several of the systems just discussed. The baseband input to be transmitted enters at the lower right corner and is frequency-modulated onto a 2-GHz oscillator. The FM signal is amplified, sampled by a directional coupler, and sent to an AFC circuit. It then passes through a load isolator to a varactor diode multiplier, where the microwave carrier with its frequency modulation sidebands is multiplied to 6 GHz. The signal then passes through a load isolator and a bandpass filter and circulator combination, which form a multiplexer to insert the signal into the transmitter feedline. Another circulator is shown in the feedline to indicate that several 20-MHz transmitter channels (up to 25, which would cover the entire 5.9–6.4 GHz band) can be multiplexed together. The transmitter is identical to the one in Figure 15.3, and the multiplexing system is identical to the one in Figure 15.4. The microwave signal is then sent to the T/R circulator and up the feed waveguide to the antenna.

The received signal from a distant location is received in this same antenna, shown on the left, and comes down the antenna feedline to the T/R circulator and into the receiver line. The receiver line contains several RF multiplexers, one for each received channel. These multiplexers consist of a circulator and a bandpass filter. The received signal passes through the multiplexer and through a load isolator into a mixer, where the signal with its modulation sidebands is converted to an IF frequency. This receiver is the same as the one in Figure 15.3. The IF signal is divided, and part is taken out through a demodulator to provide a baseband output. With this type of output, the receiver serves as a baseband receiver. The remainder of the IF signal is

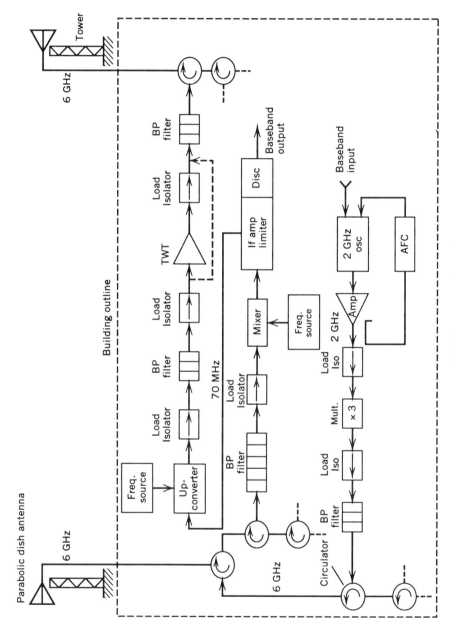

Figure 15.6 Microwave relay system.

sent to an up-converter where the frequency at 70 MHz with its modulation sidebands is converted back to 6 GHz. An up-converter uses varactor diodes to add the 70-MHz IF to the microwave transmitter frequency. The up-converted signal then passes through a bandpass filter, another load isolator, and may be amplified up to the 10-W level in a traveling wave tube so that it can be transmitted over long distances. Long-distance transmission is possible if the antenna can be located high enough to overcome earth curvature effects. The microwave signal then passes through a bandpass filter and circulator combination, which serves as a multiplexer to connect this frequency channel with other frequency channels into the transmitter feedline, which runs up to the right-hand antenna.

Figures 15.3 through 15.6 were for analog FM microwave relay. In these systems, the baseband consisted of FDM telephone channels or video signals where the amplitude of the baseband voltage carried the information; that is, the baseband was an analog signal. This analog signal was then frequency-modulated onto the microwave carrier. Noise was added to the microwave carrier during transmission, and with each repeating of the signal, whether it was kept at the IF frequency in a heterodyne repeater or converted to baseband and remodulated onto the transmitter carrier, the noise was additive. Consequently, at each repetition, the signal to noise ratio becomes worse.

In a digital microwave relay the digital signal is regenerated at each repeater station so that the noise is not additive. A block diagram of a regenerative digital repeater is shown in Figure 15.7. The received signal represents a 1101 digital signal. Note that noise has distorted the signal, but the 1101 characteristic can still be distinguished. Rather than simply transmitting this signal and having it further distorted by noise so that bit errors might occur,

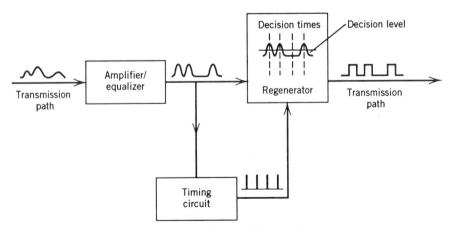

Figure 15.7 Regenerative repeater.

the digital signal is regenerated in the repeater so that it is sent out as a perfect digital signal. This regeneration is accomplished as follows: The noisy digital signal is first amplified in a limiting-type amplifier so that the pulses are sharpened. The pulses are still distorted and are not the perfect square pulses originally transmitted. The signal is then sent to a timing circuit that regenerates a timing reference, and the regenerative timing signal is then used to sample the distorted pulse waveform. At each sampling time, a decision is made as to whether the signal is a 1 or a 0. Based on this information, a new digital signal is regenerated with perfect spacing in time and a uniform amplitude. Note that the original received signal could be read as a 1101 signal, but if further noise was added it would be difficult to determine exactly what the digits were. Before that noise can be added in an additional relay, the signal is reconstructed in the regenerative repeater and transmitted as a perfect 1101 pulse series.

A block diagram of a digital microwave relay system is the same as that of the FM analog microwave systems in Figures 15.3 through 15.6 except for the modulator and demodulator blocks. Techniques for modulating digital information onto a microwave carrier were discussed in Chapter 14 and included single-level or multilevel amplitude modulation and biphase, quadra-phase, or higher-level phase modulation techniques. The more complex the modulation technique, the more information that can be transmitted in a given bandwidth, but the poorer is the bit error rate.

15.3 ANTENNAS

Microwave relay antennas are shown in Figures 15.8 and 15.9. The most common microwave antenna is the parabolic dish in the upper left. The dish operates in a single microwave frequency band with a single polarization.

Simply by a change of antenna feed to the one shown in the center left, the same antenna can be used in a dual-polarized system. With two polarizations, two microwave signals can be transmitted over the same frequency band without interference, so the information-carrying capacity of the microwave relay is doubled. Each microwave relay transmitter and receiver is identical to those already discussed, and each is connected separately to the antenna feed.

The antenna feed in the lower left allows dual-frequency operation from the same antenna. The feed contains separate waveguides with single polarizations for the 6- and 11-GHz frequency band.

The parabolic antenna on the right side of Figure 15.8 has a shielding ring around its outer diameter to reduce the sidelobe level so that transmissions will not interfere with receivers located in other directions. The antenna and shield combination are covered with a radome to protect the feed horn from weather.

Horn-reflector antennas are shown in Figure 15.9. On the left, the antenna

Dual-Polarized
Waveguide Feed

Dual-Frequency
6 and 11 GHz
Waveguide Feed

Figure 15.8   Microwave relay antennas. (Photos courtesy of Andrew Corporation.)

is mounted on a tower. The microwave signal enters the lower part of the antenna through a horn and is reflected from a parabolic surface out the front of the antenna. The horn-reflector antenna has the lowest sidelobe level of any microwave antenna. Although it is the most expensive microwave relay antenna, because of its low sidelobes it is used in congested metropolitan areas to prevent interference.

The mounting accessories of an antenna are illustrated on the right of Figure 15.9. They are shown used with a horn-reflector antenna, but the same type of accessories are also used with parabolic dish antennas. A dual-polarization system is shown, with two antenna feeds, each operating in the same frequency range but with different polarizations. The microwave signals are transmitted up the tower to the antenna through two elliptical waveguides. The waveguide is pressurized so that moisture does not collect inside to cause increased losses. Note the use of flexible waveguide to compensate for thermal expansion at the connection between the elliptical waveguide and the antenna. Also note the connection of the dual-polarization unit at the bottom of the antenna, where the separate signals are connected to a circular

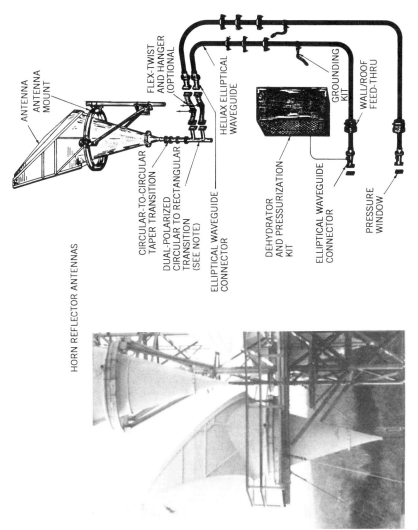

ANTENNA

ANTENNA MOUNT

FLEX-TWIST AND HANGER (OPTIONAL)

CIRCULAR-TO-CIRCULAR TAPER TRANSITION

DUAL-POLARIZED CIRCULAR TO RECTANGULAR TRANSITION (SEE NOTE)

ELLIPTICAL WAVEGUIDE CONNECTOR

HELIAX ELLIPTICAL WAVEGUIDE

GROUNDING KIT

WALL/ROOF FEED-THRU

DEHYDRATOR AND PRESSURIZATION KIT

ELLIPTICAL WAVEGUIDE CONNECTOR

PRESSURE WINDOW

HORN REFLECTOR ANTENNAS

Figure 15.9 Horn reflector antennas. (Photos courtesy of Andrew Corporation.)

waveguide section in a dual-polarized rectangular-to-circular waveguide transition.

15.4 PATH LOSS CALCULATIONS

The received microwave carrier must be 10–20 dB above the receiver noise. This SNR is not satisfactory for reception of telephone, video or data, but the FM demodulation process or the digital coding improves the SNR up to the required 40–50 dB. The received carrier power is determined by the transmitter power, the transmitter antenna gain, the receiver antenna gain, the microwave frequency, the transmitter to receiver separation, and path attenuation effects. The calculation of the received power from these parameters is called a path loss calculation.

The path loss equation is derived as follows.

Derivation of Path Loss Equation

1. Microwave power is radiated from the transmitter. If the transmitter used an isotropic antenna (Figure 15.10)—one that radiates power equally in all directions—then

$$PD = \frac{P_t}{4\pi R^2}$$

where PD = power density
P_t = transmitted power
R = distance from transmitter to receiver

This equation states that the transmitter power, as it radiates uniformly in all directions from the transmitter, will be spread at a distance R from the transmitter over a spherical surface, and the area of the spherical surface is $4\pi R^2$.

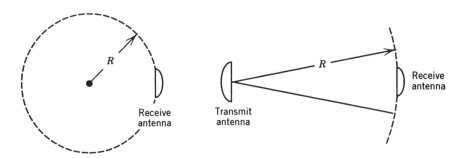

Figure 15.10 Derivation of the path loss equation.

2. With a directional antenna (Figure 15.10), power density at the receiving antenna is increased by the gain of the transmitter antenna. Note that the transmitter antenna gain depends on the wavelength and the antenna diameter:

$$PD = \frac{P_t}{4\pi R^2} \times G_t$$

$$G_t = \text{Antenna Gain} \approx 6\left(\frac{D}{\lambda}\right)^2$$

where D = antenna width
     λ = wavelength

3. The power density at the receiving antenna is actually not quite as large, as calculated in Step 2 because it is reduced by the path attenuation (α):

$$PD = \frac{P_t}{4\pi R^2} G_t \, \alpha$$

4. Since the transmitted power is spread over a spherical surface of many square miles, only a tiny fraction of the power is picked up by the receiving antenna. The amount is equal to the area of the receiving antenna times the power density at the receiving antenna:

$$C = \frac{P_t G_t \, \alpha \, A}{4\pi R^2}$$

where C = received carrier power
     A = area of receiving antenna

5. Since $G_R = 4\pi \, A/\lambda^2$, then

$$A = \frac{G_R \lambda^2}{4\pi}$$

6. The path loss equation can now be written as

$$C = P_t G_t G_R \, \alpha \, \frac{\lambda^2}{(4\pi R)^2}$$

where $\lambda^2/(4\pi R)^2$ is called the *free-space path loss.(FSPL)*

7. The path loss calculation is usually done in dBs.

$$C \; (\text{dBm}) = P_t \; (\text{dBm}) + G_t \; (\text{dB}) + G_R \; (\text{dB}) + \alpha \; (\text{dB}) + \text{FSPL} \; (\text{dB})$$

where

$$\text{FSPL} \; (\text{dB}) = -97 - 20 \log R \; (\text{mi}) - 20 \log F \; (\text{GHz})$$

See Figure 15.11.

Physically, the free-space path loss is the ratio of the area of a one-square-wavelength antenna (that is, an antenna of dimensions one wavelength by one wavelength) to the area over which the transmitter power has been spread (which is a sphere with a radius equal to the transmitter-receiver separation). The free-space path loss depends on the microwave frequency and the transmitter-to-receiver separation.

8. The noise power equation:

$$N \; (\text{dBm}) = -114 \, \text{dBm} + \text{NF} \; (\text{dB}) + 10 \log B \; (\text{MHz})$$

Example 15.1   Calculate the free-space path loss for an antenna operating at 6 GHz that is 25 mi from the transmitting antenna.

Solution

$$
\begin{aligned}
\text{FSPL} &= -97 - 20 \log R - 20 \log F \\
&= -97 - 20 \log (25 \, \text{mi}) - 20 \log (6 \, \text{GHz}) \\
&= -140 \, \text{dB} \\
&= 10^{-14}
\end{aligned}
$$

Thus, at a distance of 25 mi the transmitter power has been spread over a spherical surface whose radius is 25 mi, and the area of a 1-square-wavelength antenna is 10^{-14} times smaller than the spherical surface. This is the significance of the free-space path loss. If omnidirectional antennas were used, then the received power would equal the transmitter power times this ratio of antenna area to spherical surface area. The area of the receiving antenna is greater than 1 square wavelength, and the increase is given by the gain of the antenna. The transmitter power is not spread uniformly through space, but is concentrated in the direction of the receiver, and the gain of the transmitter antenna describes how much the power is concentrated.

Table 15.1 lists the results of a path loss calculation for 6 GHz at 25 mi. The value -67 dBm is typical of the received power in a terrestrial microwave relay system. The C/N value of 28 dB is satisfactory, because the signal-to-noise ratio is further improved by the detection process.

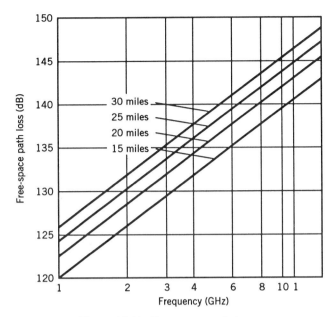

Figure 15.11 Free-space path loss.

All the factors in the pathloss equation can be determined exactly, except the path attenuation. It depends on propagation conditions that vary with time and local weather conditions, and thus can only be estimated. Major factors contributing to path attenuation are shown in Figures 15.12 and 15.13. The atmospheric attenuation due to absorption of microwaves by air, water vapor in the air, and rain are shown in Figure 15.12. The graph shows the attenuation of the microwave signal in dB per kilometer as a function of the

Table 15.1 Example path loss calculation

Frequency	6 GHz	
Transmitter-receiver separation	25 miles	
Transmitter power		30 dBm (1 W)
Transmitter antenna gain		34 dB
Free-space path loss		−140 dB
Path attenuation		−25 dB
Receiver antenna gain		34 dB
Received carrier power (C)		−67 dBm
Noise figure of mixer		6 dB
Bandwidth		13 dB (20 MHz)
Noise (N)		−95 dBm
Carrier-to-noise ratio (C/N)		28 dB

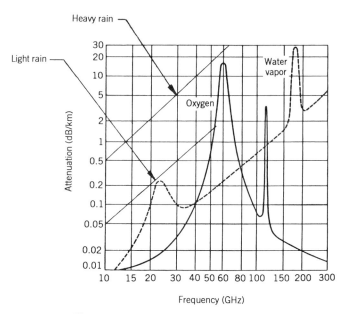

Figure 15.12 Atmospheric attenuation.

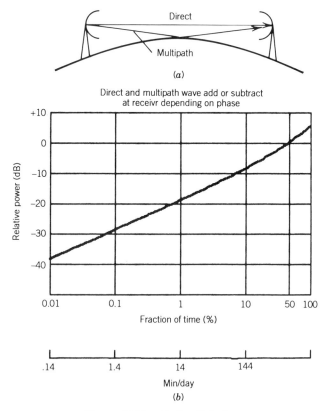

Figure 15.13. Multipath fading.

microwave frequency. Attenuation is very low below 10 GHz. The solid curve shows the attenuation of microwaves due to oxygen in the air. Note that at 60 GHz there is an extreme attenuation reaching 15 dB/km. There is a molecular resonance of oxygen at 60 GHz, and microwaves react strongly with the oxygen at that frequency. Water vapor has molecular resonances at 24 GHz and 170 GHz and, as shown by the dashed curve, high attenuation occurs at these frequencies. Thus, microwave windows exist at 35 GHz and at 90 GHz, where the attenuation due to these effects is a minimum. Note that this is in addition to the free-space path loss. The high attenuation of microwaves due to the effects of water vapor and oxygen create serious problems for the use of the microwave spectrum above 20 GHz for terrestrial microwave relay. The straight lines of Figure 15.12 show the attenuation of light rain and heavy rain. Other levels of rainfall will fall between the two curves. Note that at 20 GHz a heavy rain adds 2 dB of attenuation per kilometer of microwave path.

The most serious atmospheric attenuation effect is *multipath fading* (see Figure 15.13). In Figure 15.13*a* there are two paths by which the microwave signal from the transmitter can reach the receiver. The desired path is called the direct path, where microwaves travel in approximately a straight line from the transmitter to the receiver. However, there are "multi" paths by which the power can reach the receiver. They may occur because the microwaves are reflected from the surface of the earth, or because of atmospheric bending of the microwave signal. As it reaches the receiving antenna, the multipath signal may add or subtract from the direct signal, depending on its phase. Statistically, it adds half of the time, so the microwave signal at the receiver is greater than the direct signal. However, the other half of the time the multipath signal subtracts from the direct signal. For short periods the multipath signal may be almost equal to the direct signal and exactly 180° out of phase, so the multipath signal almost completely cancels the direct signal and hence there is no received signal.

The relative power due to multipath fading as a percentage of time is shown in Figure 15.13*b*. The 0 reference on the power scale is the power level of the direct wave, which has been calculated from the path loss equations. Note that 50% of the time the relative power is greater than the path loss calculation, and for the 50% of the time the signal is less. For 1% of the time the multipath signal is almost equal to the direct signal and in exactly the wrong phase, so the relative power is reduced by 18 dB. For $\frac{1}{10}$ of a percent of the time the cancellation is even worse, so the relative power at the receiver is 28 dB below the free-space path loss calculated value. One percent of the time is 14 min/day, so for 14 min/day, the signal is approximately 18 dB below the free-space value, and for 1.4 min/day the signal is almost 28 dB below the free-space value.

Multipath fading can be overcome by providing 20–30 dB more transmitter power or more antenna gain in the microwave relay system or by using a diversity system.

15.5 DIVERSITY SYSTEMS

A frequency diversity microwave relay uses two transmitters and two receivers, with each pair operating at a different frequency. The same baseband signal is transmitted on each frequency. If a fade occurs at one frequency, it is extremely unlikely that a fade will occur at the other frequency at the same instant in time; that is, it is not likely that multipath signals at both frequencies will be exactly 180° out of phase with the direct signals at the same instant in time. The received signal in both receivers is analyzed, and the strongest signal is switched to provide the baseband output.

A frequency diversity microwave relay protects against fading and equipment failure. If one transmitter or one receiver should fail, the diversity fading protection would be lost, but the system would still continue to operate. The disadvantage of this system is that it wastes frequency allocation. Actually, two independent sets of information could be transmitted with the same two assigned microwave frequencies. A frequency diversity system is also expensive, since it requires two transmitters and two receivers.

A space diversity microwave relay system, on the other hand, uses only one transmitter and one frequency, but has two receivers and two receiving antennas. These antennas must be approximately 100 wavelengths apart, which makes it unlikely that fading of the multipath signal will be exactly 180° out of phase to cancel the direct signal in both antennas. The space diversity system eliminates the need for two transmitters and does not waste the frequency allocation.

15.6 DIFFRACTION AND TROPOSCATTER SYSTEMS

It is possible for microwaves to be transmitted around the curvature of the earth up to several hundred miles by using diffraction or troposcatter. In Figure 15.14 a microwave signal is diffracted by a hill. The direct microwave signal travels in a straight line, hits the top of the hill, and most of it continues in a straight line. However, the small fraction of the signal diffracted by the hill is directed down the other side of the hill to the receiver.

In the lower sketch of Figure 15.14 the microwave signal is transmitted in a straight line and is reflected or scattered off clouds or air masses in the troposphere, located approximately 10 mi above the earth's surface. Most of the microwaves travel in a direct line into the sky and are lost, but a tiny fraction of the microwave power is scattered by the clouds and returns to a receiver on earth. With troposcatter communication, distances up to several hundred miles can be achieved using the very tiny fraction of transmitted power scattered back to the earth. Since the atmospheric conditions in the troposphere are continually changing, the exact amount of power received at the earth's surface varies from hour to hour, so a sufficient margin must be put into a system to account for this.

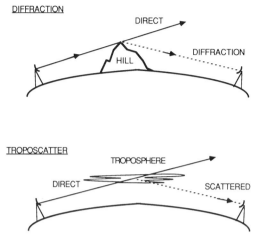

Figure 15.14 Diffraction and troposcatter systems.

The microwave path loss as a function of distance between the transmitter and the receiver, including the effect of diffraction and troposcatter, is illustrated in Figure 15.15. The direct propagation path is a dashed line. This path is simply the free-space path loss. When the radio horizon is reached, no more signal should be received. Actually, the signal is less than the direct signal at distances more than 10 mi from the transmitter, which is due to the multipath effects. The received signal at the radio horizon is reduced by 30 dB by the multipath effect. But beyond the radio horizon, the diffraction effects still provide a weak microwave signal. Finally, at about 100 mi, the diffracted signal is too weak to be useful, but a microwave signal scattered from the troposphere can be received. At 200 mi, the scattered signal is 90 dB less than the free-space path loss signal. The scattered signal is 1 billion times less than the direct signal would be. However, with sufficient transmitter power, large enough antennas, and a low-noise receiver, communication can still be obtained with this weak troposcattered signal.

Table 15.2 compares path loss calculations, for a line-of-sight microwave relay and a troposcatter system. The values for the line-of-sight system were given in Table 15.1.

If a FET preamplifier is used in the troposcatter system before the mixer, the receiver noise figure can be reduced to 1 dB. The noise power in the receiver is then -100 dBm, so the CNR is 15 dB. To achieve a S/N-50 dB, a 35-dB improvement due to the FM demodulation process is needed. More FM improvement is required for the troposcatter system than for the line-of-sight system, which means more deviation must be used. Consequently, a narrower baseband must be used, so troposystem channel capacity must be less, perhaps only 12 telephone channels, instead of 1000.

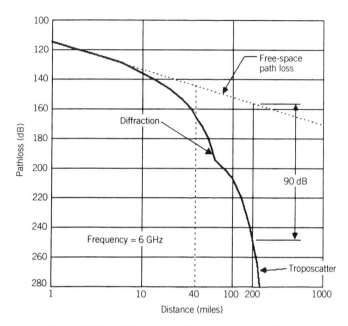

Figure 15.15 Diffraction and troposcatter path loss.

Table 15.2 Path loss calculation for troposcatter and line-of-sight microwave relays

	Line of Sight	Tropo
Frequency (GHz)	6	4
Distance (miles)	25	200
Transmitted power (dBm)	30	60 (1 kW)
Transmitter antenna gain (dB)	34	50
Free-space path loss (dB)	−140	−155
Path attenuation (dB)	−25	−90
Receiver antenna gain (dB)	34	50
Received carrier (dBm)	−67	−85
Receiver noise figure (dB)	6	1
Noise power in receiver (dBm)	−95	−100
C/N (dB)	28	15
Required FM improvement (dB)	22	35
Antenna diameter (feet)	3	30

15.7 WIRELESS LOCAL AREA NETWORKS

Another important application of microwave relay is short distance communication between buildings in a local area network (LAN) (see Figure 15.16). Local area networks allow transmission of multiplexed telephone and data between two buildings located within a few hundred feet or, at most, a few miles. Desktop data terminals in one location may be interconnected to a large computer in the other location. Often the two buildings are connected by wire or coax or fiber optic cable, but this is a problem in a congested metropolitan area. If cable is used for the LAN, it has to be run from the upper floors of the building, under the streets, and back up to the upper floor of the other building. It is much easier to simply mount the microwave relay system on the roof, or even in a window of one building and use microwaves to transmit the telephone, video, and computer information.

Popular frequencies for microwave LANs are the 18- and 23-GHz bands. These frequency bands are uncrowded, but are severely affected by rain attenuation, which limits their performance at long distances. For LAN use the transmission distances are short, so these uncrowded frequency ranges are ideal.

Microwave wireless LANs can also be used inside buildings to connect individual telephones to their private branch exchange (PBX) and individual computers to Ethernet or other LAN networks. Whenever offices are rearranged, the desks, cabinets, and partitions can be easily moved, but changing the phone and computer wiring is much more complicated. With wireless microwave connections, these instruments can be instantly moved to any location.

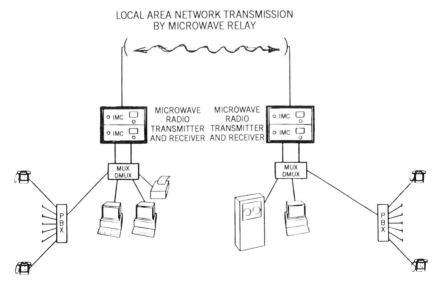

Figure 15.16 Local area network transmission by microwaves.

ANNOTATED BIBLIOGRAPHY

1. R. L. Freeman, *Telecommunication Transmission Handbook,* 2d ed. Wiley, New York, NY, 1985, pp. 177–294.
2. F. Ivanek, *Terrestrial Digital Microwave Communications,* Artech House, Dedham, MA, 1989.

Reference 1 provides a detailed explanation of microwave relay and troposcatter communication systems. Reference 2 provides a thorough explanation of digital microwave relay.

EXERCISES

15.1. Define the microwave relay terms in the following list:
1. Baseband microwave radio
2. RF multiplexing
3. Baseband repeater
4. Heterodyne repeater
5. RF repeater
6. Regenerative repeater
7. Free-space path loss
8. Multipath fading
9. Frequency diversity
10. Space diversity
11. Diffraction microwave relay system
12. Troposcatter microwave relay system
13. Local area network
14. Common carrier frequency band
15. Studio transmitter link
16. Channelization of bandwidth
17. Dual polarization

15.2. Why is a microwave relay often called a terrestrial microwave relay?

15.3. Why is a microwave relay often called a line-of-sight microwave relay?

15.4. What are the frequency ranges of the three most commonly used common carrier frequency bands?

15.5. What is the frequency range of the most common industrial microwave relay band?

15.6. What two frequency bands are most often used for local area microwave networks?

15.7. What are the typical transmitter-receiver spacings for microwave relay?

15.8. What limits the transmitter-receiver spacing for a microwave relay?

15.9. What are two advantages of a digital microwave relay system?

15.10. What is a radome?

15.11. What is a horn-reflector antenna?

15.12. What signal-to-noise ratio is required of the baseband signal when it has been demodulated from the received microwave carrier?

15.13. What carrier-to-noise ratio is required for microwave relay systems?

15.14. What are two methods for improving the received carrier-to-noise ratio up to the required baseband signal-to-noise ratio?

Exercises 15.15 through 15.19 refer to a microwave relay system with the following characteristics.

Frequency	11 GHz
Transmitter-receiver spacing	20 mi
Transmitter power	10 W
Transmitter antenna gain	30 dB
Receiver antenna gain	30 dB
Path attenuation	30 dB
Receiver noise figure	3 dB
Bandwidth	20 MHz

15.15. What is the free-space path loss?

15.16. What is the received carrier power?

15.17. What is the noise power in the receiver?

15.18. What is the carrier-to-noise ratio?

15.19. For how many minutes per day will multipath fading be greater than 18 db?

16

SATELLITE COMMUNICATIONS

The chapter reviews types of satellite communication systems and what they do.

The path loss formulas will be applied to satellite communications, and ERP and G/T, which are commonly used to specify satellite communication system performance, are defined.

There are four basic types of satellite communication systems:

1. *International Satellites* for continent-to-continent communication services
2. *Domestic or Regional Satellites* for communication over limited geographic areas.
3. *Ship-to-Shore Satellites* for mobile communication from and to ships
4. *Direct Broadcast Satellite (DBS)* for television broadcasts directly from a satellite to a home TV

Remote sensing satellites, where television pictures and other data are transmitted by microwaves back to the earth from weather satellites and deepspace probes, are also described.

16.1 INTRODUCTION

Figure 16.1 illustrates a satellite communication system. Almost every modern satellite communication system uses a *synchronous satellite*, a satellite in orbit 22,300 mi above the equator. The satellite takes 24 hours to orbit the

396

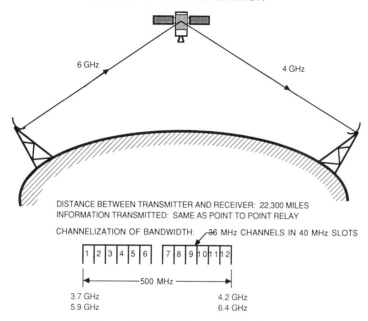

Figure 16.1   Satellite communication.

earth; since the earth turns on its axis once every 24 hours, the satellite appears fixed in space relative to the earth. Therefore, the earth station antenna need not be moved to track the satellite, since it always appears in the same relative position in the sky.

Information transmitted by satellite is the same as information transmitted by microwave relay, such as multiplex telephone, video, and data. Many of the frequency bands used for microwave relay are used for satellite communication. This is possible because the transmission is in different directions—vertically for satellite communications and horizontally for microwave relay—so there is no interference between systems as long as the antennas have low sidelobes or are shielded by their surroundings. Microwave relay antennas are mounted on high towers, mounted on buildings built on hills, to obtain the greatest transmission distance around the curvature of the earth. Satellite antennas, which point toward the sky, are located in valleys or between buildings to shield them from terrestrial microwave relay transmission.

A common set of frequency bands for satellite communication is the 6-GHz band (from 5.925 to 6.425 GHz) for transmission to the satellite and the 4-GHz band (from 3.7 to 4.2 GHz) for transmission from the satellite. The full 500-MHz bandwidth is used for up and down transmission. It is divided into twelve 40-MHz channels. The remaining 20 MHz of bandwidth in the center is used for sending control signals to, and receiving telemetering signals from, the satellite. Into each of the 40-MHz channels 900 FDM telephone calls or a TV program can be fitted.

A typical microwave relay system transmits the same amount of information in a 20-MHz bandwidth. A 40-MHz bandwidth is required for satellite transmission because more carrier deviation must be used to improve the signal to noise ratio, due to the great transmission distance between the satellite and the earth terminal.

Some earth stations in remote areas use only part of the bandwidth of each channel because they do not have 900 telephone calls at one time to fill up the channel. Techniques for dividing the 40-MHz frequency channel for use by several earth stations are discussed later.

Types of communication satellites are

1. *International (INTELSAT)* provides continent-to-continent communication
2. *Domestic or regional (DOMSAT)* provides communication over a limited geographic area such as a country or a continent
3. *Ship-to-Shore (MARISAT)* provides mobile communication between ships and the shore
4. *Direct Broadcast (DBS)* provides television transmission directly from the satellite to home TV sets

INTELSAT satellites are used for worldwide communication. The system consists of one satellite over the Atlantic Ocean, one over the Pacific Ocean, and one over the Indian Ocean. Each satellite thus provides communication over a third of the earth's surface.

DOMSAT satellites provide communication over limited geographical areas, such as the United States.

MARISAT satellites provide ship-to-shore communication. MARISAT is a mobile communication system because the earth stations are on moving ships. Until the advent of MARISAT, reliable ship-to-shore communication was not possible.

Direct broadcast satellites (DBS) broadcast TV programs directly from the satellite to home TV sets.

The limiting path is the downlink, because the size of the satellite antenna and the amount of satellite transmitter power are limited. The uplink is not a limiting path, because as much power as necessary can be generated on the ground to make the received signal at the satellite as large as is needed.

A satellite communication uplink is shown in Figure 16.2. The top half shows the equipment in the earth station, the lower half the equipment in the satellite. The equipment is for just one of the available channels in the satellite. Channel 3, for example, has an uplink frequency at approximately 6 GHz. The block diagram starts in the upper left with the baseband, which

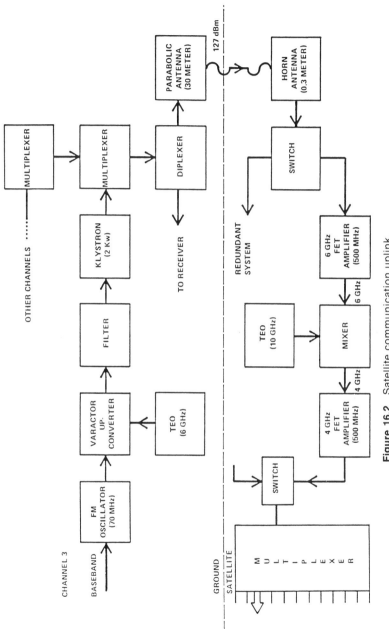

Figure 16.2 Satellite communication uplink.

may be the 900 FDM phone channels, a TV program, or digital data. The baseband signal is modulated onto an FM oscillator, which operates at an IF frequency of 70 MHz. This signal with its modulation sidebands is up-converted by using a varactor diode up-converter to channel 3 of the communication satellite band, which is at approximately 6 GHz.

The up-converted signal at the 0.1-W power level is sent through a band-pass filter, which passes only the channel 3 signal within a 40-MHz bandwidth at 6 GHz. The microwave signal is then amplified to the 2-kW level by a klystron. The klystron has enough gain to amplify the 0.1-W signal to the full 2-kW level, and it has good efficiency, low spurious signal levels, and adequate bandwidth for the 40-MHz-wide channel.

The 2 kW of power is then multiplexed into the transmission line that runs to the antenna. The multiplexed signals are passed through a diplexer, which consists of a circulator and filter that allow the 6-GHz transmitted signal to pass to the parabolic antenna and block the transmitted signal from entering the receiver. The diplexer also allows the received signal, which comes back from the satellite in the 4-GHz band, to enter the receiver and blocks it from the transmitter. After passing through the diplexer, the transmitter power goes to a parabolic dish antenna.

The power transmitted from the earth station travels to the satellite, where a tiny fraction of the transmitted power is received at the satellite by a small horn antenna. The antenna pattern of the horn antenna covers the one third of the earth's surface seen by the satellite. The satellite can therefore receive signals from all ground station locations. This power is passed through a mechanical switch, which allows it to be routed to one receiver or through a redundant receiver. The redundant receiver is used for reliability. The signal is amplified at 6 GHz by a FET amplifier. This amplifer needs a 500-MHz bandwidth because it must amplify all 12 channels being received by the satellite. The amplified 6-GHz signal is then sent to a mixer, where it is mixed with a Schottky diode mixer using a transferred electron local oscillator down to the 4-GHz frequency range at which the microwave signal will be retransmitted from the satellite to the ground. The 4-GHz signal is further amplified by a FET, which has a bandwidth of 500 MHz in order to pass all 12 channels. The amplified signal is then routed through a mechanical redundancy switch into a multiplexer, which separates the 12 channels so that each one can be amplified before retransmission.

As the 12 channels leave the multiplexer, they are ready for amplification and retransmission to an earth station. The satellite communication downlink is shown in Figure 16.3. The top half shows the transmitter in the satellite, and the lower half shows the receiver at the earth station. For simplicity, just channel 3 is shown, but all 12 channels are present in the satellite, and some or all are present in the earth station.

Each of the 12 channels has a separate amplifier in the satellite, and the channel amplifier components are called a *transponder*. The signal from each channel comes from the multiplexer in the satellite and is sent through a filter

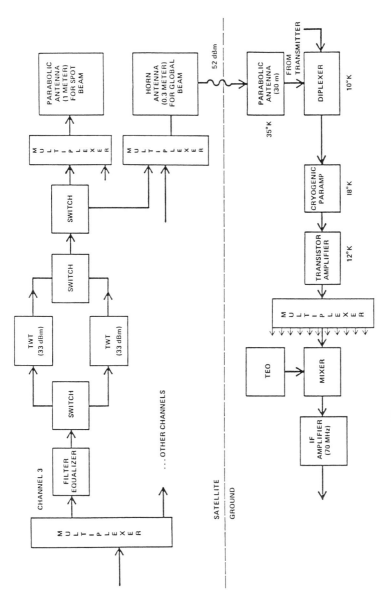

Figure 16.3 Satellite communication downlink.

and an equalizer. The filter ensures that only the 40-MHz bandwidth of channel 3 is passed to the amplifier, and the equalizer adjusts the phase delay so that signal distortion is minimized.

The amplifier contains mechanical switches that provide a redundant path, so if one amplifier should fail the signal for that channel can be switched to the redundant amplifier. The amplifiers are traveling wave tubes (TWTs), which are periodically permanent magnets focused to reduce their size and weight, and they use multistage depressed collectors to increase their efficiency. The amplifiers provide an output of 33 dBm, or about 2 W.

There are 12 transponders, each with a redundant TWT, so 24 TWTs are used. They have 25% efficiency, and 12 are on at any one time. Each requires 8 W of power to operate, so the total power supplied to the satellite transmitter is about 100 W. This power comes from the satellite's solar panels. The amplified signal from each TWT is then switched to one of two possible satellite transmitter antennas. The antennas are preceded by multiplexers so that some or all 12 channels can be combined into the antennas. One antenna is a parabolic dish. Its beam is narrow and does not cover one third of the earth's surface, but only an area such as the United States or Europe, where the major telephone traffic exists. The other antenna is a horn antenna, which has a global beam, so the transmitted signal can be received at any earth station on the third of the earth's surface covered by the satellite.

The satellite signal is received in the same type of earth station antenna that was used for the transmitter. The received signal is sent through the diplexer and is amplified by a cryogenically cooled paramp. The received signal is further amplified by a transistor amplifier and is sent through a multiplexer, where each of the 12 channels is separated for further amplification. Each channel is then mixed with a local oscillator down to an IF frequency of 70 MHz, and amplified. The 70-MHz can be demodulated and the original baseband signal recovered, or it signal can be sent directly to a terrestrial microwave relay for tranmission to its ultimate destination.

A spin-stabilized international satellite is shown in Figure 16.4. The long axis of the satellite is maintained parallel to the axis of the earth by the rotation of the outer satellite structure, which spins at about 50 revolutions per minute. A drawback of this scheme is that fewer than half of the solar cells are in sunlight at a given instant. The antenna mast and the rest of the communication package are "despun," so that they point steadily at the earth. Despinning is accomplished by a small electric motor in the bearing and power transfer assembly. Hydrazine fuel is used in the small jets that control satellite orientation and maintain its desired orbital location. Control signals are radioed from the earth as needed, usually every month or so, to hold the satellite within $0.1°$ (60 km) of its assigned station. The fuel supply places the main limitation on satellite life. The apogee motor is used only once, at launching, to provide the final "kick" that propels the satellite from its transfer orbit into its circular equatorial orbit.

The microwave communication system is mounted in the upper part of

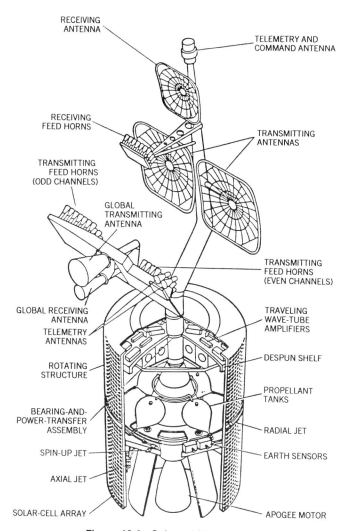

Figure 16.4 Spin stabilized satellite.

the satellite body. The solar cells are used to charge batteries, which provide continuous power for the satellite system. The microwave components and the antenna are despun together. The antenna mast contains a global receiving antenna that receives signals from any location on the third of the earth's surface seen by the satellite. Just above the global receiving antenna is a global transmitting antenna that transmits to any point on the one third of the earth's surface. The upper part of the antenna mast contains three parabolic dishes and multiple feed horns feed each dish. The uppermost parabolic antenna is a receiving antenna. Note its array of feed horns. The lower two antennas are transmitting antennas, with one containing the even transmitter channels and the other the odd transmitter channels. These higher-gain antennas concentrate the transmitted power separately at North America and Europe, the areas of the largest communication traffic. The multiple horns shape the antenna beam so that the power is concentrated on land areas, and none of it is wasted over water areas.

The spin-stabilized satellite in Figure 16.3 is approximately 6 ft in diameter. The satellite body is approximately 8 ft high, and the antenna mast extends another 12 ft above the body, so the total height is 20 ft. The satellite weighs 1700 lb. By comparison, a mid-size passenger car is about 18 ft long and weighs about 3000 lb.

A three-axis-stabilized satellite is shown in Figure 16.5. The satellite is maintained in a stationary orientation relative to the earth. Small jets of hydrazine fuel are fired periodically to keep the satellite properly oriented in all three axes. The satellite's antennas always point toward the earth, and the solar panels always point toward the sun. The satellite is approximately 50 ft long with its solar panels extended. The solar panels are retracted when the satellite is launched. When the satellite reaches its orbital position, the panels and the microwave antennas are then extended.

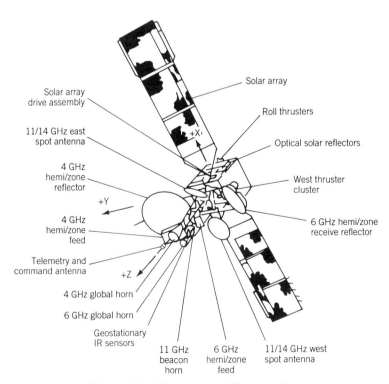

Figure 16.5   Three axis stabilized satellite.

16.2 PATH LOSS CALCULATIONS (ERP and *G/T*)

The same path loss calculation is used as was used for microwave relay systems in Chapter 15. However, the equation is simplified because the transmitter-receiver distance is always 22,300 mi. Also, atmospheric attenuation is negligible, because except for the last 2 mi in which the signal travels through the earth's atmosphere, the microwave signal travels from the satellite to the earth through free space.

The downlink is the most critical path, since the power in the satellite is limited, and the satellite antenna gain is limited by the area on the surface of the earth that the satellite must cover. Therefore only downlink calculations will be discussed.

The received power C is

$$C = P_t G_t G_R \text{ FSPL}$$

where P_t = transmitter power
G_t = transmitter antenna gain
G_R = receiver antenna gain
FSLP = free-space path loss

The receiver noise N is

$$N = kTB$$

The bandwidth of all the satellite channels is 40 MHz. The noise temperature of the receiver is the antenna noise temperature plus the receiver amplifier noise temperature. Satellite communication is the one case where the antenna temperature is *not* 290 K, because the satellite is looking into the cool sky. The antenna temperature of satellite earth stations is 25 to 50 K, depending on the antenna elevation angle.

The microwave carrier-to-noise ratio is

$$\frac{C}{N} = \frac{P_t G_t G_R \text{ (FSPL)}}{kTB}$$

$$= (P_t G_t)\left(\frac{G_R}{T}\right)\frac{\text{FSPL}}{kB}$$

$$= (\text{ERP})\left(\frac{G}{T}\right)(\text{Constant})$$

where ERP is the effective radiated power and completely characterizes the satellite transmitter.

The performance of the satellite system is the same whether a large transmitter power and a small antenna gain or a small transmitter power and a large antenna gain is used, as long as the ERP is the same.

G/T is the gain of the receiving antenna divided by the total noise temperature of the receiver, including the antenna noise temperature and the receiver noise temperature. G/T completely specifies the earth station receiver.

In dB format,

$$\frac{C}{N} \text{ (dB)} = \text{ERP (dBm)} + \frac{G}{T} \text{ (dB)} + \text{FSPL (dB)} - 10 \log kB \text{ (dBm)}$$

The $10 \log kB$ term is the noise in a bandwidth B for a 1 K noise temperature. The actual receiving system noise is greater than this by the noise temperature of the receiving system and this term has been included in G/T. For a 40-MHz channel, $10 \log kB = -123$ dBm. At 4 GHz, the free-space path loss is -196 dB and FSPL $= -205$ dB at 11 GHz, for the satellite-to-earth distance of 22,300 mi. Inserting these values into the above equation gives at 4 GHz.

$$\frac{C}{N} \text{ (dB)} = \text{ERP (dBm)} + \frac{G}{T} \text{ (dB)} - 73$$

for C/N to be 20 dB, we must have

$$\text{ERP} + \frac{G}{T} = 93 \text{ dBm} \qquad \text{at 4 GHz}$$
$$= 102 \text{ dBm} \qquad \text{at 11 GHz}$$

ERP and G/T are normally used to specify satellite communication systems.

A comparison of the path loss calculations for the microwave relay system used as an example in Chapter 15 and an INTELSAT communication satellite is shown in Table 16.1. In spite of the great distance of 22,300 mi between the transmitter and receiver, the received power from the satellite is maintained high by the 100-ft-diameter receiving antenna. Note that the total noise temperature of the microwave relay system is 1160 K, which corresponds to a 6-dB noise figure in the receiver. In contrast, because of paramp receiver and the fact that the satellite antenna is looking into the cool sky, the total noise temperature of the satellite communication earth station receiver is 75 K. Consequently the noise is -104 dBm in the satellite system, or 9 dB less than in the microwave relay. The carrier-to-noise ratio in the satellite system is therefore 20 dB, which is poorer than in the microwave relay system, so additional carrier deviation is required to improve the signal to noise ratio.

Table 16.1 Comparison of path loss calculations for microwave relay and communication satellite

	Microwave Relay	INTELSAT
Frequency (GHz)	6	4
Range (MI)	25	22,300
Voice circuit capacity	1000	900
Transmit antenna diameter (ft)	3	1
Receive antenna diameter (ft)	3	100
P_t (dBm)	30	32
G_t (dBm)	34	20
ERP (dBm)	64	52
Free-space path loss (dB)	-140	-196
α (dB)	-25	0
G_R (dB)	34	60
C (dBm)	67	-84
T_A (K)	290	45
T_R (K)	870	30
B (MHz)	20	40
N (dBm)	-95	-104
C/N (dB)	28	20
G/T	3	41

16.3 INTERNATIONAL SATELLITES

International satellites is to provide worldwide communication. For obvious reasons, microwave relay cannot be used to communicate across oceans. Submarine cables using coaxial cables or fiber optics can provide high-capacity tranmission between fixed locations, but with satellites all countries can communicate with one another.

The international satellite communication system is called INTELSAT. It operates with three satellites, one over the Atlantic Ocean, one over the Pacific Ocean, and one over the Indian Ocean. Each satellite covers one third of the world's surface. The first international satellite, *INTELSAT I*, was launched in 1965. It had a capacity of 240 telephone circuits and a design lifetime of 1.5 years. The investment per circuit year was $32,500 for each telephone circuit. *INTELSAT V* was put into operation in the mid 80s. Its capacity is 13,400 two-way telephone circuits and it has a lifetime of seven years. Although the satellite cost is $75 million, the investment cost per circuit year is $800, less than 5% of a circuit in *INTELSAT I*.

The capacity of *INTELSAT V* was increased by using

The same frequency in different directions
The same frequency with different polarizations
Different frequency bands

A new set of satellites, *INTELSAT VI*, with even greater capacity, will be operational in the 1990s.

An INTELSAT earth station is shown in Figure 16.6. The various signals to be transmitted (voice, video, data) come in through the microwave relay antenna and are received in the microwave relay link equipment. The signals must be demodulated, separated into individual channels, and remultiplexed in a suitable format for international transmission. Remember that the FDM or TDM hierarchies in United States differ from other countries. The baseband signal is modulated onto the microwave carrier frequency, which is transmitted through the waveguide to the satellite transmitter. The signal is transmitted by the 100-ft-diameter Cassegrainian antenna to the satellite.

The downlink signal, repeated from the satellite in the 4-GHz band, is received by the antenna and amplified in the low-noise cryogenically cooled paramp behind the antenna. The resulting received signal is further amplified, demodulated, remultiplexed, and transmitted from the earth station by microwave relay for distribution as individual telephone calls or TV programs.

16.4   DOMESTIC SATELLITES

Domestic satellites are used for communication within a limited geographic area. Usually it competes with other types of communication systems, such as microwave relay, which can be implemented over the land area covered by the satellite.

The antenna on a domestic satellite is simpler than the INTELSAT antenna, since the domestic satellite antenna must radiate the transmitted power over a single geographic area. The antenna footprint of the *WESTAR V* is shown in Figure 16.7. The satellite antenna is a parabolic dish with a special feed horn to give the pattern shown. The antenna beam is not perfectly sharp, so the power varies from the center to the edge of the beam. The contour lines are lines of equal effective radiated power from the satellite. The ERP is plotted in dBw, as is often done. (Note that 30 dBw = 60 dBm.)

Most domestic satellite have 24 transponders, obtained by using vertical polarization for 12 of the channels and horizontal polarization for the other 12.

The domestic satellites operating in the 6-GHz uplink and 4-GHz downlink bands over North America are shown in Figure 16.8. (Not all the satellite systems are shown.) Both spin-stabilized and three-axis-stabilized satellites are used. Two satellites are shown in some orbital positions, actually there are two satellites in all orbital positions. One is in use, and the second is a spare.

Since the satellites are all in a synchronous orbit, they appear stationary to a receiving antenna on the earth, so the signals from each satellite can be received simply by pointing the earth station antenna at the satellite. The

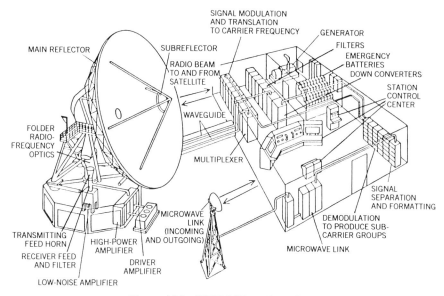

Figure 16.6   INTELSAT earth station.

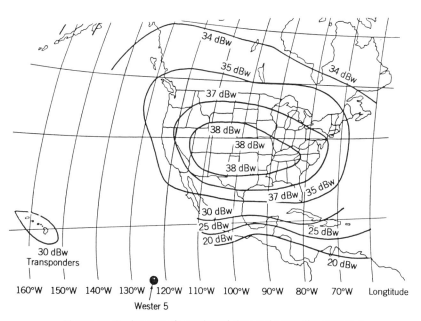

Figure 16.7   Antenna footprint of domestic satellite antenna.

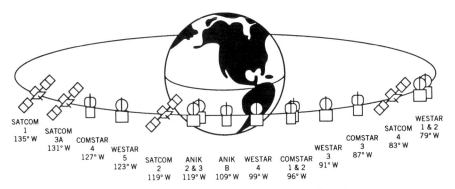

Figure 16.8 Domestic satellites over North America.

satellites are located 4° apart, which in their orbit 22,300 mi above the earth puts them about 1700 mi apart.

The ERP of the U.S. domestic satellites is 64 dBm, 12 dB greater than the ERP from the INTELSAT global beam, because the power is spread over a much smaller area of the earth's surface. Consequently, the area of the receiving antennas can be made approximately 10 times smaller. Therefore a 30-ft-diameter antenna can be used for U.S. domestic satellite systems rather than the 100-ft-diameter antenna required for the INTELSAT system. The receiving system of a domestic satellite earth station is simpler than that of an INTELSAT earth station, using an uncooled paramp or a FET as the low-noise receiver. Consequently the receiver noise temperature is higher. With the lower antenna gain and the higher noise temperature, the G/T of a domestic satellite earth terminal is only 29 dB, whereas the INTELSAT earth terminal G/T is 41 dB. However, the domestic satellite ERP is higher, so both types of 4-GHz satellite systems have a combined ERP plus G/T equal to the required 93 dBm to give a carrier-to-noise ratio of 20 dB.

Many of the U.S. domestic satellite transponders are used for the distribution of cable TV programs, such as HBO. The program is distributed from a single satellite transponder to all cable distribution centers throughout the United States. A cable distributor normally uses a 30-ft-diameter antenna and achieves a 20-dB carrier-to-noise ratio, and a 50-dB signal-to-noise ratio after demodulation. This high signal-to-noise ratio allows for further degradation as the television signal is transmitted through the cable system to the individual subscribers, where at the television set, the signal-to-noise ratio must be 40 dB for satisfactory television reception.

It is also possible to receive the television signal directly from the satellite with a 10-ft-diameter antenna. This antenna has one-tenth the area of the 30-ft-diameter antenna, so the signal-to-noise is degraded to 40 dB, but since the TV reception only (TVRO) system is connected directly to the TV set, the smaller antenna and poorer carrier-to-noise ratio are acceptable.

Domestic satellites are also used for private business communication. For example, major department store chains with locations in several cities can send all their interstore telephone conversations and data such as inventory control, payroll, and sales information from one location to another via satellite. Because the data rate from digitized phone calls and the inventory and sales information is small, small-diameter antennas can be used. Because the antennas are small, these systems are called very small aperture terminals (VSATs). Most VSAT systems use antennas with diameters of 6 to 10 ft; they operate in the 14-GHz uplink and 12-GHz downlink bands. The digital information occupies the small frequency range from 50 kHZ to 1 MHz; therefore each VSAT terminal uses only a small fraction of the total 40-MHz satellite transponder bandwidth. Many VSAT terminals can be serviced by a single satellite transponder of the type used in the Satellite Business Systems (SBS) satellite.

A block diagram of a VSAT ground station is shown in Figure 16.9. The digital baseband signal, which may be digital data or PCM telephone signals, enters from the right side. A stable 200-MHz crystal oscillator generates a subcarrier, and onto this subcarrier the digital signal is phase-modulated with QPSK.

The phase modulated subcarrier is up-converted to the 14-GHz transmitter frequency in two steps, as shown in Figure 16.9. These steps allow the same oscillators used for up-conversion in the transmitter to be used as local oscillators in the receiver. As shown in the middle right of the figure, the 1-GHz stable frequency is generated from a 33-MHz crystal oscillator and a two-stage frequency multiplier, which multiples the frequency by 30 times. The signal is used to up-convert the 200-MHz phase modulated subcarrier to 1.2 GHz.

In the center left of Figure 16.9, a second stable frequency is generated at approximately 2 GHz by using a crystal reference and a phase-locked oscillator. It is multiplied by six times to 12.8 GHz. The phase modulated carrier at 1.2 GHz is up-converted with this 12.8-GHz signal to give a 14-GHz transmission signal. This signal is amplified in a power amplifier to the 2-W level and passes through the diplexer to the small-aperture antenna. (A diplexer is a two-frequency multiplexer that routes the 14-GHz transmitted signal to the antenna and routes the 12-GHz received signal from the antenna to the receiver.)

The received signal, 11.73 GHz, comes in from the antenna and passes through the diplexer, through a bandpass filter, and through the low-noise amplifier chain to the first mixer. This mixer uses the stable 12.8-GHz source that was also used for up-conversion in the transmitter as its local oscillator. The IF from this first mixer is 1.07 GHz. The IF signal is amplified with an automatic gain control (AGC) circuit and goes to a second mixer. The second mixer uses the other stable source used in the transmitter up-conversion to generate a standard IF of 70 MHz. This signal is amplified and demodulated to remove the digital information from the phase modulated subcarrier.

412

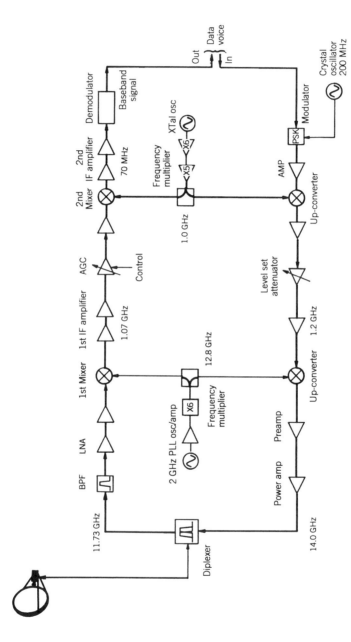

Figure 16.9 Very small aperture terminal (VSAT).

16.5 SHIP-TO-SHORE COMMUNICATION BY SATELLITE

Figure 16.10 shows ship-to-shore communications using the MARISAT satellite, which is spin-stablized. Transmission from the shore to the satellite is at 6 GHz. The signal from the shore station is received in the satellite with a global beam receiving antenna, so the satellite can receive from all shore stations. Transmission from the satellite to the ship is in the frequency range of 1.534 to 1.599 GHz. An array of four helix antennas is used for the satellite downlink antenna. This helical array provides a global beam with a smaller antenna than could be achieved with a horn or parabolic reflector at the downlink frequency of 1.5 GHz. The ERP is 50 dBm.

A 1.3-m (4-ft) parabolic dish is used for the shipboard downlink receiving antenna. The satellite is in a synchronous orbit and appears stationary relative to a fixed location on the earth, but the ship is continually moving, so the shipboard antenna must track the satellite. To permit tracking, the shipboard antenna must be small, so it can be easily moved. The small shipboard antenna and the use of a bipolar transistor low-noise receiver give a G/T of -4 dB for the MARISAT downlink. ERP $+ G/T$ is only 46 dBm, well below the value required for a 20-dB CNR with a 40-MHz channel. However, these values are entirely satisfactory for the MARISAT system because it does not use a 40-MHz-wide channel but a 25-kHz-wide channel to receive a single phone message or an equivalent amount of data. The narrow bandwidth greatly reduces noise, so the required 20-dB carrier-to-noise ratio can be achieved with the small antenna, which is required so that the antenna can track the satellite.

The signal from the ship back to the satellite is transmitted from the same shipboard antenna in the 1.625–1.660 GHz band. The signal is then retransmitted from the satellite to the shore station in the 4-GHz band.

MARISAT satellites are located over the Atlantic, the Pacific, and the Indian oceans to provide worldwide communications.

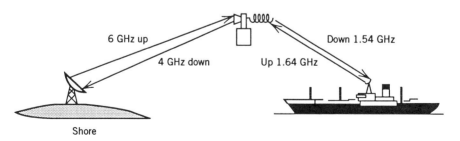

Figure 16.10 Ship-to-shore communication system (MARISAT).

16.6 DIRECT BROADCAST SATELLITES

Direct broadcast satellites (DBS) are illustrated in Figure 16.11. They are intended for broadcast directly from the satellite to home TV receivers. Domestic satellites can be used, but they require a 10-ft-diameter antenna for the home installation, which is expensive. The DBS uses a special satellite with a high power transmitter and high-gain antennas so that the receiving system at the individual homes can use a smaller antenna and an inexpensive receiver.

Figure 16.11 shows a proposed DBS satellite system for the United States. The satellite would have four beams—one for each time zone, which would allow the power to be concentrated into four regions of the United States rather than be spread over its entire surface. The antenna could thus have approximately 40-dB gain rather than the 30-dB gain of a domestic satellite. The DBS satellite would use a 100-W transmitter for each beam rather than the 2-W transmitter used in domestic satellites. Therefore, the DBS satellite would have an ERP of 90 dBm rather than the 62 dBm of U.S. domestic satellites. This nearly 1000-fold increase in ERP would allow the G/T of the home earth station to be reduced by 1000 times, which would significantly reduce the cost of the earth stations.

The frequency band to be used for the DBS is 12.3 to 12.7 GHz. An ERP + G/T = 106 dB is needed to achieve a C/N = 20 dB. Since the ERP = 90 dBm for the DBS satellite, the G/T of the home earth terminal need only be 16 dB.

Antenna diameter	1.4 m
Antenna gain	38 dB
Receiver	FET
Receiver noise	300°K
G/T	16 dB
Total antenna/receiver cost	$250

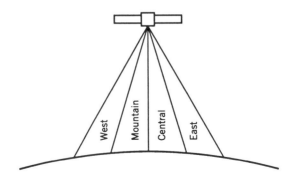

Figure 16.11 Direct broadcast satellite (DBS).

16.7 COMPARISON OF COMMUNICATION SATELLITES

Several communication satellite systems are compared in Table 16.2. The *INTELSAT V* 27-transponder capability is obtained by frequency reuse and dual polarization in the 6 GHz–4 GHz band, and by frequency reuse in the 14 GHz–11 GHz band. The transmitting antennas in the satellite are parabolic dishes with complex feed horn arrays to distribute the power into various antenna patterns covering only land areas on the earth's surface. The earth station antenna is the standard INTELSAT 30-m parabolic dish. The ERP for the global beam is 52 dBm, the G/T is 41 dB, and the sum is 93 dBm, for a 20-dB carrier-to-noise ratio. If zone or spot beams are used, the ERP is greater, the received carrier-to-noise ratio is greater, less FM deviation is required, so more channels can be fitted into the 40-MHz transponder bandwidth.

In the RCA SATCOM twenty-four transponders are obtained in the 6-GHz uplink–4-GHz downlink band by using horizontal and vertical polarizations. The satellite is three-axis-stabilized and uses a transmitter antenna dish in the satellite. A 2-W transmitter is used in each SATCOM transponder, just as in INTELSAT satellites, but the SATCOM ERP is greater because the power is spread over the United States instead of over one third of the earth's surface. For that reason a 10-m (30-ft) diameter antenna can be used for the SATCOM earth terminal, with a FET low-noise amplifier. The G/T of the SATCOM earth terminal is 29 dB, which is less than that of the INTELSAT earth terminal, because of the smaller antenna and less expensive preamp. However, because of the higher ERP, the sum of the ERP and the G/T adds up to the required 93 dBm.

The SBS satellite is used for private business purposes, where the business organizations provide their own VSAT earth station transmitter and receiver, and the SBS company leases transponder space on its satellite. This satellite is spin-stabilized, and no effort has been made to increase capacity by frequency reuse or polarization diversity. A parabolic dish is used for the satellite transmitter. The receiving antenna is a 2-m (6-ft) parabolic dish, which can be mounted on roofs of office buildings. The high ERP is obtained by using a satellite antenna pattern that covers only part of the United States. The G/T value is obtained by using a FET low-noise receiver. The sum of the ERP and G/T is 103 dBm, which is required for C/N 20 dB at the 11-GHz downlink frequency.

The MARISAT satellite is spin-stabilized and uses four helical antennas for the satellite-to-ship up- and downlinks. The G/T of the shipboard terminal is -4 dB, which is satisfactory since the channel bandwidth is so small that the noise in the receiver is greatly reduced, and the required 20-dB carrier-to-noise ratio is achieved.

The DBS satellite may be three-axis-stabilized by using a parabolic antenna dish with multiple feed horns to shape the beams to cover limited

Table 16.2 Comparison of satellite systems

System	Usage	Transponders	Up/Down Frequency (GHz)	Stabilization	Receiver Antenna Diameter (m)	ERP (dBm)	G/T (dB)
INTELSAT V	International	27	6/4 14/11	3-Axis	30	52	41
RCA SATCOM	U.S. domestic	24	6/4	3-Axis	10	64	29
SBS	U.S. domestic	10	14/11	Spin	5	73	30
MARISAT	Ship to shore	25 kHz	1.6/1.5	Spin	1.3	50	−4
DBS	Direct broadcast TV	4	12/12	3-Axis	1	90	13

geographic regions. The ERP of the satellite must be high, perhaps 90 dBm, to reduce the G/T of the earth station, to reduce the earth station cost.

16.8 REMOTE-SENSING SATELLITES

Remote sensing is the technology of gaining knowledge of the earth by sensing conditions on its surface through orbiting satellites, or about distant planets through deep-space probes. Sensors (TV cameras, infrared sensors, radar, magnetic probes) are mounted in the satellites. The information from these sensors is transmitted to the earth by microwave communication systems. These systems are similar to those discussed in this chapter, with the following exceptions:

1. The satellite is not always in a synchronous orbit, so the earth antenna may need to track the satellite.
2. The transmission is one way from the satellite to the earth, not two way as for communication satellites.

Typical remote-sensing satellites include weather satellites, which provide television pictures of cloud movement and other weather conditions over the earth, and the LANDSAT satellite, which measures conditions on the earth's surface to assess water polution, highway construction, urban sprawl, deforestation, and potential mineral deposits.

Another example of remote-sensing satellites is the *Voyager* spacecraft, which explored Jupiter and Saturn and the interplanetary region between the earth and Saturn. The *Voyager* contained a TV camera and other sensors, whose signals were transmitted to earth by microwaves. At the time *Voyager* passed Saturn, it was 1 billion miles from earth.

The design of the *Voyager* microwave communication system is compared with the INTELSAT communication satellite downlink in Table 16.3. The numbers for the INTELSAT satellite were taken from Table 16.1. Both the INTELSAT system and the *Voyager* transmitted a TV picture to the earth. The main difference in the systems is that the communication satellite is located at 22,300 mi above the earth, whereas the *Voyager* is 1 billion miles above the earth. Consequently, the free-space path loss is 100 times greater for the *Voyager*. To prevent the *Voyager*'s received signal from being lost in the noise, its TV picture was not transmitted in real time. It was transmitted instead as a slow scan, with the total transmission time for one frame taking about 5 min. This lowered the data rate of the video signal so that only 20 kHz was required. This significantly reduced the noise in the receiver, and the TV pictures were 9 dB above the noise. The signal-to-noise ratio is not very good, but the received video signal was high enough above the noise so that the resulting pictures could be received and then optically enhanced.

Table 16.3 Comparison of path loss calculations for communication satellite and Voyager deep-space probe

System	INTELSAT	Voyager Deep-Space Probe
Frequency (GHz)	4	8.4
Transmitter-receiver spacing (mi)	22,300	1 billion
TV capacity	1 broadcast	1 slow scan
Transmit antenna diameter (m)	0.3	3.7
Receive antenna diameter (m)	30	60
P_T (dBm)	32	43
G_T (dB)	20	48
ERP (dBm)	52	93
Free-space path loss (dB)	-192	-296
α (dB)	0	0
G_R (dB)	60	72
C (dBm)	-84	-132
T_A (K)	45	20
T_R (K)	30	10
B (MHz)	40	.020
N (dBm)	-104	-141
C/N (dB)	20	9

ANNOTATED BIBLIOGRAPHY

1. B. R. Elbert, *Introduction to Satellite Communication*, Artech House, Dedham, MA, 1987.
2. S. Prentiss, *Satellite Communications*, TAB Books, Blue Ridge Summit, PA, 1983.
3. W. S. Cheung and F. H. Levien, *Microwaves Made Simple; Principles and Applications*, Artech House, Dedham, MA, 1985, pp. 269–288.
4. L. Y. Kantor, *Handbook of Satellite Telecommunication and Broadcasting*, Artech House, Dedham, MA, 1986.

References 1, 2, and 3 present simple descriptions and explanations of satellite communication systems. Reference 4 gives a complete technical discussion of satellite communication with design formulas and tables.

EXERCISES

16.1. Define the communication satellite terms in the following list:
  1. International satellite
  2. Domestic satellite
  3. MARISAT

   4. Direct broadcast satellite

   5. Synchronous orbit

   6. ERP

   7. G/T

   8. Spin-stabilized satellite

   9. Three-axis stabilized satellite

   10. Transponder

   11. Antenna footprint

   12. Global beam

   13. Spot beam

16.2. What is the synchronous orbit for a communication satellite in miles?

16.3. Why is the downlink the most difficult path in a communication satellite system?

Exercises 16.4 through 16.10 refer to a domestic communication satellite downlink with the following characteristics:

Frequency	11 GHz
Satellite transmitter power	1 W
Satellite transmitter antenna gain	34 dB
Earth station antenna gain	63 dB
Noise temperature of receiving system	300°K
Transponder bandwidth	40 MHz

16.4. What is the free-space path loss?

16.5. What is the ERP?

16.6. What is the received carrier power?

16.7. What is the total receiver noise power?

16.8. What is C/N?

16.9. What is G/T?

16.10. What is the sum of ERP and G/T?

17

RADAR SYSTEMS

17.1 INTRODUCTION

The purpose of radar is to

Detect the presence of a target
Determine where the target is, including its range, azimuth, and elevation
Determine how fast the target is moving
Determine (maybe) what the target is

A radar must do all of the above in the presence of large interfering reflections, multiple targets, and jamming.

In Figure 17.1 a ground radar is tracking an aircraft, but the same principles apply to all radar situations–for example, an aircraft tracking a ship or a ship tracking another ship. The radar transmits a microwave signal that is reflected from the target and returns to the radar. The radar compares the received echo signal with the original signal to determine velocity, range, azimuth, and elevation. Azimuth and elevation are the technical terms for the angular location of the target: azimuth is the angular location in the horizontal plane, elevation in the vertical plane.

Types of radar and their functions are shown in Table 17.1 Several of the radars in the table must be used on a ship or an aircraft. Consequently, many radars are multifunction.

Radar frequency bands are illustrated in Figure 17.2. The lettered band designations are shown above the frequency scale. The original lettered bands, which were assigned during World War II, are shown just above

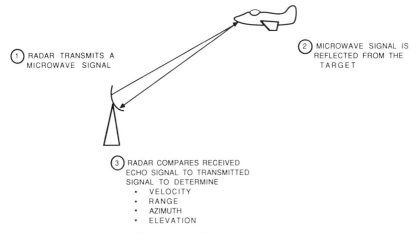

Figure 17.1 How radar works.

the frequency numbers. These bands are still in common use. The newest designation of radar bands, shown above the original lettered bands, classifies the bands from A through M, with A occurring below the microwave band.

17.2 VELOCITY MEASUREMENT

Figure 17.3 shows the principle of radar velocity measurement. The reflected signal is shifted in frequency by the movement of the target. This is the Doppler shift, and so velocity-measuring radars are often called Doppler radars. Figure 17.3a shows the signal transmitted from the radar to the target. The curved lines represent peaks of the microwave signal. The number of peaks leaving the transmitter or being received at the target is the signal

Table 17.1 Types of radar

Radar	Function
Police	Determines vehicle speed
Air traffic control	Determines location of aircraft coming into an airport
Marine	Determines location of ships and shoreline
Aircraft weather	Determines cloud formations in aircraft flight path
Air defense	Determines location of target aircraft and directs weapons against them
Missile guidance	Controls flight of an attack missile to the target
Docking	Guides vehicle to a docking position
Terrain guidance	Determines location of mountains, etc., in the path of aircraft
Ground mapping	Makes a radar map of the ground from aircraft or satellite

422

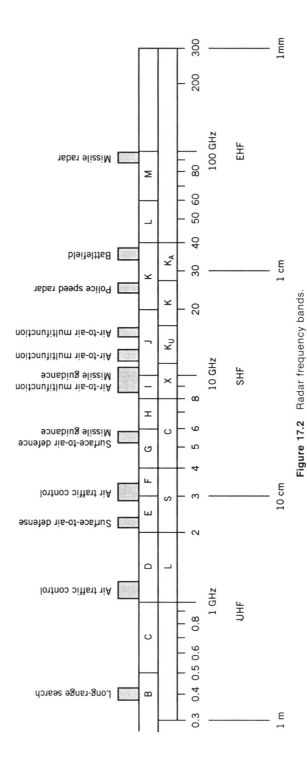

Figure 17.2 Radar frequency bands.

Reflected signal is shifted in frequency by
velocity of target

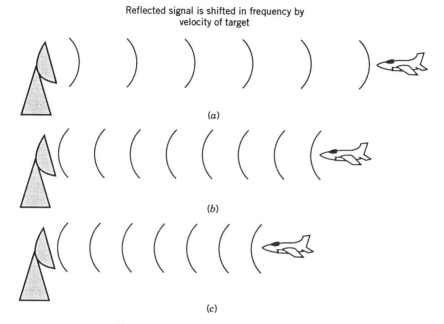

(a)

(b)

(c)

Figure 17.3 Radar velocity measurement.

frequency. The microwave signal is reflected from the target and returns to the radar; if the target is stationary, the number of waves reaching the radar occurs at the same frequency as the number of waves that are transmitted. However, if the target is moving toward the radar, each successive wave peak travels a shorter distance in going from the target and back to the radar, because the distance between the transmitter and the target is continuously decreasing. Since each wave has a shorter distance to travel, the waves return to the radar faster then they were transmitted. This is the Doppler shift. The change in frequency is very small (in the kilohertz range for a gigahertz microwave signal), but it can be easily measured by comparing the transmitted signal to the received signal, and the velocity of the target can be accurately determined. If the target is moving away from the radar, then the number of waves per second received back from the target is less than the original transmitted frequency because each successive wave has further to go as the target moves away from the radar.

Figure 17.4 is a block diagram of a simple Doppler radar (a police radar). A microwave signal is generated in an oscillator (usually a Gunn) and directed into a circulator. Most of the oscillator power is transmitted through the circulator to the horn antenna. However, some of it leaks through the circulator, and this power is used for comparison with the reflected signal. The main part of the oscillator power is transmitted through the horn antenna to the target, where it is reflected and returns to the horn antenna and is routed by

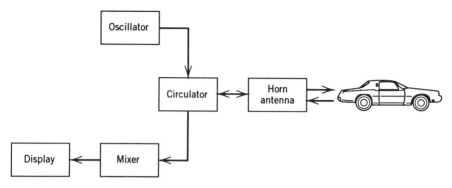

Figure 17.4 Simple Doppler radar.

the circulator into the mixer. The reflected signal is mixed with the sample of the transmitted signal in the mixer, which produces the sum and difference frequencies. The sum frequency is ignored, but the difference frequency is the change in frequency between the transmitted signal and the received signal, due to the Doppler shift. This Doppler signal is then sent to a display or detector where, since the frequency difference is known, the target velocity can be determined.

The following formulas are used to determine target velocity from the Doppler shift:

$$\text{Velocity (mi/h)} = 0.26 \times \frac{\text{Doppler shift (Hz)}}{\text{Microwave frequency (GHz)}}$$

$$\text{Velocity (knots)} = 0.29 \times \frac{\text{Doppler shift (Hz)}}{\text{Microwave frequency (GHz)}}$$

$$\text{Doppler frequency (Hz)} = 3.9 \times \text{Velocity (mi/h)} \times \text{Microwave frequency (GHz)}$$
$$= 3.4 \times \text{Velocity (knots)} \times \text{Microwave frequency (GHz)}$$

The Doppler frequency shift measures the relative velocity between the target and the radar. It only measures the absolute velocity if the target is moving directly toward or away from the radar. A target moving in a circle about the radar gives a zero Doppler shift; even though the target is moving, it is not moving relative to the location of the radar. If the target is not moving directly toward or away from the radar, the Doppler frequency shift measures only the component of target velocity in the direction directly toward or away from the radar. With a 10-GHz radar and a 25-mi/h velocity, the Doppler shift is approximately 1 kHz. If the velocity is 2500 mi/h, say for a supersonic missile and supersonic aircraft on a collision course, then the Doppler shift is 100 kHz. Consequently, Doppler frequency shifts range from a few hundred hertz to a few hundred kilohertz. In this range frequency is extremely easy to measure. It is also easy to tell if one or several frequencies are present, and to measure the frequencies of the several signals, which allows multiple targets to be easily resolved. Thus, velocity measurement with radar is extremely accurate.

17.3 RANGE MEASUREMENT

Figure 17.5 shows the principle of range measurement. The radar transmits pulses of microwave power rather than a continuous signal. The pulses travel at the speed of light (300 meters per microsecond or approximately 1000 feet per microsecond, the actual velocity of the pulses is 984 feet per microsecond). The radar measures the time for the signal to go out to the target and return, and then the range of the target is simply this time multiplied by the velocity of light divided by 2:

$$\text{Range} = \frac{\text{Time} \times \text{Velocity}}{2}$$

The factor 2 takes into account the round-trip transmission path from the radar to the target and back. For example, if it takes a microwave pulse 10 microsec to go out to the target and return, then

$$\text{Range} = \frac{10\ \mu s \times 1000\ \text{ft}/\mu s}{2}$$
$$= 5000\ \text{ft}$$

Figure 17.6 shows a radar "A scope" presentation, which is an oscilloscope display of the returned radar signal as a function of time. The transmitted signal is shown at the beginning of the display and then a noise background, which is the noise in the radar system. In approximately the middle, an echo, which is the microwave signal reflected from the target, is shown. The time required for the microwave signal to travel to the target and return is also shown and by measuring this time the range of the target can be determined.

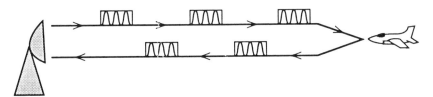

Figure 17.5 Radar range measurement.

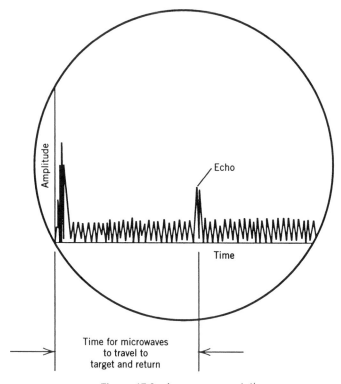

Figure 17.6 A scope presentation.

17.4   COMBINED RANGE AND VELOCITY MEASUREMENT

Range and velocity measurement problems are clutter, ambiguity, and range resolution.

Figure 17.7 defines the clutter problem. A ground-based radar is tracking an aircraft flying over a mountain. The mountain and the aircraft are in the radar beam. The mountain is much larger than the aircraft and reflects a much greater microwave signal back to the radar. Thus, the echo from the target cannot be distinguished from the mountain. The echo from the mountain is called *clutter*. Another cause of clutter is the echo from ocean waves, which masks the radar signal from low-flying aircraft. The surface of the ocean is so large compared with the size of the aircraft that the echo from the ocean waves is much greater than that of the aircraft.

The clutter and the target are at the same range, so the echoes from both return to the radar at the same time, and the target is hidden by the large echo from the clutter. However, the aircraft is moving, and the clutter is not, so a solution to the clutter problem is to distinguish between moving and

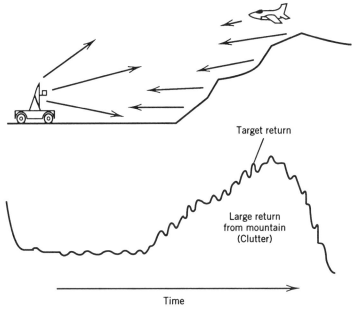

Target return

Large return
from mountain
(Clutter)

Time

Figure 17.7   Clutter.

stationary targets by using Doppler shift principles. Two such techniques are
moving-target indicator (MTI) and pulse Doppler.

Clutter suppression using MTI is illustrated in Figure 17.8. MTI is useful
when the radar is stationary, such as for an air traffic control radar at an
airport. Figure 17.8*a* shows the radar echo from two successive pulses. The
transmitted pulse is shown, and then the echo from this pulse, then a second
transmitted pulse, and the echo from the second pulse. The echo from the
moving target is indicated. All the rest of the echoes are from stationary
targets. The actual radar display would not indicate which was the moving
target and which were the stationary targets, and this is the clutter problem.
The echo from the second pulse is identical to the echo from the first pulse
for all of the clutter returns, because these are from stationary targets.
However, the echo from the second pulse is different from the echo from the
first pulse for the moving target, because in the time between the transmission
of the first and second pulses the target has moved, so the phases of the
echoes are different. Therefore, by delaying the first pulse by one pulse
repetition frequency period and subtracting it from the second pulse, all of
the echoes from stationary targets that are identical can be made to cancel
and only the moving-target echo remains. The principle of the MTI radar is
to store the echo from the first pulse and subtract it from the second, then
store the echo from the second and subtract it from the third, store the echo
from the third and subtract it from the fourth, and so on. Sometimes the

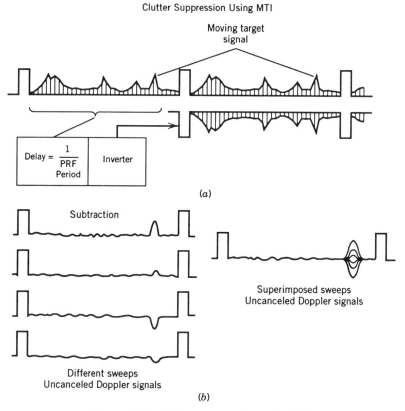

Figure 17.8 Clutter suppression using MTI.

difference between the two pulses will give a large indication for the moving target. In other cases, as shown in Figure 17.8b, the target echo is almost canceled because the target has moved approximately a full wavelength during the time between transmission of the first and second pulses, so the echo from the second pulse looks almost exactly the same as from the first pulse. Five successive pulses are usually required to completely cancel the clutter and obtain a good indication of the moving target.

MTI works for stationary radars (it is used in every air traffic control radar) or on slow-moving radars mounted on ships or land vehicles. However, MTI will not work if the radar itself is moving because even stationary targets appear to the radar to be moving, so their clutter echo changes from pulse to pulse. Consequently, for a moving radar, a pulse Doppler radar, which measures the range and velocity of the target, must be used. Echoes that appear to be moving at the velocity of the aircraft are coming from stationary objects and can be eliminated. Echoes that appear to be moving at velocities other than the ground speed of the aircraft are from the desired moving targets.

A block diagram of a simplified pulse Doppler radar is shown in Figure 17.9. The radar starts with a timer and a stable microwave oscillator. The timer controls a pulse modulator, which turns the microwave amplifier on and off. The stable microwave signal from the oscillator is amplified by the amplifier, but only for short periods when the amplifier is on. The resulting pulses are radiated from the antenna, reflected from the target, and return to the radar. In the radar receiver, the returned echo is mixed with the stable oscillator frequency, and the Doppler shift frequency is obtained. The Doppler frequency is passed through a set of Doppler filters to determine target velocity. The returned signal is also compared with the reference time in a time comparer, such as the A scope, to determine the target range. The pulse Doppler radar therefore measures target range and velocity and allows echoes from stationary ground targets to be eliminated because they are at different velocities from actual moving targets.

The pulse Doppler radar would appear to be a perfect radar, but ambiguity has been introduced. Range ambiguity is shown in Figure 17.10. Three transmitted pulses, A, B, and C, and the echoes from three targets are shown.

Echo 1 is the echo from pulse A. By measuring the time between the transmission of pulse A and the echo return, the radar can determine the target range. Echo 2 is also from pulse A, but it occurs at the same time pulse B is being transmitted. Consequently, echo 2 is eclipsed by the transmitted pulse B. Echo 3 is the echo from pulse A, but it occurs during the listen time for pulse B. Echo 3 is a distant echo from pulse A, but whether it belongs to pulse A or B cannot be determined. Consequently echo 3 could represent two possible ranges as far as the radar information is concerned. Echo 4 is from the same target as echo 1 and is from pulse B. It is delayed

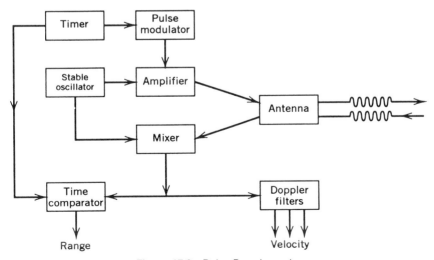

Figure 17.9 Pulse Doppler radar.

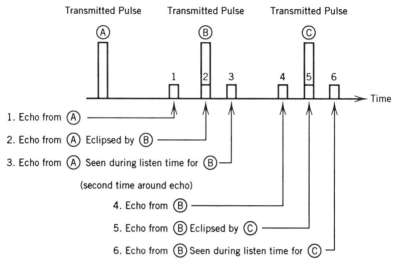

Figure 17.10 Range ambiguity.

in time from pulse B by the same time that echo 1 was delayed from pulse A. Echo 5 is the echo from the second target from pulse B, and it occurs exactly when pulse C is being transmitted. Finally, echo 6 is the echo from the third target from the transmitted pulse B, but it occurs during the listen time for pulse C. The range ambiguity arises when the echo pulse occurs after a second pulse, or even a third or fourth, has been transmitted. A solution is to space the pulses very far apart. However, this eliminates the possibility of obtaining velocity information.

Pulse repetition frequency (PRF) is the number of times per second that the pulse is repeated. Note that the time between pulses, from the start of one pulse to the start of the next, is the reciprocal of the pulse repetition frequency.

Figure 17.11 illustrates the problem of range and velocity ambiguities. Any electrical signal can be analyzed in terms of single-frequency components. Figure 17.11 shows the radar signals as functions of time on the left side, and the frequency spectra of each of these signals on the right side. The upper-signal is continuous with time. The envelope of the signal rather than the microwave waveform is shown. The envelope of a continuous or CW signal is constant with time, and the frequency spectra of this signal is just a single line at the microwave frequency. This CW signal has no range measurement capability because it is constant with time. Its frequency spectra is just a single line with no sidebands. A Doppler-shifted echo can be easily compared to the transmitted frequency.

The lowest signal is a pulsed microwave signal with a low PRF. The pulses are spaced far enough apart to avoid range ambiguities. The spectrum of this

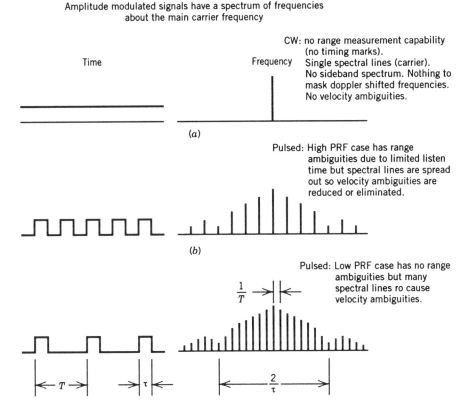

Figure 17.11 Velocity ambiguities.

signal contains many sideband lines, and they are combined to form the pulsed waveform.

The frequency spectrum contains frequency lines spaced by the reciprocal of the pulse repetition time; that is, they are spaced by the pulse repetition frequency. Most of the lines occur within a frequency range of 2 divided by the time duration of the pulse.

The many spectral lines cause velocity ambiguities. A Doppler-shifted return signal will fall between the spectral lines, and it cannot be determined whether the Doppler-shifted echo signal goes with the first PRF line or the second PRF line or the third PRF line. Consequently, several target velocities can be calculated, but it is not known which one is the actual target velocity. Also, targets whose Doppler shifts fall at the same frequencies as the PRF line are not visible to the radar. These are called blind speeds.

Figure 17.11b is a pulsed radar signal with a high PRF. It has range ambiguities due to the limited listening time, but the spectral lines are spread out so that there will be no velocity ambiguity. Thus, by choosing the right

PRF, we can have no range ambiguities, or no velocity ambiguities, or some compromise between them.

Figure 17.12 shows an example of the range-velocity ambiguity trade-off. The radar has a pulse repetition frequency of 10 kHz, which means that the time between pulses is 1/10 kHz, or 100 microseconds. Looking at the microwave signal envelope as a function of time, the pulses are 100 microseconds apart. The frequency spectra of this signal consists of a series of frequency lines 10 kHz apart about the carrier. An echo is received, and its characteristics are shown. The microwave frequency is 10 GHz, and the echo returns 50 microseconds after transmission, with a 5-kHz Doppler frequency shift. This echoes are shown as a function of time in Figure 17.12a. Note that they occur 50 microsec after each transmitted pulse, but it is not known whether the echo belongs to the first, second, or third pulse. Consequently, the time could be 50 microseconds, or it could be 150 microseconds if the echo belongs to the previous pulse, or 250 microseconds if the echo belongs to the second previous pulse. The range could be 25,000, 75,000, or 125,000 ft (using the relationship that the range is equal to 500 ft times the time in microseconds).

The possible velocity shift can be calculated from the Figure 17.12b. The echo is shifted 5 kHz away from each spectra line, but it is not known which line of the transmitted signal that the echo is associated with. Consequently, the Doppler shift could be 5, 15, or 25 kHz, and so forth, and the velocity could be 130, 390, or 650 mi/h. Spacing the pulses farther apart would reduce the range ambiguity, but the velocity ambiguity problem would be worse, and conversely.

A solution to the range ambiguity problem is shown in Figure 17.13. It involves changing the PRF from time to time. A series of pulses at two PRFs and the echo from a single target are given. The echo actually is in the listen time of the second pulse and should be related to the previous pulse. Looking at the top line only with PRF 1, it is not known which pulse the echo was related to. Looking at the lower line with PRF 2, it is still not known which pulse the echo was related to. By looking at the complete drawing with the two PRFs, one common target range can be determined.

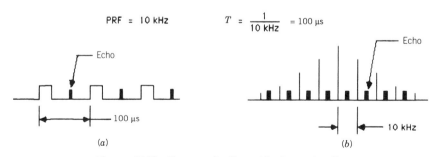

Figure 17.12   Range-velocity ambiguity trade-offs.

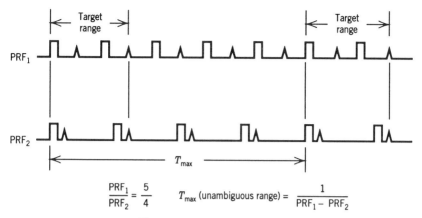

$$\frac{PRF_1}{PRF_2} = \frac{5}{4} \qquad T_{max} \text{ (unambiguous range)} = \frac{1}{PRF_1 - PRF_2}$$

Figure 17.13   PRF switching.

The range resolution problem is illustrated in Figure 17.14. Range resolution is the ability to distinguish between two targets closely spaced in range. Figure 17.14 shows the echo returning from targets A and B as a function of time. The transmitted microwave pulse must last for some time. It is not possible to turn the radar transmitter on and off in an infinitesimal amount of time. Typical radar pulses are one to several microseconds long. Therefore, the echo from target A occupies a period of time equal to the pulse length. The beginning of the pulse is reflected from the target and, as time continues, the remainder of the pulse is reflected from the target. The echo from the beginning of the pulse begins to arrive from target B while the end of the pulse from the target A is still arriving. The pulses thus overlap. Unfortunately, the radar display does not show one pulse as solid and the other as dotted. The pulses actually overlap, and it is not possible to tell whether there is one target or two targets. Thus, the individual targets cannot be resolved. The resolution capability of a radar is determined from the relation

$$\text{Resolution Capability (ft)} = 500 \times \text{Pulse Length } (\mu s)$$

The range resolution problem can be overcome by pulse compression, illustrated in Figure 17.15. With pulse compression the radar transmits a long

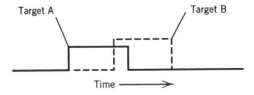

Figure 17.14   Range resolution.

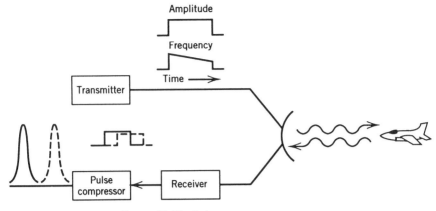

Figure 17.15 Pulse-compression radar.

pulse, but the range resolution is independent of pulse length. During pulse transmission, the transmitter frequency is varied. For example, for a 10-GHz radar, the transmitter frequency might be varied from 10.1 to 9.9 GHz during the duration of the transmitted pulse. In this way the pulse is marked at every time during its transmission because the frequency is different. This feature is used to compress the received echo pulse to an extremely narrow width.

The long transmitted pulse with its varying frequency during transmission is transmitted from the radar, reflected from the target, returned to the radar, and amplified in the radar receiver. At the receiver output the pulses from two targets overlap in time, so it is not possible at this point to resolve the targets. However, at a given instant of time where the pulses overlap, the frequency of one pulse is different from the frequency of the other pulse. For example, at the time where the two pulses just begin to overlap, the frequency of the echo from the first target (shown solid) is at 10.0 GHz, since it is in the middle of the pulse, but the echo from the more distant target, which is at the leading edge of the pulse, is 10.1 GHz. The two targets could be distinguished if there were some way of distinguishing the frequency difference between the echos at this instant of time.

The frequencies can be distinguished by using a pulse compressor. The pulse compressor allows the low frequencies, which occur at the end of the pulse, to catch up with the high frequencies, which occur at the beginning, so that the pulse is squeezed in time, or compressed. As the overlapping pulses pass through the pulse compressor, they are compressed into two extremely short duration pulses. The time length of the compressed pulses is

$$\frac{\text{Time Length of}}{\text{Compressed Pulse}} = \frac{2}{\text{Frequency variation during transmitted pulse}}$$

The time length of the compressed pulse is not related to the time length of the original transmitted pulse.

The most common type of pulse compressor is the surface acoustic wave (SAW) delay line, shown in Figure 17.16. The SAW delay line uses a piezoelectric material, lithium niobate. A piezoelectric material is a material that vibrates mechanically if an alternating voltage is applied. In Figure 17.16 the microwave signal is applied to the crystal at the source and causes vibrations in the crystal at the microwave frequency. The vibrations travel through the crystal at a velocity about 1 millionth the velocity of light. Consequently, it takes many microseconds for the mechanical vibration wave in the crystal to travel the length of the crystal. The mechanical vibration wave is reflected at the right end of the crystal by the grating and travels back to the load. When the mechanical vibrations reach the load, they generate an electrical signal. Note that low frequency signals are reflected in the near end of the diagonal gratings, mid-frequencies are reflected at the middle, and high frequencies are reflected at the far end, where the grating lines are very close together. Consequently, the high frequency signals travel farther then the low frequency signals and get delayed. The high frequency signals were transmitted at the beginning of the pulse, whereas the low frequency signals were transmitted at the end of the pulse. Thus, as the microwave signal passes through the SAW delay line, the high frequencies are delayed more than the low frequencies, and this compresses the pulse.

The SAW device does not work in the microwave frequency range, but in the 100-MHz range. This is no problem because the microwave signal with its frequency variation is simply shifted down to 100 MHz by the mixer in the receiver.

Pulse compression allows extremely good range resolution, down to 5 ft. A related technique, which uses a frequency variation of the transmitted signal, is FM CW radar (Figure 17.17). FM CW radar is used for targets at

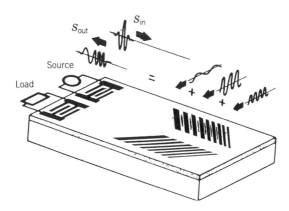

Figure 17.16 Pulse compressor. Surface acoustic wave (SAW) delay line.

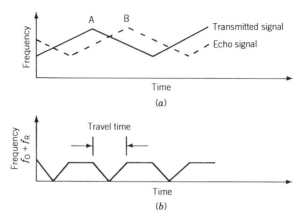

Figure 17.17 FM-CW radar.

very short ranges, ranges so short that the echo would be received while the transmitted signal was still being transmitted. In FM CW radar, the radar signal is transmitted continuously, as with Doppler radar, but as shown by curve A of 17.17a the transmitter frequency is continually varied during transmission. The frequency shift of the echo signal, shown by curve B, occurs at a different time. The time difference is related to the time that it takes the microwave signal to travel out to the target and return. By measuring the time difference between the occurrence of the same frequency in the transmitted and received signals, the target range can be determined. FM CW radar can measure range down to a foot or even less.

17.5 ANGLE MEASUREMENT

Figure 17.18 shows how a radar measures the angular location of a target. The microwave beam from the radar antenna is moved through space. The direction from which an echo returns to the radar is the angular location of the target. Figure 17.18 shows a PPI scope presentation, a common radar presentation used in air traffic control radars. As the antenna rotates to move the beam through space, the oscilloscope display shows the returned echo. The angular location around the scope corresponds to the angular location of the targets and is controlled by the direction in which the antenna points. The distance from the center of the display is the range. Both azimuth and elevation coordinates must be measured. The PPI scope normally is used to show the azimuth and range of targets, with a separate presentation for elevation.

The problem with angle measurement is that the beam is so broad that the angular location of a target cannot be accurately determined, and multiple targets cannot be resolved.

PPI Scope

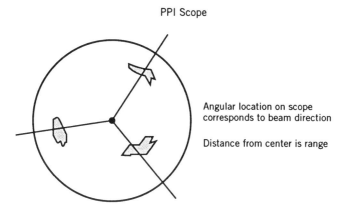

Angular location on scope
corresponds to beam direction

Distance from center is range

Both angle coordinates (azimuth and elevation) must be measured

Figure 17.18 Radar angular measurement.

This problem is illustrated in Figure 17.19, which shows that two or more targets cannot be individually distinguished if they are within the antenna beamwidth. If pulse compression is used, a radar can resolve targets in range if they are only 5 ft apart. But if the targets are 50 miles from the radar, the radar cannot resolve the targets in angle if they are closer than 2 miles. The angular resolution of a radar is much poorer than its velocity or its range resolution.

The angular resolution of a radar can be calculated from the following formulas:

$$\theta \sim \frac{\lambda}{D} \quad (\text{rad}) \sim 60° \frac{\lambda}{D}$$
$$L = \theta R = \frac{\lambda R}{D} \quad (\text{rad})$$

Two or more targets cannot be individually distinguished
if they are within the antenna beamwidth

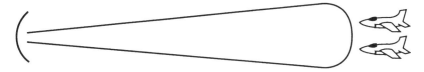

The angular resolution of radar is much poorer than
its velocity or range resolution

Figure 17.19 Angular resolution.

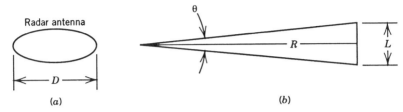

Figure 17.20 Calculation of angular resolution.

Figure 17.20a shows a radar antenna, and the azimuth resolution is determined by its width D. Figure 17.20b shows the geometry of the situation: R is the range from the radar to the target, θ is the antenna beamwidth between the half-power points, and L is the lateral resolution, or the minimum spacing between targets that permits them to be resolved.

17.6 TECHNIQUES TO IMPROVE ANGULAR RESOLUTION

Solutions to the angular resolution problem include

Larger antenna
Higher frequency
Tracking mode
Radar closer to the target
Synthetic aperture radar.

Making the antenna larger reduces the beamwidth. However, since the angular location of the target is determined by moving the antenna beam, using a larger antenna makes it more difficult to rapidly move the beam. The solution in most military radars is to move the beam without moving the antenna by using a phased array antenna. This approach makes the radar complicated and expensive, but it is essential, because in most military applications the beam must be moved in a fraction of a second.

Increasing the frequency to reduce the wavelength is equivalent to increasing the size of the antenna at a given frequency. For example, a 3-ft-diameter antenna operating at 100 GHz has the same beamwidth as a 100-ft antenna at 3 GHz. This clearly shows the advantage of operating at high frequency, but it is not usually possible to select the radar frequency arbitrarily.

A tracking radar mode is illustrated in Figures 17.21 through 17.23. In Figure 17.21, a tracking radar uses two offset beams. The received power of a radar changes very little as the angular position of the target changes around the center of the antenna beam. However, a much larger change in received power occurs as the target moves around the edge of the beam. A tracking

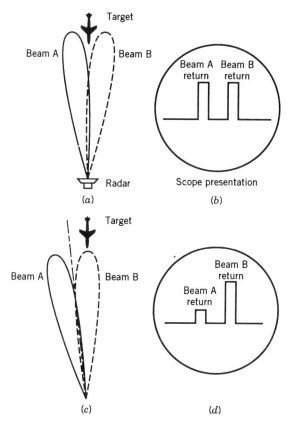

Figure 17.21   Radar tracking mode.

radar uses two offset beams and positions the antenna so that the target lies midway between the two beams (Figure 17.21a). In Figure 17.21b an equal return would be received from the target in beams A and B. If the target is off the axis of the antenna (Figure 17.21c), a greater return will be received from beam B than from beam A. This difference signal can be used to reposition the antenna so that equal echo returns are received in both beams. The angular accuracy of a search radar using a single beam and of a tracking radar using two beams are compared in Figure 17.22 which shows the signal amplitude as a function of the angle between the antenna boresight or axis and the target. The signal amplitude changes very little as the target moves off the antenna axis for the single-beam search radar. However, the difference in the received signal between the two tracking beams changes rapidly as the target moves in either direction off the antenna axis. Approximately 10-fold greater accuracy can be obtained from the two-beam tracking mode.

    The two tracking beams are obtained by using either conical scanning (which uses a single rotating feed horn) or monopulse (which uses two pairs of

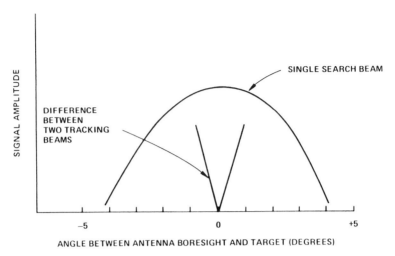

Figure 17.22 Comparison of search and tracking.

feed horns—one pair for azimuth and one pair for elevation). The monopulse tracking radar requires four feed horns and four receivers. Complete angular information in azimuth and in elevation is obtained on each pulse, hence the name *monopulse*. Conical-scan tracking is further illustrated in Figure 17.23. As shown in Figure 17.23*a*, the antenna beam is rotated at an angle with respect to the antenna axis by rotating the feed horn. The result is a conelike antenna pattern transmitted toward the target. The echo returns from the target and contains an amplitude modulation due to this rotation. The amplitudes of the successive echoes are equal when the target is exactly on the antenna axis. In Figure 17.23*b,c*, the target is not located at the center on the antenna axis, so the return when the beam is at position 1 is the greatest and the return when the beam is at position 5 is the lowest. Comparing this amplitude modulation of the echo returns allows the target's angular position can be determined. A conical-scan tracking radar is simpler to implement than a monopulse radar because only one feed horn and only one receiver are required. Unfortunately, it is extremely easy to jam because the target can repeat back the echo returns with a different modulation pattern and confuse the radar. This is discussed further in Chapter 18.

The lateral resolution of a radar can obviously be improved if the radar is moved closer to the target. This is accomplished in radar weapons control systems by putting the radar into a guided missile (see Figure 17.24 and 17.25). One means of missile guidance, called *command guidance* is shown in Figure 17.24. A radar tracks the target, and the same radar or another one tracks the missile. The positions of the target and missile are then fed into a computer, which generates a control signal, to command the missile to fly toward the target. The problem with command guidance is the poor angular accuracy of the radar. If the target is 50 mi away and tracking mode is used

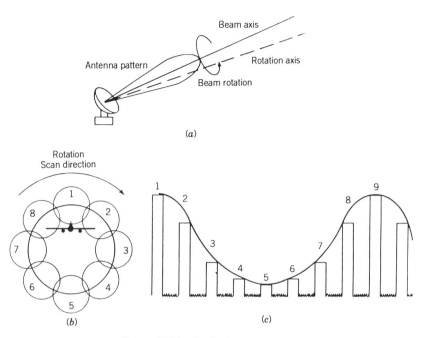

Figure 17.23 Conical scan tracking.

(so that a 10-fold improvement in angular accuracy is obtained), a 3-GHz radar has a lateral accuracy of 1250 ft, not close enough for the missile to hit the target. When the missile is close to the target, the missile's radar takes over. Now with 3-GHz frequency and the same beamwidth, the lateral resolution is 25 ft when the missile is within 1 mi of the target, good enough angular accuracy to achieve a hit.

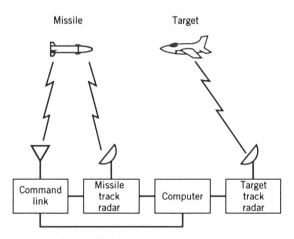

Figure 17.24 Command guidance.

Two types of missile-homing guidance are shown in Figure 17.25: active and semiactive. In active homing the missile contains a radar transmitter and a receiver. The disadvantage is that the transmitter is large, which makes it difficult to mount in the missile, and expensive. The transmitter and receiver are destroyed every time a missile is used.

In semi-active homing the transmitter is mounted on the launch vehicle (in this example a ship). The transmitter illuminates the target, and the missile receives the signal reflected from the target and the signal transmitted from the radar on the launch vehicle. Therefore, it must have two receivers. Receivers are much smaller and much less expensive than transmitters. Therefore, with semiactive homing, only the small receivers must be mounted in the missile; the large expensive transmitter remains on the launch vehicle.

Synthetic aperture radar is shown in Figure 17.26. The only way to achieve good angular resolution is to use a large antenna, if the radar frequency and the range between target and radar are fixed. The synthetic aperture radar achieves an antenna several miles in dimension, synthetically. It must have a stationary target and a moving radar, and is therefore used in airborne and satellite radars for ground mapping.

The synthetic aperture radar forms a linear antenna array by moving a single antenna element. As the aircraft or satellite moves, the antenna element "looks" at the target, then moves as the aircraft or satellite moves along its

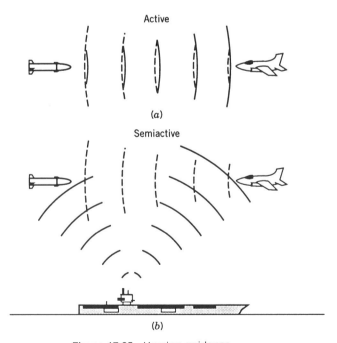

Figure 17.25 Homing guidance.

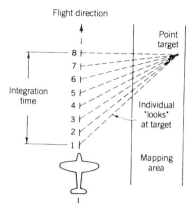

Figure 17.26   Synthetic aperture radar (SAR).

flight path, looks again, moves, and so forth. An extremely large effective antenna aperture is achieved, not with a large antenna made up of many elements but with a single element moved from position to position to position. For a 5-ft lateral resolution at 100 miles with a 10-GHz radar, the integration path is

$$D = \frac{\lambda R}{L} = 1.95 \text{ mi}$$

Since the synthetic aperture radar antenna must look at the target every half wavelength, for the 5 ft resolution at 100 miles, the radar must make 211,500 looks at each target. The synthetic aperture radar in Figure 17.26 is achieving 5-foot angular resolution at 100 miles, which is comparable to the range resolution obtained with pulse compression, three orders of magnitude better than can be obtained with a conventional radar. At present, adequate computer capacity cannot be mounted on the aircraft to process the information. A normal synthetic aperture radar works by flying the plane with the radar over the target area, then relaying the raw radar information by microwave relay to the high-altitude aircraft, which repeats the radar information to a ground location that has adequate computer capacity to analyze the data.

The obtainable lateral resolution with a conventional antenna and a synthetic aperture radar are compared in Figure 17.27 as a function of the range between the radar and the target. With a conventional antenna, the resolution worsens as the range between the radar and the target increases. With a synthetic aperture radar, the lateral resolution is independent of range. See Figure 17.27, which shows the aircraft flying along its flight path and two targets indicated by the dots, to see why. When the aircraft is in the lowest position in the figure, the beam of the single antenna element picks up both

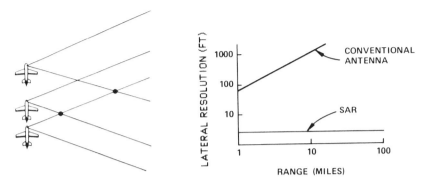

Figure 17.27 Resolution of synthetic aperture radar.

targets. When the aircraft is flown to the middle position, the closest target is just leaving the antenna beam. However, the aircraft can fly to the upper position before the more distant target drops out of the antenna beam. Consequently, with more distant targets, the aircraft flies over a longer flight path and more integrations are made. The net result is that since more looks are taken at distant targets than at close targets, the lateral resolution is independent of target range.

Synthetic aperture radar is used for ground mapping and for satellite surveillance of objects on the earth's surface. When combined with pulse-compression techniques to give good resolution in range, a synthetic aperture radar can achieve 5-ft resolution in range and azimuth.

17.7 PHASED ARRAY RADAR

For most military radars, the antenna beam must be moved very rapidly in order to update the angular location of many targets. This is usually accomplished with a phased array antenna (see Chapter 13).

Figure 17.28a is a reflective array type of phased array antenna. In this case, a horn feed illuminates the antenna array. The radiation pattern of the horn is adjusted so that the total power is divided uniformly among the 64 elements of the array. At each array element, part of the power from the feed horn is received by the small dipole antenna and travels through a phase shifter, is reflected from the reflecting surface, travels again through the phase shifter, and then is radiated from the dipole. The phase of the reflected signal from each element of the array is adjusted electronically by the phase shifter, and the total beam from all the elements of the antenna can be made to point in any direction simply by controlling the phase of the radiating elements.

The number of radiating elements used in the phased array antenna should be minimized. However, the elements must be no further than half a wave-

Reflective Array   Active Array

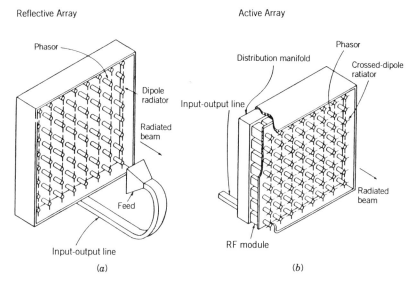

Figure 17.28   Reflective and active phased array antennas.

length apart. If farther apart, the antenna pattern will not be smooth but will have holes in it.

The reflective array can be modified to be a "lens" array. In this case, the surface is fed by the feed horn, but instead of reflecting the power after it is received by the individual dipoles at each element and passed through the phase shifter, the power is transmitted, with the correct phase relationships, out the other side of the array.

An active array is shown in Figure 17.28b. In this case, a low-level micro-wave signal is divided among the various elements of the array. This low-level signal can be supplied to each element by radiating the low-level power from a feed horn as with the lens array. Alternatively, the power can be distributed by a distribution manifold. Each array element contains the phase shifter to provide the beam steering and an amplifier to amplify the power in each element to the required transmitter power. In this way, the total radiated power is not supplied by a single high power transmitter but is provided by the sum of the many low power elements. Some long-range search radars have as many as 10,000 individual transmitting elements. The transmitter power of each element is about 100 W, which can be obtained at frequencies below 1 GHz from a microwave transistor, and the total transmitted power from the array is 10,000 elements times 100 W per element, or 1 MW.

The active phased array offers the advantage of "graceful degradation," which means that if one active transmitter element of the array fails the array continues to perform with just slightly degraded performance. This differs from the reflective or lens array in which the whole microwave system stops working if the transmitter fails.

17.8 BLOCK DIAGRAMS

A pulse radar is shown in Figure 17.29. The process begins with the timer. The timer signal turns on a modulator, which turns on the transmitter for a short period of time, and the transmitter microwave signal passes through the circulator and out the antenna. The received signal enters the antenna and is routed by the circulator to the mixer, where the incoming radar signal is mixed with the LO signal to obtain the IF signal, which is amplified in the IF amplifier and then detected. The resulting video pulse is amplified by a video amplifier and sent to an indicator, such as the A scope (see Figure 17.6). The timer provides a time reference for the indicator; and from the time between the transmission of the microwave pulse and the return of the echo from the target the target range can be determined.

A pulse-compression radar is shown in Figure 17.30. A timer, not shown, is used to start the process and turn on the transmitter. In addition, the transmitter is frequency modulated during the pulse transmission, and the frequency modulated transmitter microwave signal passes through the circulator to the antenna. The returned signal enters the antenna and is routed by the circulator to the mixer, where the return signal with its frequency modulation is shifted down to the IF and amplified in the IF amplifier. The pulse then passes through the pulse-compression filter, usually a SAW device, where the low frequency, which occurs at the end of the pulse, catches up with the high frequency at the beginning of the pulse to achieve pulse compression. The compressed pulse then passes to the detector, video amplifier, and indicator.

A Doppler radar is shown in Figure 17.31. This block diagram is more complicated than that of the simple Doppler radar shown in Figure 17.4 because a low-noise receiver is used. The radar begins with a stable microwave oscillator generating a microwave frequency and a coherent oscillator generating a stable IF. The coherent oscillator frequency is added to the stable oscillator frequency in a modulator, which is a varactor up-converter, and the sum signal is amplified by a driver amplifier, further amplified by the power amplifier, and routed through the circulator to the antenna. The Doppler-shifted frequency is reflected from the target and passes through

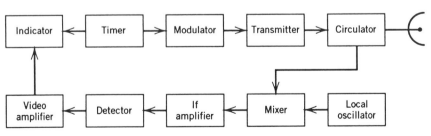

Figure 17.29 Pulse radar.

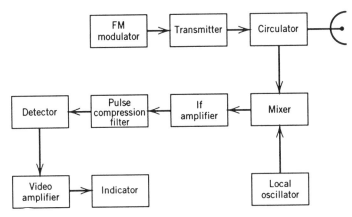

Figure 17.30 Pulse-compression radar.

the circulator to the mixer. The original microwave signal from the stable oscillator mixes with the returned signal from the target. Note that the transmitted signal had been shifted up by the IF in the modulator, so the signal leaving the mixer is the IF plus the Doppler shift. The signal is amplified in an IF amplifier and mixed with the signal from the coherent oscillator in a second mixer. The information from the second mixer is the Doppler-shifted signal.

A block diagram of a pulse Doppler radar is shown in Figure 17.32. This block diagram is more complicated than the simple pulse Doppler radar in Figure 17.9 because of the need to use a low-noise receiver with an IF amplifier. The process begins with a timer to give the time reference so that the range can be determined. In this case, the signal from the stable microwave oscillator is amplified by an amplifier, which is pulsed on and off by

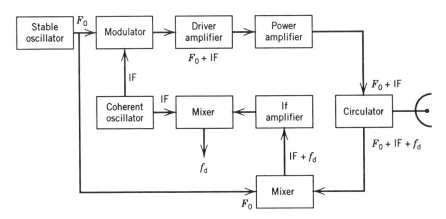

Figure 17.31 Doppler radar.

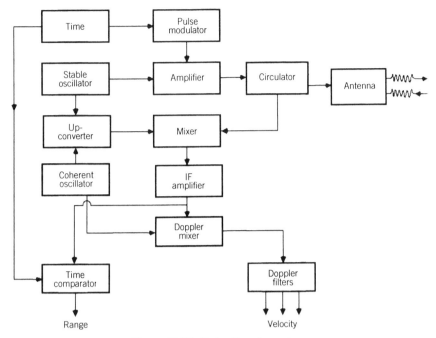

Figure 17.32   Pulse Doppler radar.

the pulse modulator controlled by the timer. The pulsed microwave signal is routed by the circulator to the antenna. The reflected microwave signal from the target comes back through the antenna and is routed by the circulator to the mixer. The mixer must have a stable LO related to the transmitter microwave frequency but changed in frequency by the IF. This is done by up-converting the microwave signal from the stable oscillator by the frequency of a coherent oscillator, which is at the IF. This up-converted signal forms the LO. (In Figure 17.31, the stable microwave signal was upshifted in frequency before transmission. This same technique could be used for the pulse Doppler radar.) The output of the mixer contains the pulsed information for range and the Doppler-shifted information for velocity. The amplified IF signal is sent to a time comparator where the range information is obtained. The signal is also mixed with the coherent oscillator signal in the Doppler mixer to obtain the Doppler-shifted signal, which is passed through Doppler filters to determine the target velocity.

A block diagram of a monopulse radar is shown in Figure 17.33. Only the azimuth channel is shown, a similar channel for elevation must also be used. The microwave signal from the transmitter passes through a circulator and into the hybrid junction. The signal from the hybrid feeds both feed horns, and the sum antenna pattern is transmitted from the antenna hybrid junction. The sum of the antenna patterns passes back through the circulator into a

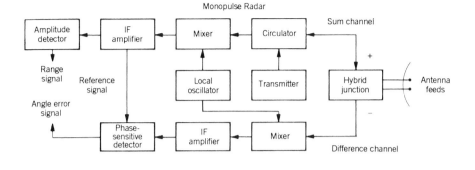

Monopulse Radar

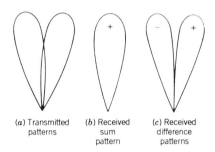

(a) Transmitted patterns

(b) Received sum pattern

(c) Received difference patterns

Figure 17.33 Monopulse radar.

mixer, an IF amplifier, and an amplitude detector in the upper channel, and this provides the range signal. This sum signal could also be used for obtaining Doppler or pulse-compression information by combining the pulse-Doppler and the pulse-compression block diagrams with the monopulse radar block diagram.

The difference signal from the two antennas, which gives the azimuth information, passes into the difference channel, is mixed and IF amplified, and the angle error signal is obtained.

17.9 THE RADAR EQUATION

The requirements on transmitter power, antenna size, and receiver noise figure for a radar are determined by the need to achieve a satisfactory signal to noise ratio at the radar receiver.

A signal to noise ratio of 20 dB is required to ensure satisfactory range, velocity, and angle-tracking accuracy and to ensure a satisfactorily low false-alarm rate. False-alarm rate is the time interval between noise signals that are large enough to be mistaken for targets.

The formula for the received power in a radar system is called the radar equation.

$$P_R = \frac{P_t G \sigma A}{(4\pi)^2 R^4 L}$$

where P_R = received power
G = transmitter antenna gain
σ = radar cross section
A = receiver antenna area
R = range from radar to target
L = atmospheric losses

$$N \ (\text{dBm}) = -114 + \text{NF} + 10 \log B$$

where N = receiver and antenna noise
NF = receiver noise figure
B = receiver bandwidth (MHz)

The equation is derived in the same way as the equation for received power in a communication system (see Chapter 15) except that the microwave signal must go out from the transmitter to the target, be reflected, and then return to the radar. The amount of microwave signal reflected depends on the size of the target and is called the target's radar cross section. Because the microwave signal must travel from the transmitter to the target and then be reflected back to the radar, the received signal varies as the fourth power of the distance between the radar and the target. Note that for a communication system, since the microwave power has to go only in one direction from the transmitter to the receiver, the received power varies as the square of the transmitter receiving spacing.

The radar equation derivation is given in the books listed in the bibliography.

All of the terms in the radar equation are familiar, except for the radar cross section, which is the effective area of the target as seen by the radar's antenna beam. The radar cross section of a large jet aircraft as seen from the front is approximately 100 m^2. Seen from the side, the same aircraft has a much larger radar cross section because its area as seen by the radar beam is much larger.

The expression for the noise received in the radar receiver when the radar antenna is looking along the surface of the earth is also given. For a radar transmitting a pulsed microwave signal, the bandwidth is normally required to be the reciprocal of the pulse length so that most of the spectral lines of the pulsed signal get into the receiver.

Example 17.1   A radar has a transmitter power of 10 kW, an area of 10 m², a gain of 40 dB, and the range between the radar and the target is 100 km. The target cross section is 100 m², the total atmospheric and system losses are 4 dB, (2.5 times) the receiver noise figure is 2 dB, and the bandwidth is 1 MHz.

$$
\begin{aligned}
P_R &= \frac{P_t G \sigma A}{(4\pi)^2 R^4 L} \\
&= \frac{(10^4 \text{ W})(10^4)(10^2 \text{ m}^2)(10 \text{ m}^2)}{(4\pi)^2 (10^5 \text{ m})^4 (2.5)} \\
&= 2.5 \text{ pW} \\
&= -86 \text{ dBm}
\end{aligned}
$$

$$
N = -114 + 2 + 2.5 = -112
$$

$$
\therefore \frac{P_R}{N} = 26 \text{ dB}
$$

ANNOTATED BIBLIOGRAPHY

1. P. A. Lynn, *Radar Systems*, Van Nostrand Reinhold, New York, 1989.
2. S. A. Hovanessian, *Radar System Design and Analysis*, Artech House, Dedham, MA, 1984.
3. E. Brookner, *Aspects of Modern Radar*, Artech House, Dedham, MA, 1988.
4. M. I. Skolnick, *Radar Handbook*, New York, 1990.

References 1 and 2 are basic radar textbooks with simple explanations of radar principles and systems. References 3 and 4 cover radar theory and equipment in complete technical detail.

EXERCISES

17.1. Define the radar system terms in the following list:
   1. Range-measuring radar
   2. Velocity-measuring radar
   3. Pulse Doppler radar
   4. Moving-target-indicator (MTI) radar
   5. Pulse-compression radar
   6. Search radar
   7. Tracking radar
   8. Monopulse radar
   9. Conical scan radar

10. Phased array radar

11. Synthetic aperture radar

17.2. What three things can a radar measure?

17.3. Which of the above can radar measure with the best accuracy?

17.4. Which of the above are the most difficult for a radar to measure accurately?

17.5. Complete the table below.

Range (ft)	Time for pulse to go out and return
	50 μs
10,000 ft	
5 ft	

17.6. Complete the table.

Microwave Frequency (GHz)	Doppler Shift (Hz)	Velocity (mph)
10.1	1500	
10.1		50
24	1500	
14	50,000 (50 kHz)	

17.7. What range resolution can be obtained with a 2-μs pulse?

17.8. What range resolution can be obtained with the following pulse compression radar? Pulse length is 100 μs, Frequency sweep during pulse is 50 MHz.

17.9. What is the beamwidth of a 5-meter antenna at 3 GHz?

17.10. What lateral resolution in feet can be obtained with the antenna of Exercise 17.9 at a range of 25 miles?

17.11. What is the beamwidth of a 10-cm-diameter missile-seeking antenna at 10 GHz?

17.12. What lateral resolution in feet can be achieved with the antenna of Exercise 17.11. at a range of $\frac{1}{2}$ mi?

17.13. How long must the flight path be for a 5-GHz synthetic aperture radar to achieve 10-ft resolution at 50 mi?

17.14. How many ''looks'' at the target must the synthetic aperture radar of Problem 17.13 make?

18

ELECTRONIC WARFARE

Electronic warfare comprises the techniques and equipment for reducing the effectiveness of an enemy's radar. Electronic warfare terminology is defined, and various techniques for rendering a radar ineffective are discussed, including stealth, antiradiation missiles, chaff, decoys, noise jamming, and deceptive jamming.

The chapter also discusses techniques used by the radar to counteract the enemy's electronic warfare, known as electronic counter-countermeasures.

18.1 INTRODUCTION

Electronic warfare comprises the techniques, equipment, and tactics to render electronic target location and weapons control ineffective. It includes techniques to render infrared heat-seeking sensors ineffective, to jam weapons control communications systems, tactics, such as flight plans, to avoid detection, and electronic surveillance and spying techniques to determine how enemy weapons systems work and how they are deployed. This chapter concentrates only on radar aspects. The following definitions are used.

Electronic Warfare (EW) The techniques, equipment, and tactics used to render electronic target location and electronic weapons control ineffective.

Electronic Countermeasures (ECM) The electronic techniques and equipment used to render electronic target location and electronic weapons control ineffective.

Electronic Counter-Countermeasures (ECCM) Techniques and equipment used by the radar to render ECM ineffective.

Electronic Support Measures (ESM) The equipment used to determine the characteristics of the enemy's electronic target location and electronic weapons control systems.

The terms *electronic warfare* and *electronic countermeasures* are often used interchangeably. The distinction between them is that EW refers to the *total* function of rendering the enemy's electronic target location and weapons control system ineffective, whereas ECM refers only to the electronic aspects of EW.

Radar can determine target range to within a few feet, target velocity to within a fraction of a mile per hour, and angular location to within a few feet (if the radar is mounted in a missile and gets close to the target). Therefore, in a military engagement, the enemy's radar must be rendered ineffective. This is easy to accomplish against a simple enemy radar, but quite difficult against a sophisticated radar. Therefore, there is a continuing battle between making radars that cannot be jammed and better ECM equipment to jam the radars, and still better radars that cannot be jammed by this more sophisticated equipment, and on and on.

EW techniques used to render a radar ineffective include

Stealth Aircraft are specially designed to reduce the amount of microwave power they reflect so that they cannot be seen by radar.

Antiradiation missile A missile is launched by the target to home on the microwave signal transmitted by the radar. This forces the radar to shut down or be destroyed.

Chaff Chaff is metal foil, dropped from an aircraft or launched from a ship or land vehicle, that reflects a microwave signal back to the radar to create false targets.

Decoys Decoys are small missiles with amplifiers in them. They receive and amplify the radar signal and transmit it back to the radar, so that they look like targets.

Noise Jamming Noise-modulated microwave signals are transmitted to raise the noise level in the radar receiver so that the radar cannot accurately determine the range, velocity, or angular location of the target. The technique is dangerous because the enemy can launch missiles that home on the jamming signal.

Deceptive Jamming The jammer repeats an amplified version of the radar signal, but with false information. The radar determines the range, velocity, and angular location of the target with this false information.

ECM has a power advantage over radar. A small amount of jamming power can put a large jamming signal into the radar receiver. This happens because the jamming signal must go from the target to the radar, whereas, the radar signal must go out to the target, be reflected, and return to the radar. Since the jammer signal only goes in one direction, a small jammer transmitter can put more power into the radar receiver than the desired reflected signal from the target, even though the radar is using a large transmitter with thousands of watts of power.

Although the jammer has a power advantage, the radar's major advantage is that it knows the characteristics of the signal it is transmitting. The jammer must measure the enemy radar signal and then decide what to do to counter it. This measurement and decision can be accomplished in fractions of a second. However, if the radar changes its transmission frequency with every transmitted pulse, the jammer does not have time to analyze the pulse and respond with a false signal, so it will not be effective. However, since the jammer need only transmit a small amount of power to exceed the reflected radar signal in the radar receiver, the jammer can be located in an "off-board" missile, which is ahead of the target that it is protecting, and then it will have adequate time to deceptively jam the radar.

In most tactical military situations, the various ECM techniques are used together. For example, for an air strike against a ground location protected by ground-based radars, decoys would first be used to force the enemy to turn on all its radars. Once the radars were tracking the decoys, antiradiation missiles would be launched to home on them. Then noise jamming would be used to confuse the enemy radars, and, finally, the remaining radar sites would be attacked by aircraft protected with deceptive jamming.

18.2 STEALTH

An effective countermeasure is to design aircraft to reflect no microwave signal so that enemy radar cannot detect them. This technique is used for the Stealth aircraft (Figure 18.1). An aircraft cannot be made completely nonreflecting to microwaves, but its radar cross section can be significantly reduced. For example, the radar cross section of a fighter-bomber is approximately 100 square meters. This aircraft can be detected up to 100 miles. If the radar cross section could be reduced by a factor of 100 to only 1 square meter, the detection range could be reduced from 100 miles to 3 miles. At this short range, although the radar could finally detect the stealth aircraft, it would be too late to take any action against it.

Techniques to reduce the radar cross section include redesigning the aircraft structure so that there are no sharp corners, which can act as "corner reflectors" and enhance the reflected microwave signal. The major sharp corners in a conventional jet aircraft are where the wings join to the body and where the engines join the wings.

Another technique is to cover the jet engine turbine intakes, so that the radar cannot "see" the rotating engine blades.

Reduce radar reflections from aircraft

- No sharp corners
  (wings to body)
- Cover exposed jet turbines
- Single-frequency radome
- Absorptive coating on
  aircraft skin

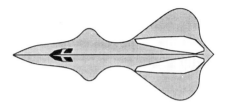

Figure 18.1 Stealth aircraft.

The nose of most military aircraft contains the aircraft's radar and, from the outside, is sharp and pointed for aerodynamic purposes and would seem to present a very small radar cross section. However, the radar signal travels through the dielectric radome on the nose of the aircraft and is reflected by the large radar dish behind the pointed radome. Therefore the radome for stealth aircraft must be designed to pass only the microwave frequency used by the aircraft's own radar, and to reflect all other radar signals so that they cannot be reflected by the antenna. Then the cross section of the stealth aircraft will approach the physical cross section of its nose, which is very small.

Finally, an absorptive coating can be used on the aircraft skin. One way to do this is to use a dielectric material that is a quarter-wavelength thick at the radar frequency. Part of the microwave signal is reflected from the front surface of this coating, and part travels through the coating and is reflected off the aircraft metal skin. However, when the two signals combine they are 180° out of phase and thus cancel, so no signal is reflected back to the radar.

18.3 ANTIRADIATION MISSILES

A very effective EW technique is to use an antiradiation missile (ARM) against an enemy radar. The ARM is a radar-controlled guided missile. It has a receiver (but no transmitter) that homes in on the microwave signal transmitted by the enemy radar. The ARM has no reference signal to obtain range and velocity information, but it can easily determine angle, and the angular resolution gets better and better as the missile gets closer and closer to the target.

The radar is providing a beacon for the ARM to home on, and the ARM can actually fly right through the center of the radar dish. When attacked by an ARM, the radar can shut down or be blown up. Either way, it is rendered ineffective.

18.4 CHAFF AND DECOYS

Chaff is metal foil used to reflect radar signals. It is a simple and inexpensive jamming technique. The chaff can be launched from an aircraft, a ship, or a land vehicle. It is designed to be dispersed over a wide area. The echo from the metal foil creates false targets that mask the real target. The disadvantage of chaff is that it remains stationary. It therefore can be distinguished from the actual targets if the radar applies MTI or Doppler techniques to distinguish moving and stationary targets.

Decoys are small unarmed missiles launched from aircraft, or buoys dropped from ships. They contain a microwave amplifier that amplifies the received radar signal so that when it is received back by the radar it is as large as the actual echo from a target. The decoy has a major advantage over chaff in that an electronically generated Doppler shift can be added to its repeated signal so it appears to be moving just like the real target.

18.5 NOISE JAMMING

Noise jamming involves transmitting high levels of noise-modulated microwave power to raise the noise level in the radar receiver so that the radar cannot function properly. Noise jamming is illustrated in Figures 18.2 and 18.3. One danger with noise jamming is that enemy missiles can be made to home on the jammer. For this reason, noise jammers usually stay beyond missile range. For this reason noise jamming is often called *stand-off* jamming. It is also called *barrage* jamming, since the jammer barrages the radar with power.

The use of noise jamming to destroy the enemy's range and velocity information is shown in Figure 18.2. The upper drawing shows the echo pulse, which would give the range of the target, buried in the large noise background. The actual pulse cannot be distinguished from the noise, so the radar can no longer determine range. The lower drawing shows the transmitted frequency and the Doppler-shifted echo; From the difference between these frequencies, the radar could determine target velocity. However, if noise is transmitted at all frequencies, including the transmitter frequency and the Doppler-shifted frequency, the Doppler-shifted echo cannot be found in the noise, so the target velocity cannot be determined.

The use of noise jamming to destroy the radar's angular information is shown in Figure 18.3. Figure 18.3a shows the antenna pattern of the radar, with the target located in the direction of one of the antenna sidelobes. The radar determines the angular location of the target by determining if an echo returns from the direction that the antenna is pointing. Since the antenna sidelobe level is 20 dB below the main-lobe level, the antenna pointing in the

TRANSMIT AMPLIFIED NOISE TO DESTROY RANGE AND VELOCITY INFORMATION

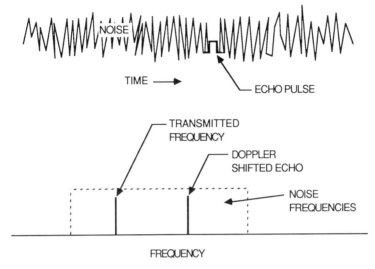

Figure 18.2   Noise jamming.

direction shown of Figure 18.3 would indicate that no target was present in the direction of the main lobe. However, since the noise jammer can put large amounts of power into the sidelobes, the received radar signal, as displayed on a PPI scope, would appear as shown Figure 18.3*b*, and the target could be at any one of several angular locations.

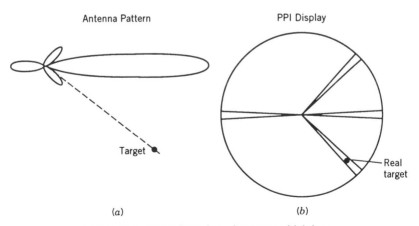

Figure 18.3   Noise jamming of antenna sidelobes.

18.6 DECEPTIVE JAMMING

With deceptive jamming, the target deliberately amplifies the radar signal so that the radar has a strong signal above its own receiver noise to work with. However, the amplified radar signal has false information—for example, arriving at the wrong time, so that the radar calculates the wrong range, or arriving with a false Doppler shift, so that the radar calculates the wrong target velocity.

With deceptive jamming, the radar doesn't even know that it is being jammed, and directs its weapons to where it has calculated the target to be.

There are two phases to the deceptive jamming process:

1. Deciding what radar should be jammed
2. Jamming the radar

The decision as to what should be jammed is the most difficult of these two requirements in a modern military environment, because various other systems are transmitting microwave signals, including fighter aircraft, expendable jammers, missile seekers launched from the ground, various air defense units on the ground, stand-off jamming aircraft, and communication systems from other aircraft. Out of all these emitters, the jammer must decide which radar is the enemy radar that should be jammed.

The receiver used to detect and analyze all the radar signals in order to decide what should be jammed is called an ESM receiver. Possible types of ESM receivers are crystal video, superheterodyne, channelized, and Bragg cell. The first two are not suitable for the scenario in Figure 18.4. These were the first types of ESM receivers developed and were suitable during the Southeast Asia conflict of the early 1970s, when there were only a few enemy radars in operation. The channelized receiver and the Bragg cell receiver are more suitable in a modern military environment.

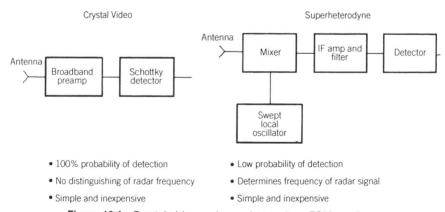

Figure 18.4 Crystal video and superheterodyne ESM receivers.

Crystal video and superheterodyne ESM receivers are shown in Figure 18.4. The crystal video receiver consists of a broadband preamp and a Schottky detector. The radar signals come through the antenna and are amplified by the broadband preamp, so any signal over a wide frequency range at the -90-dBm level is brought up to the -60-dBm sensitivity of the Schottky diode. The signal is then detected in the Schottky diode, which indicates that a target is present. The crystal video receiver has a 100% probability of detection, and it is simple and inexpensive. The significant disadvantage is that it provides no determination of the radar frequency. It simply tells that a radar signal is being transmitted.

The superheterodyne ESM receiver consists of a swept local oscillator, mixer, IF amplifier, and detector. At a given instant of time, the LO is at some freuency, and only at that instant of time can a radar frequency shifted from the LO frequency by the IF frequency be amplified by the IF amplifier and passed to the detector. A detector output at a given instant of time means that a certain radar frequency has been received by the antenna.

The superheterodyne receiver is simple and inexpensive. It does not require a broadband preamp, because the mixer itself has enough sensitivity for a -90-dBm received signal. The superheterodyne receiver can determine the frequency of the incoming signal, but it has a very low probability of detection, because, while the local oscillator is adjusted so that the superheterodyne receiver is checking for a 2-GHz signal, signals entering the receiver at this instant of time at all other frequencies do not get through the IF amplifier and so are not detected.

A channelized ESM receiver is shown in Figure 18.5. The radar signals to be analyzed pass through a broadband amplifier and are divided and put through a parallel combination of perhaps 100 superheterodyne receivers. Each one has a fixed LO, operating at a different frequency. There are 100 outputs from the channelized receiver, one for each small frequency range across the total microwave frequency range being analyzed. The channelized ESM receiver has 100% probability of detection, and it can determine the frequency of every radar signal. However, it is basically 100 times larger than the superheterodyne or the crystal video receiver. Even with miniaturization using microwave integrated circuits, the channelized ESM receiver is still large and expensive. Right now, however, it is the only way to effectively solve the problem of deciding what signal to jam in a modern military engagement.

The Bragg cell ESM receiver is shown schematically in Figure 18.6. It uses a variety of physical effects to obtain an instantaneous measurement of the time and frequency of all received microwave signals. The Bragg cell consists of an acousto-optic crystal. Microwaves are applied to the crystal and set up mechanical vibrations. The Bragg cell crystal is piezoelectric, which means that the electrical microwave signal applied to the crystal sets up mechanical vibrations at the microwave frequency that travel through the crystal.

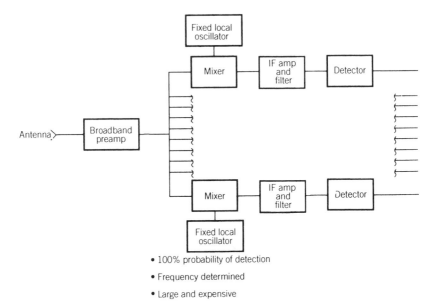

- 100% probability of detection
- Frequency determined
- Large and expensive

Figure 18.5 Channelized ESM receiver.

Laser light is directed through the crystal. The light is bent by the crystal, and the amount of bending depends on the crystal structure, which is continually changed by the mechanical vibrations set up by the microwaves. Therefore the direction in which the light leaves the Bragg cell crystal depends on the microwave frequency. The light beam is scanned in and out of the plane of the paper with time. Therefore the position of the light beam, as it leaves the Bragg cell crystal, varies in two dimensions. The position in one dimension depends on the microwave frequency, and the position in the other dimension depends on the microwave signal arrival time at the receiver.

An array of photodetector diodes is located at the Bragg cell output, shown on the right side of the Bragg cell in Figure 18.6. Each diode detector in the two-dimensional output array represents a particular frequency and time of arrival of the microwave signal. If a signal is present in a photodetector, the receiver knows that a microwave signal at a particular frequency and time was received.

Figure 18.7 shows deceptive range jamming. The signal at the radar receiver is shown as a function of time at three different intervals. The upper sketch shows the real echo and the noise, and this is what the radar would normally receive. To provide false range information, the radar signal is received by the jammer, amplified, and retransmitted to the radar. Note that there is a slight delay, which corresponds to the time required for the jammer to respond and amplify the signal. The radar is now actually receiving a larger signal than usual because the jammer is amplifying the radar signal and retransmitting it. At this point, the jammer is helping the radar.

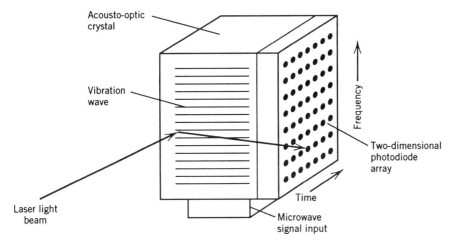

Figure 18.6 Bragg cell ESM receiver.

Once the radar has determined the range position of the target, it tracks the range position in a *range gate*. The range gate measures the amount of signal during the first half of the pulse and during the second half of the pulse. By adjusting its time position so that these signals are equal, it then determines the target range. The deceptive jammer technique designed to

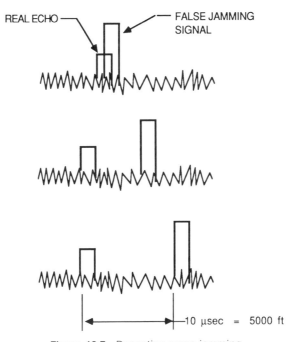

Figure 18.7 Deceptive range jamming.

pull this range gate off the actual target is shown in the second and third sketches.

In the middle figure the real echo and noise are shown. The false jamming signal has been delayed, so it occurs at a different instant in time than the real echo. The false jamming signal is continuously moved from its original time location, which was coincident with the real echo, to later and later times, until it gets to the position shown in the middle of Figure 18.7. The radar cannot track the smallest signal, but the largest; otherwise it would be continuously tracking on false-alarm noise. Consequently, the range gate follows the larger signal and is pulled off the real target. In the bottom figure the jamming signal has been delayed 10 μs behind the real target, and the radar has followed the larger signal away from the target so that it thinks the target is 5000 ft behind where it is. This technique is called range-gate pull-off (RPGO) and is very effective against a simple range-tracking radar.

The radar may suspect that it is being jammed by range-gate pull-off because this technique is common. If the radar is more sophisticated, it can track the first pulse. Note that it cannot track the smaller pulse because it would always be tracking the false-alarm noise, but it could track the first pulse and thus completely avoid the deceptive jamming pulse.

If the jammer suspects that the radar is tracking the first pulse, and if it is more sophisticated, it can estimate when the pulse will be received and generate and transmit a jamming pulse just before the real echo pulse. It can then progressively move this jamming pulse ahead of the real echo. Therefore, a radar tracking the first pulse thinks the target is ahead of where it actually is. This technique is called range-gate pull-in (RGPI). Note that the use of range-gate pull-in requires that the jammer estimate the microwave frequency and the pulse repetition frequency of the radar from the previous pulses that it has obtained. If the radar changes its frequency on each pulse and uses a random PRF, the range-gate pull-in will not work.

In a similar way a Doppler radar can be jammed by having the jammer repeat the received radar signal with a false Doppler frequency shift. This technique is called velocity-gate pull-off (VGPO).

A block diagram of a deceptive jammer, which uses RGPO, RGPI, and VGPO, is shown in Figure 18.8. The microwave signal is received by a horn antenna and is amplifed in a transistor amplifier. A sample of the radar signal is taken through the directional coupler to a detector, and then the radar threat is analyzed. The remainder of the received microwave signal is amplified in a second transistor amplifier and sent to a variable delay line, shown in the lower right. If the jammer decides to use range-gate pull-off, the threat analysis and control computer causes a delay to occur, and the delayed received signal is then amplified in a low power amplifer and a pulsed high power traveling wave tube. It is then transmitted through the horn antenna as a range-shifted jamming signal. This signal is an amplified version of the received signal, but is delayed in time, so the radar thinks that the target is behind its actual position.

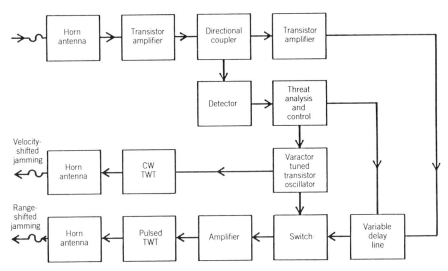

Figure 18.8   Deceptive jammer.

A deceptive jammer must operate over a wide frequency range to jam any radar. Consequently, all the components in the jamming system must cover wide frequency ranges. Modern jamming systems cover the frequency range from 2 to 18 GHz in two bands, with the lower band covering the frequency range from 2 to 8 GHz and the upper band from 8 to 18 GHz. The traveling wave tube amplifiers, transistor amplifiers, and all other components shown in Figure 18.8 must cover one of these two wide ranges.

If the jammer decides to use range-gate pull-in and transmit a false signal ahead of the received signal, the threat analysis and control computer sets the frequency of the varactor-tuned transistor oscillator and the time that its signal will be transmitted. The internally generated signal, which has been estimated to be at the frequency and time of the transmitted radar signal, is then switched through the chain of transistor and traveling wave tube amplifiers and is transmitted ahead of the actual echo pulse.

If the jammer decides to use velocity-gate pull-off, the received signal is measured and a signal at a false Doppler-shifted frequency is generated by the varactor-tuned transistor amplifier, amplified by the CW TWT, and transmitted through a separate antenna as a velocity-shifted jamming signal.

Deceptive angle jamming against a conical-scan radar is shown in Figure 18.9. With conical-scan tracking, the angular location of the target is determined in azimuth (or in elevation) by scanning the antenna beam progressively from one position to another. The antenna is then moved so that the target is located midway between the two beams. In Figure 18.9a the target is not on the antenna axis but in beam B. Consequently, the radar will receive a larger signal when the beam is pointing in position B than in position A. This signal variation is labeled "actual signal." Note that as the beam is

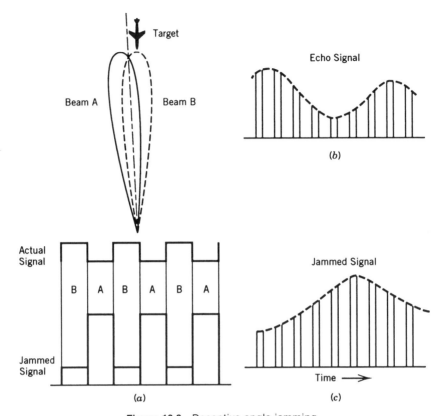

Figure 18.9 Deceptive angle jamming.

scanned in time, the return signal is larger at times when the beam is in direction B than in direction A. The returned echo signal thus appears as in Figure 18.9*b* as a series of pulses whose amplitude varies as the beam is scanned from one position to another.

It is extremely easy to deceptively jam a conical-scanning radar simply by repeating the signal back at the right time in range and with the right Doppler velocity, but with an amplitude modulation that is out of phase with the real conical-scanning information. As in Figure 18.9*c*, when the return echo should be small, because the beam is in position A and not pointing at the target the jammer amplifies this signal, so the return signal is very large. When the beam is pointing in position B, the jammer provides no amplification. Consequently, as in Figure 18.9*a*, a much larger signal is received by the radar when the beam is pointing in position A. Consequently, the radar thinks that the target is located in beam A rather than in beam B.

Note that the angle jamming technique in Figure 18.9 is only effective against conical-scanning radars. If the radar obtains its angular information with monopulse tracking, where all the angular information is obtained on a single pulse, this type of jamming is not effective. Monopulse tracking radars are very difficult to deceive in angle. Jamming a monopulse radar requires two jammers separated by a large distance, either in the opposite wing tips of an aircraft or in two aircraft.

The jammer has a power advantage over the radar. A jammer usually uses an omnidirectional antenna, which radiates the jamming power in all directions. It would be more effective to concentrate all the jamming power on the radar being jammed. This would not only increase the jamming-to-signal ratio at the radar but would reduce the problem of jamming other friendly radars. However, this approach requires a high-gain antenna with a narrow beam, which would have to be moved instantaneously to point at the location of the radar being jammed. The jammer antenna cannot be moved fast enough mechanically, but the beam can be moved rapidly by using a phased array.

A phased array jammer is shown in Figure 18.10. The phased array technique provides a high jammer effective radiated power (ERP) for protecting targets like ships, which have large radar cross sections. The phased array can automatically transmit the jamming power in the direction of the target, using the lens array shown, which consists of a dielectric lens. On the receiving side of the lens are separate antennas, each pointing in a different direction. When a signal is received in one of these antennas, the lens divides the received power and directs it with the right phase shifts to each of the transmitting locations on the other side of the lens, where the signals are amplified and transmitted from the output antennas. The lens array automatically provides the correct phase shift to each output antenna, so the transmitted signal is radiated in exactly the same direction from which the received signal was received.

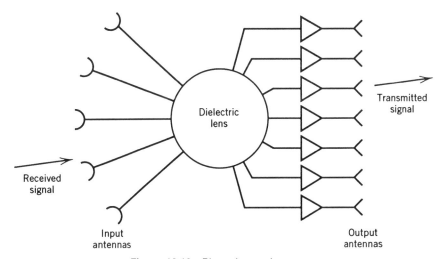

Figure 18.10 Phased array jammers.

Table 18.1   Electronic counter-countermeasures

ECCM	Effective Against
Frequency hopping	Barrage jamming
Leading edge tracking	Range-gate pull-off
Variable PRF	Range-gate pull-in
Monopulse	Angle-tracking deception

18.7   ELECTRONIC COUNTER-COUNTERMEASURES

Electronic counter-countermeasures (ECCM) are techniques that the radar can use to render the jamming ineffective. Some ECCM techniques are shown in Table 18.1. The radar has an advantage over the jammer in that the radar knows the characteristics of the signal it is transmitting. The jammer must measure the radar signal and then respond. If the radar can make the analysis of its signal difficult for the jammer, the jammer will not be able to respond fast enough to confuse the radar.

The table shows the ECCM techniques that the radar can use, and what type of jamming they are effective against. If the radar uses frequency hopping, then a barrage jammer must spread its jamming power over a wide frequency range, and, consequently, less power can be put into the radar bandwidth.

If the radar uses leading-edge tracking and determines the range of the target by looking just at the beginning of the return echo pulse, then range-gate pull-off will not be effective because the radar will have determined the range to the target before the false jamming pulse, which is delayed in time, will be received. If the radar uses pulse compression, a similar jamming immunity to range gate pull off will be achieved.

If the radar uses a variable pulse repetition frequency, then the jamming techniques of range gate pull-in will not be effective. The range gate pull-in technique requires that the radar estimate when the actual radar pulse will be received and send out a jamming pulse ahead of this. If the pulse repetition rate is continually varied, the jammer has no way of knowing when the actual radar pulse will be received.

If the radar uses monopulse tracking for determining the angular location of the target, then angle detection jamming is made very difficult compared with the case if the radar had used conical scanning.

ANNOTATED BIBLIOGRAPHY

1. D. C. Schleher, *Introduction to Electronic Warfare*, Artech House, Dedham, MA, 1986.

2. E. J. Chrzanowski, *Active Radar Electronic Countermeasures,* Artech House, Dedham, MA, 1990.

Reference 1 presents a thorough technical presentation of electronic warfare. Reference 2 provides a detailed and clear explanation of electronic jamming systems.

EXERCISES

18.1 Define the electronic warfare terms in the following list:
  1. Electronic Warfare (EW)
  2. Electronic countermeasures (ECM)
  3. Electronic counter-countermeasures (ECCM)
  4. Electronic support measures (ESM)
  5. Stealth
  6. Antiradiation missile (ARM)
  7. Chaff
  8. Decoy
  9. Barrage jamming
  10. Stand-off jamming
  11. Deceptive jamming
  12. Range-gate pull-off (RGPO)
  13. Range-gate pull-in (RGPI)
  14. Velocity-gate pull-off (VGPO)
  15. Inverse conical-scan jamming
  16. Home-on jammer
  17. Channelized receiver
  18. Bragg cell

18.2. List four stealth techniques that reduce the radar reflections from an aircraft.

18.3. What is the advantage of decoys compared with chaff for creating false targets?

18.4. How does noise jamming prevent the radar from determining target range and velocity?

18.5. What are four types of ESM receivers?

18.6. Why is the crystal video receiver ineffective in a modern electronic warfare scenario?

18.7. What is the disadvantage of a superheterodyne receiver in a modern electronic warfare scenario?

18.8. How does range-gate pull-off work?

18.9. How does range-gate pull-in work?

18.10. How does velocity-gate pull-off work?

18.11. How does deceptive angle jamming of a conical-scan radar work?

18.12. What is the basic advantage that a jammer has over a radar?

18.13. What is the basic advantage that a radar has over a jammer?

19

NAVIGATION AND OTHER MICROWAVE SYSTEMS

The global positioning system (GPS) is discussed first. This fantastic navigation system uses 24 orbiting satellites and gives the position of receivers located anywhere in the world with an accuracy of 2 meters.

Cellular mobile telephones systems are discussed next. These systems, located in the microwave band around 900 MHz, provide a major increase in mobile radio and portable telephone channel capacity.

Medical applications of microwaves, including the use of microwaves for x-ray generation using linear accelerators and the use of microwaves for hyperthermia treatment of cancer, and scientific applications of microwaves are also reviewed.

19.1 GLOBAL POSITIONING SYSTEM

The global positioning system is illustrated in Figures 19.1 and 19.2. The GPS consists of 24 orbiting satellites (Figure 19.1) that transmit positioning signals back to the earth. By processing the signals from four of the satellites, a GPS receiver on the earth can determine the receiver's location, including its latitude, longitude, and altitude above the earth to an absolute accuracy of 2 meters (6 feet). The 24 GPS satellites are in half-synchronous orbits; that is, they make two revolutions about the earth each day. The orbits are inclined at 55° to the equator; and four satellites will be equally spaced in each of six equatorial planes.

GPS was originally developed for defense applications to provide navigation and position information for military aircraft, ships, and personnel, and for weapons guidance. However, since the GPS signals are available, they

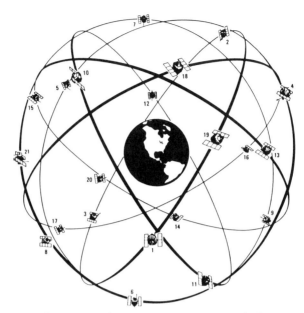

Figure 19.1 Global positioning system (GPS).

can be used for many commercial applications, such as surveying, navigation for aircraft, ships, trucks, and automobiles, and timing. The GPS navigation system can replace all the lower frequency navigation and landing systems for commercial aircraft. Auto manufacturers worldwide are considering GPS navigation systems to provide a dashboard display of the vehicle's location. Because the GPS system provides accurate timing information, the GPS signal can replace the time reference signal currently being broadcast by each nation's bureau of standards.

To determine position, the GPS receiver must receive signals from four satellites. The pulse code signal from each satellite tells the satellite's position. The satellite position is given to within 200 meters for civilian applications, and anyone in the world can receive the signal from the satellites and determine their position to within 200 meters. The satellite position is given to within 2 meters for military applications. In this case the pulsed information from the satellite is coded so that only the military will know the satellite position to within the 2 meter accuracy. During peacetime, the 2-meter accuracy signal is also made available for civilian use.

The GPS receiver receives position information from each of three satellites. The signal from the fourth satellite gives a time reference. The GPS receiver knows signal transmission time from the satellites and measures the reception time. It then determines its range from each satellite. The range and known position of a satellite define a spherical surface. The intersection of the three spherical surfaces gives the receiver location.

The location of a civilian receiver can be determined to even greater accuracy by using an additional ground transmitter at a known location. Consequently, within a metropolitan area a ground station, whose position is known, can also transmit a signal, and by using the civilian signals from the GPS satellite and the signal from the known ground location the position can be determined to even better than a 2-meter accuracy.

A block diagram of a GPS receiver is shown in Figure 19.2. The signal is transmitted from the satellite at 1575 MHz. The satellites can be in any location relative to the receiver, so an omnidirectional antenna must be used. The antenna for the GPS system is a helical antenna fitted inside a radome. With a helical antenna, the length of the antenna is shortened so that the radome and antenna are only a few inches high. The recieved signal passes through a bandpass filter and then through a low-noise transistor amplifier to the first mixer. The local oscillator for the first mixer is a phase-locked loop. A crystal oscillator supplies a reference signal for the phase detector. A VCO generates the 1.530-GHz LO signal. Rather than multiplying the reference signal up to compare with the LO signal in the phase detector, the LO signal is divided down by frequency dividers (scalers) and this signal is compared with the crystal reference in the phase detector. The IF frequency from the first mixer at 45 MHz is amplified in the first IF amplifier. The IF amplifier includes an automatic gain control (AGC) circuit. The signal then goes to a second mixer, which has a fixed crystal-controlled LO. After amplification, the signal is analyzed in a signal processor to give latitude, longitude, and altitude, or it can be further processed to give other information such as the location of the receiver on a city map.

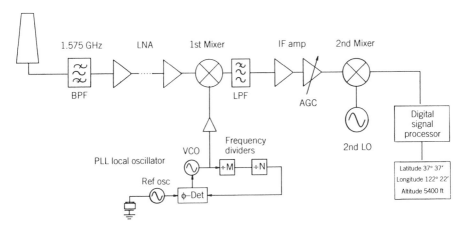

Figure 19.2 Global positioning system (GPS) receiver.

19.2   CELLULAR TELEPHONES

Cellular mobile radio is illustrated in Figure 19.3. The mobile radios used by police, fire, and other emergency services, taxicabs and truck lines, and for mobile telephones have operated in assigned frequency ranges in the VHF band around 50, 150, and 450 MHz. Each mobile radio has a 25-MHz bandwidth assigned for the exclusive use of the mobile-radio owner. Into this bandwidth a single 5-kHz voice signal is frequency modulated. VHF mobile radio capacity, however, is severely limited in metropolitan areas.

To relieve this lack of capacity, the microwave band from 806 to 947 MHz, originally to be allocated for UHF TV, has now been assigned for microwave mobile radio usage. Over 600 channels can be fitted into this frequency range. The channel capacity has been further increased to several times this value by dividing the metropolitan areas into cells, as shown in Figure 19.3, and reusing frequencies in nonadjacent cells. Each user is given a mobile radio, but is not assigned a frequency. The frequency of transmission and reception in the mobile radio is controlled from a central switching location. As the mobile radio leaves one cell and enters another cell, its transmitting and receiving frequency is switched, leaving the original frequency available to another customer. This type of ''cellular'' mobile communication system is much more complicated than the VHF mobile radio systems, because of the frequency switching required and the need for a range of transmitter-receiver

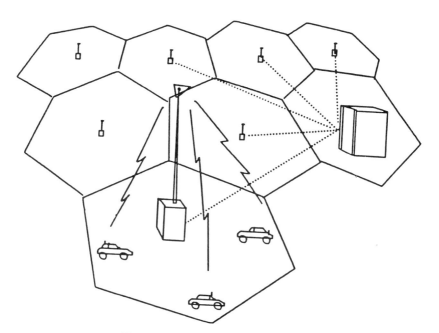

Figure 19.3   Cellular mobile telephone.

frequencies available in each mobile radio. It provides the necessary capacity for mobile radio in congested metropolitan areas.

A block diagram of a cellular telephone set is shown in Figure 19.4. The cellular phone, in the middle right, contains a crystal oscillator capable of generating 650 different subcarriers in the 138–to 141 MHz frequency range. The subcarrier being used is controlled by the base station. When a call is to be made and the cellular phone is turned on, a request for a channel assignment is automatically sent to the base station. The base station sends back a signal that selects the channel. Actually, a pair of channels is assigned, one for transmission and one for reception. The assigned frequency from the crystal oscillator is used both for the transmitter and as a local oscillator for the receiver. The subcarrier frequency assigned in the 138–141 MHz frequency band is amplified and sent to an FM modulator where the telephone voice signal is frequency-modulated on to the subcarrier. The FM modulated subcarrier is amplified, multiplied up in frequency by 3 with a frequency multiplier, amplified again and multiplied up in frequency by 2, amplified, and passed through a bandpass filter that filters out the undesired harmonics of the multiplication process. Now the signal with its frequency modulation of the telephone voice information is located in a 20 MHz frequency band in the range between 825 and 845 MHz. This 20-MHz band is the frequency range assigned for cellular mobile telephone transmission. The signal is amplified through several stages of amplification up to a power level of approximately 1 W and then passes through the diplexer to the transmitting antenna. The diplexer is a two-frequency multiplexer that allows the transmitted frequency to pass from the transmitter into the antenna and allows the received signal, which is at a different frequency, to come in from the antenna and pass through to the receiver.

If the mobile telephone is very close to the base station, the transmitted signal from the cellular telephone is very strong and could saturate the base station receiver, which is receiving not only this signal but signals from all other telephones in the cell. Therefore the base station continually sends a message back to the individual phone to control the power level of its transmitted signal. The signal level control circuitry is shown in the lower left of Figure 19.4. The control signal from the base station is sent to a power level controller. A directional coupler samples the output signal just before it is to be transmitted. The power level is compared with desired value, and an error signal is applied to control the gain of the transmitter amplifiers.

The received signal will be at an assigned frequency in the 870–890 MHz band. The signal comes in the same antenna that was used for transmitting and passes through the diplexer into the receiver through a bandpass filter. Note that the filter allows any signal within the assigned cellular telephone band to pass into the receiver because the cellular phone is using different frequencies at different times as assigned by the base station. The received signal is then amplified in a low-noise preamplifier and goes to the first mixer. The mixer uses the original subcarrier generated for the transmitter multiplied

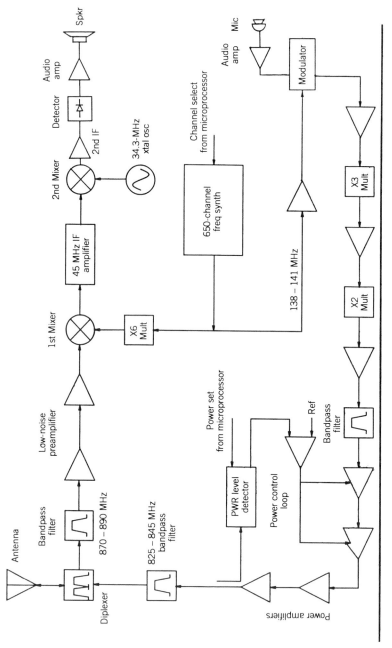

Figure 19.4 Block diagram of Cellular mobile telephone.

up 6x in frequency as its local oscillator. The difference between the received signal and the local oscillator is 45 MHz, which is the difference between the transmitted and received frequencies. After amplification, the first IF frequency is mixed in a second mixer with a fixed 34.3-MHz crystal-controlled LO and amplified in a 10.7-MHz IF amplifier. The amplified signal then goes to a demodulator where the FM modulation is removed, through an audio amplifier, to the speaker in the telephone set, as shown in the upper right of Figure 19.24.

The cellular mobile phone was first developed as a car phone. The concept has now been extended to portable phones that can be removed from the car and used in homes and offices as well. In the future the cellular phone concept will be extended to personal communications systems in which the cellular phone can be used to transmit and receive computer data.

The cellular telephone system best serves large metropolitan areas where many mobile phones are within the 25-mi-diameter cell. A cellular phone system could also be used in rural areas, but it would not be economically feasible because there would be so few customers to be served by the expensive base station. A solution to this problem, which would also make worldwide cellular telephone communication possible, is to put the base station in a satellite. Currently, plans for such a satellite cellular telephone system are underway. The system would require many orbiting satellites. The mobile or portable phone would have the same basic block diagram as in Figure 19.4. The individual mobile phones would communicate directly to the orbiting satellites, and the satellites would then transmit the phone signal back to a base station on the ground.

19.3 MICROWAVE OVENS

Microwaves are used for industrial heating and for heating and cooking food in a microwave oven. Most heating and cooking is done with infrared radiation, which is commonly called "heat" and occurs at short wavelengths in the frequency range between microwaves and light. The short wavelengths of infrared radiation cannot penetrate materials such as food, so the infrared radiation is absorbed on the surface and conducted into the inner part of the material. Normally the material to be heated is a poor thermal conductor, so the heating is a slow process.

In contrast, the longer wavelength of microwaves can penetrate the entire material and heat it from the inside out. In a microwave oven the microwaves are absorbed by water in the food, and the food is uniformly heated throughout its cross section.

The frequency for microwave cooking and heating is 2450 MHz. The wavelength is long enough to penetrate the material, yet short enough so that the microwaves can fit into a moderate size enclosure such as a home microwave oven.

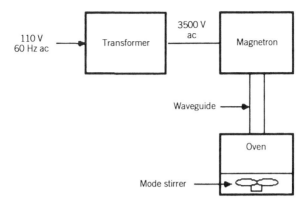

Figure 19.5 Microwave oven.

A block diagram of the microwave oven electronics is shown in Figure 19.5. The 110 V, 60 Hz, ac power enters the oven and is transformed to 3500 V ac, which operates the magnetron (see Chapter 12). The magnetron operates from the raw ac power and does its own rectification. The microwave power at the microwave oven frequency of 2.45 GHz is transmitted through a waveguide, which is simply a metal duct above the oven, into the oven enclosure. Somewhere in the oven enclosure is a mode stirrer, slowly rotating fan blades that change the microwave field configuration inside the oven to distribute the microwave power uniformly into the food. The complete microwave electronics package, shown in Figure 19.5, costs less than $50 in large quantities, including $20 for magnetron transmitter tube.

The assigned frequency for microwave oven operation is 2.450 GHz. The microwave power in the oven is 600 W which is obtained even when the SWR that the oven presents to the magnetron tube is 3:1. The microwave oven cavity is deliberately loaded so that its SWR is always around 3:1 whether there is food in the oven or not.

19.4 MEDICAL APPLICATIONS OF MICROWAVES

One promising application of microwaves is for the hyperthermia treatment of cancer. This technique, just being developed, uses microwaves to differentially heat the cancer tissue to destroy it. The cancer tissue absorbs more microwaves than the adjacent healthy tissue, so the simple heating effect of microwaves can be used to destroy cancerous tissue.

The major medical use of microwaves is to generate cancer treatment x-rays (see Figure 19.6). X rays are generated by accelerating electrons to high energies and directing them into a metal target. As the electrons hit the metal target, they displace electrons in the atomic shells of the target

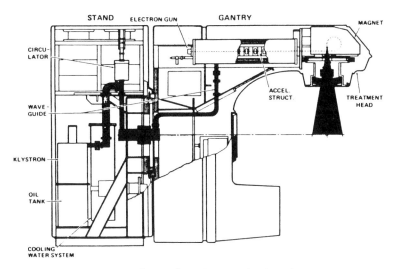

Figure 19.6 Generation of cancer treatment x-rays.

metal and the atom emits x-radiation. The greater the electron velocity, the higher the X-ray frequency and the more effective they are for cancer treatment. For most effective cancer treatment, the electrons should be accelerated through 10 million volts to give them sufficient energy to penetrate the atoms of the target and generate the high-energy x rays. A simple way of accelerating the electrons would be to accelerate them through a dc voltage of 10 million volts, but it is difficult to generate this high voltage. A more effective way, which is used in most cancer treatment x-ray machines, is to accelerate the electrons in a linear accelerator with microwaves as the power source.

In coupled-cavity TWTs a beam of electrons travels down in a microwave transmission line. Loading the transmission line causes microwaves to travel at the same velocity as the electrons, so the microwaves and the electron beam interact and generate microwave power. This same principle is used in a linear accelerator, except the phase is changed so that the electron beam absorbs energy from the traveling microwave field, so the electrons can be continuously accelerated almost up to the speed of light. These high-energy electrons are then directed to an x-ray target; upon hitting the target, they generate x rays. A linear accelerator is shown in Figure 19.6. The microwave signal is provided by a high power klystron mounted in the equipment enclosure behind the x-ray machine. The linear accelerator is at the top. The electrons are bent by the magnet in the treatment head, impinging on the x-ray target, and the x rays are emitted in a vertical direction from the treatment head.

19.5   SCIENTIFIC APPLICATIONS OF MICROWAVES

Some of the most important scientific applications of microwaves are radioastronomy, nuclear research, and triggering of thermonuclear power generators. In radioastronomy the nature of the universe is determined by studying the microwave radiation from distant galaxies. Much of our knowledge of the universe has been obtained from optical astronomy, where telescopes have detected the light radiation from distant galaxies. However, much of the matter in the distant galaxies does not emit light, but it does emit microwaves at a frequency of about 1.4 GHz, which is a natural resonance of the hydrogen atom composing this material. Consequently, a more complete picture of the universe can be obtained.

Microwaves also aid nuclear research from their use in linear accelerators. High-energy linear accelerators powered by microwaves at 2.45 GHz are used to accelerate electrons to sufficient energies to where they can penetrate the nucleus of the atoms, split the atoms, and provide information about the internal nuclear structure. To obtain sufficient energy, linear accelerators must be very long, up to several miles, and a microwave klystron must be used every 20 feet to provide the microwave power for electron beam acceleration.

Perhaps one of the most exciting scientific applications of microwaves, which is currently under investigation and which may lead to a major practical application, is the use of microwaves at 100 GHz to trigger a safe thermonuclear power generation process. All existing nuclear power generating plants use the fission process, where plutonium is split in the reactor into other radioactive elements. The energy released during the fission process is converted to heat, which boils water that drives steam turbines to generate electricity. The problem is that, if an accident should occur, the radioactive elements from the fission process are released into the surroundings. Much greater power generation capability can be obtained from the fusion process, where two hydrogen atoms are combined to form helium, and vast amounts of energy are released. This is the principle of the hydrogen bomb. The fusion process has not yet been developed for commercial power generation. Its advantage is that the helium resulting from the process is not highly radioactive, so if an accident occurs there is no radiation danger. The problem is getting the thermonuclear fusion started. Once started, it is self-sustaining, but some technique must be developed for injecting enough energy into the hydrogen plasma to start the process. Investigation is currently underway to see if microwaves at 100 GHz (a natural resonance of the hydrogen atom) can be used. To achieve the high powers necessary to make thermonuclear power generation practical on a commercial scale, thousands of watts of power at 100 GHz must be generated. This can be done by using several klystron-like transmitter tubes, called *gyrotrons*. Gyrotrons have already been developed that are capable of generating 10 kW of power at 100 GHz.

If several gyrotrons are paralleled, enough power is hopefully available to start the thermonuclear fusion process.

ANNOTATED BIBLIOGRAPHY

1. G. Calhoun, *Digital Cellular Radio,* Artech House, Dedham, MA, 1988.
2. W. S. Cheung and F. H. Levien, *Microwaves Made Simple; Principles and Applications,* Artech House, Dedham, MA, 1985, pp. 307–320.

Reference 1 describes analog and digital cellular telephone systems. Reference 2 describes medical applications of microwaves.

EXERCISES

19.1. What position accuracy can be obtained with the GPS?

19.2. How does the GPS receiver determine its location?

19.3. What is the microwave transmission frequency of GPS satellites?

19.4. What is the frequency range of cellular telephones?

19.5. In what two ways can the channel capacity of cellular mobile radio be increased as compared with VHF mobile radio?

19.6. What is the frequency of microwave ovens?

19.7. How are microwaves used to generate cancer treatment x rays?

19.8. Name three scientific applications of microwaves.

EXERCISE ANSWERS

CHAPTER 1

1.1. What is the frequency range of microwaves?
300 MHz to 300 GHz

1.2. Which of the common frequency bands (HF, VHF, UHF, SHF, etc.) are in the microwave range?
UHF, SHF, EHF

1.3. What types of communication systems lie below the microwave band?
AM broadcast radio; shortwave radio; FM broadcast radio; VHF TV; mobile radio

1.4. What is the wavelength at the low frequency end of the microwave band?
1 m

1.5. What is the wavelength at the middle of the microwave band?
3 cm

1.6. What is the wavelength at the high end of the microwave band?
1 mm

1.7. List six types of microwave communication systems, and, for each system, list a frequency at which it operates.

UHF TV	600 MHz
Microwave relay	3.9 GHz
Satellite communication	6 GHz (up)
	4 GHz (down)

Troposcatter communication	2 GHz
Mobile radio	900 MHz
Telemetry	2 GHz

1.8. List four types of radar systems and the frequencies at which they operate.

Search radar	400 MHz
Airport traffic control	3 GHz
Airborne fire control	10 GHz
Police radar	24 GHz
Missile-seeker radar	94 GHz

1.9. What two problems prevent conventional electronic equipment from working at microwave frequencies?
Lead reactance and Transit time

1.10. What six basic types of microwave devices are used in all microwave systems?
transmission lines; signal control components; semiconductor sources; low-noise receivers; tubes; antennas

1.11. What are the three types of microwave transmission lines?
coaxial cable; stripline; waveguide

1.12. Name eight signal control components.
attenuators; phase shifters; cavities; couplers; filters; loads; circulators and isolators; switches

1.13. What are the six major types of microwave semiconductor devices?
bipolar transistors; field-effect transistors; HEMTs; varactor multipliers; IMPATTs; Gunns

1.14. What is the maximum power that can be obtained from a microwave semiconductor at 10 GHz?
10 W

1.15. What is the maximum power that can be obtained from a microwave tube at 10 GHz?
500 kW

1.16. What are the five types of microwave tubes?
klystron; coupled-cavity TWT; helix TWT; Gridded tube; CFA

1.17. What are the six types of low-noise microwave receivers?
mixer; bipolar transistor; field-effect transistor; HEMT; paramp; cyrogenic paramp

CHAPTER 2

2.1 Complete the following table.

Frequency	Period
60 Hz	0.017 s
2 kHz	0.5 ms = 5×10^{-4} s
5 MHz	0.2 μs = 2×10^{-7} s
500 MHz	2 ns = 2×10^{-9} s
2 GHz	0.5 ns = 5×10^{-10} s
5 GHz	0.2 ns = 2×10^{-10} s
10 GHz	0.1 ns = 10^{-10} s
100 GHz	10 ps = 10^{-11} s
100 Hz	10 ms
50 kHz	20 μs
200 kHz	5 μs
10 MHz	100 ns
20 MHz	50 ns
2 GHz	500 ps
250 kHz	4 μs = 4×10^{-6} s
20 MHz	5 ns = 5×10^{-8} s
400 kHz	2.5 μs

2.2. Complete the following table.

Frequency	Wavelength
600 MHz	50 cm
5 GHz	6 cm
2 GHz	15 cm
15 GHz	2 cm
75 GHz	4 cm
150 GHz	2 mm
10 GHz	3 cm
3 GHz	10 mm
100 GHz	3 mm
600 MHz	50 cm
1 GHz	300 mm

Frequency	Wavelength
6.8 GHz	44 mm
10.7 GHz	28 mm
15.5 GHz	19.3 mm
1.2 GHz	25 cm
3.5 GHz	8.57 mm
1.5 GHz	20 cm
200 GHz	1.5 mm

2.3. Draw arrows whose length is equal to one wavelength at the following frequencies:

Frequency (GHz)	Wavelength (mm)	Arrow Representing Wavelength
4	75	————————————————————————→
8	37.5	———————————→
15	20	————→
60	5	→

2.4. Complete the table.

Elec Field (V/m)	Magnetic Field (A/m)	Impedance (Ω)	Power Density (W/m²)
1	.02	50	0.02
754×10^{-3}	2×10^{-3}	377	1.51×10^{-3}
0.025	5×10^{-4}	50	1.25×10^{-5}
15	0.04	377	0.6
0.22	0.0044	50	10^{-3}
0.61	0.0016	377	10^{-3}

2.5. How much microwave power is received by a 3-m² antenna if the electric field at the antenna is 10^{-5} V/m and the magnetic field is 2.6×10^{-8} A/m?

$$\text{Power} = \text{Electric Field} \times \text{Magnetic Field} \times \text{Antenna Area}$$
$$= 10^{-5} \text{ V/m} \times 2.6 \times 10^{-8} \text{ A/m} \times 3 \text{ m}^2$$
$$= 7.8 \times 10^{-13} = 0.78 \text{ pW}$$

2.6. What is the phase relationship of the following signals?

B ___in phase___ A by __−__ degrees
  leads or lags

C ____lags____ A by _90_ degrees
  leads or lags

D ____lags____ A by _180_ degrees
  leads or lags

2.7. Complete the following table.

Frequency (GHz)	Material	Skin Depth (microns)
2	Copper	1.48
6	Gold	0.60
10	Aluminum	0.80
10	Copper	0.66
30	Copper	0.38

CHAPTER 3

3.1. Convert the following numbers to dB (approximately) without using a calculator.

Numbers	dB
6	8
8	9
400	26
3000	35
0.05	−13
0.002	−27

3.2. Convert the following dB values to numbers (approximately) without using a calculator.

Numbers	dB
5	7
20	13

Numbers	dB
300	25
0.05	−13
$2 \times 10^{-6} = 0.000002$	−57

3.3. Convert the following powers to dBm (approximately) without using a calculator.

Power	dBm
1 W	30
2 kW	63
50 W	47
30 mW	15
200 nW	−37

3.4. Convert the following dBm values to power (approximately) without using a calculator.

Power	dBm
20 mW	13
250 W	24
300 W	55
0.05 mW = 50 μW	−13
100 nW	−40

3.5. Convert the following phrases to the equivalent numerical change in dB.

a. "10 times larger"   dB = ___10___
b. "one-half as large"   dB = ___−3___
c. "decreased by a factor of 12"   dB = __−11__
d. "twice as large as"   dB = ___3___
e. "3 times larger than"   dB = ___5___
f. "smaller by 700 times"   dB = __−28.5__
g. "150 times greater than"   dB = __21.8__
h. "a 60 to 1 reduction"   dB = __−17.8__
i. "8 orders of magnitude larger"   dB = ___80___
j. "3 orders of magnitude smaller"   dB = ___30___

Note: One order of magnitude means a factor of 10, two orders of magnitude means a factor of 100, and so on.

3.6. Express the following numbers in dB notation without using a caclulator.

20	13 dB
5	7 dB
500	27 dB
40	16 dB
3000	35 dB
0.25	−6 dB
0.20	−7 dB
0.05	−13 dB
6	8 dB
30	15 dB

3.7. Express the following dB values as numbers without using a calculator.

3	2
23	200
−3	0.5
35	3000
46	40,000
−13	0.05
−20	0.01
17	50

3.8. Use a calculator for these conversions.

Ratio	dB
14	11.46
560	27.48
7.2	8.57
984	29.92
0.013	−18.86
10^{-5}	−50
10^4	40

Ratio	dB
6607	38.2
29.51	14.7
1.26×10^5	51
0.0166	-17.8
3.98×10^{-4}	-34
0.00316	-25

3.9. Use a calculator for these conversions.

Power	mW	dBm
1.5 kW	1.56×10^6	61.76
250 W	2.5×10^5	53.98
18 W	1.8×10^4	42.55
4 kW	4×10^6	66.02
600 mW	6×10^2	27.78
20 μW	20×10^{-3}	-16.99
32 nW	32×10^{-6}	-44.95
15 pW	15×10^{-9}	-78.24
482 pW	482×10^{-9}	-63.17
5 mW	5	6.99

3.10. Use a calculator for these conversions.

	mW	dBm
100 mW	100	20
39.8 μW	0.0398	-14
2.5 W	2511	34
398 W	3.98×10^5	56
17.7 kW	1.77×10^7	72.5
63 mW	63	18
316 mW	316	25
50 μW	0.05	-13
3.2 nW	3.2×10^{-6}	-55
32 pW	3.2×10^{-8}	-75
125 mW	125	21

3.11. Determine from the context of each statement whether the answer should be dB or dBm. Remember, dB indicates a change of power (or a ratio), and dBm indicates an absolute power level.

   a. The lab detector is rated for a maximum power of 25 <u>dBm.</u>
   b. An attenuator has an insertion loss of 10.4 <u>dB.</u>
   c. The coupled power level of a 20-dB directional coupler is nominally 20 <u>dB</u> less than the input power.
   d. When tested at input power of 33 dBm, the return loss of the *pin* attentuator was 16.5 <u>dB.</u>
   e. The reflected power level from the circulator was −18.5 <u>dB.</u>
   f. The output power of a TM-812 amplifier is always 37 <u>dB</u> higher than the input power.

CHAPTER 4

4.1. Calculate the insertion loss.

Incident Power	Insertion Loss	Transmitted Power
6 mW	5 dB	2 mW
100 W (50 dBm)	17 dB	2 W (33 dBm)
7 dBm	20 dB	13 dBm
10 mW (10dBm)	40 dB	−30 dBm

4.2. Calculate the transmitted power.

Incident Power	Insertion Loss	Transmitted Power
0.4 mW	3 dB	0.2 mW
15 mW	6 dB	4 mW
20 W	16 dB	0.5 W
(43 dBm)		(27 dBm)
27 dBm	20 dB	7 dBm

4.3. What is the difference between insertion loss and attenuation?
Insertion loss is the difference between incident (input) power and transmitted power. It can be measured, and includes losses due to reflection and absorption inside component. Attenuation is power lost inside com-

ponent. Insertion loss is equal to attenuation only if component is perfectly matched.

4.4. Complete the table for three components in cascade.

$$\underrightarrow{P_{\text{IN}}} \boxed{\begin{array}{c}\text{Insertion}\\\text{loss 1}\end{array}} \xrightarrow{P_{\text{A}}} \boxed{\begin{array}{c}\text{Insertion}\\\text{loss 2}\end{array}} \xrightarrow{P_{\text{B}}} \boxed{\begin{array}{c}\text{Insertion}\\\text{loss 3}\end{array}} \xrightarrow{P_{\text{OUT}}}$$

P_{IN} (dBm)	Insertion Loss 1 (dB)	P_{A} (dBm)	Insertion Loss 2 (db)	P_{B} (dBm)	Insertion Loss 3 (dB)	Total Insertion Loss of Chain (dB)	P_{OUT} (dBm)
0	3	−3	10	−13	7	20	−20
25	15	10	5	5	8	28	−3
−17	14	−31	25	−56	3	42	−59
3	10	−7	8	−15	5	23	−20
−35	3	−38	3	−41	8	14	−49

4.5. Calculate the gain.

Input Power	Gain	Output Power
20 mW (13 dBm)	10 dB	200 mW (23 dBm)
−13 dBm	13 dB	1 mW (0 dBm)
500 W (57 dBm)	6 dB	2000 W (63 dBm)
−33 dBm)	43 dB	10 dBm

4.6. Calculate the output power.

Input Power	Gain	Output Power
3 μW (−25 dBm)	10 dB	−15 dBm
5 kW (67 dBm)	6 dB	73 dBm
10 dBm	3 dB	13 dBm
1 nW (−60 dBm)	20 dB	−40 dBm

4.7. Complete the table.

Input Power	Insertion Loss of Filter (dB)	Gain of Ampli- fier (dB)	Insertion Loss of Cable (dB)	Total Insertion Loss (dB)	Output Power (dBm)
1 mW (O dBm)	3	10	1	6	6
20 W (43 dBm)	6	10	1	3	46
1 μW (−30 dBm)	10	12	1	1	−29
23 dBm	10	20	1	9	32

4.8. Convert the following percent reflected powers to return loss.

Reflected Power (%)	Return Loss (dB)
1	20
5	13
10	10
20	7
50	3
100	0

4.9. Convert the following return loss values to percent reflected power.

Reflected Power (%)	Return Loss (dB)
0.1	30
1	20
2	17
5	13
10	10

4.10. Convert the following return loss values into percent reflected power, reflection coefficient, and SWR.

Return Loss (dB)	Reflected Power (%)	Reflection Coefficient	SWR
0	100	1	∞
10	10	0.32	1.92
13	5	0.22	1.56
17	2	0.14	1.33
20	1	0.10	1.22
30	0.1	0.03	1.06

4.11. Rank the following mismatches from least reflected power (1) to most reflected power (6).

Return loss	= 0 dB	6
Return loss	= 20 dB	2
Reflected power	= 10%	5
Reflected power	= 5%	4
SWR	= 1.4	3
SWR	= 1.1	1

CHAPTER 5

5.1. Plot the following mismatch points on a Smith chart.
See Figure A1.

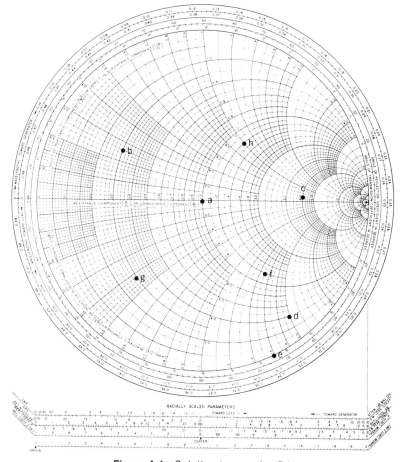

Figure A.1 Solution to exercise 5.1

5.2. The amplitude and phase of the reflection coefficient of a mismatch are 0.57 and $-144°$ respectively.
Match out the mismatch using a series capacitor. (See Figure A2.)

a. How many wavelengths toward the generator from the mismatch must the capacitor be placed?
0.224

b. How much capacitive reactance must be added?
70 Ω

c. If the frequency is 3 GHz, what is the value of the capacitor?
0.8 pF

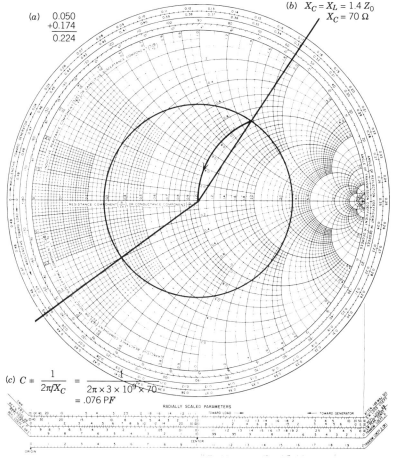

Figure A.2   Solution to exercise 5.2

5.3. If the matching capacitor of Problem 2 is located at the correct position, but has a capacitive reactance of 1.0 (instead of the correct value) what is the resulting return loss?
14 dB
(See Figure A3.)

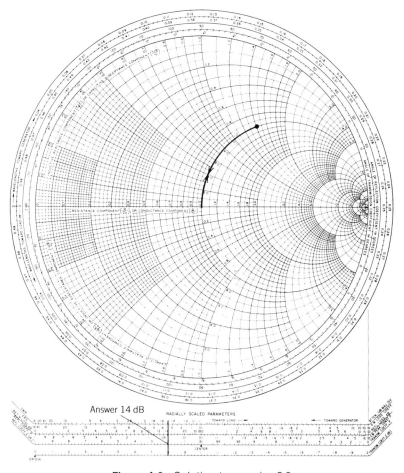

Answer 14 dB

Figure A.3   Solution to exercise 5.3

Match the following mismatch in Exercises 5.4 through 5.6

$$p = 0.34/120°$$

5.4. Match out the mismatch using a capacitive stub. (See Figure A4.)

    a. How many wavelengths toward the generator from the mismatch must the stub be placed?

    0.015

    b. What must the normalized susceptance of the stub be?

    $b_c = 0.72$

    c. How long (in wavelengths) must an open stub be to provide the required capacitive susceptance?

    0.10

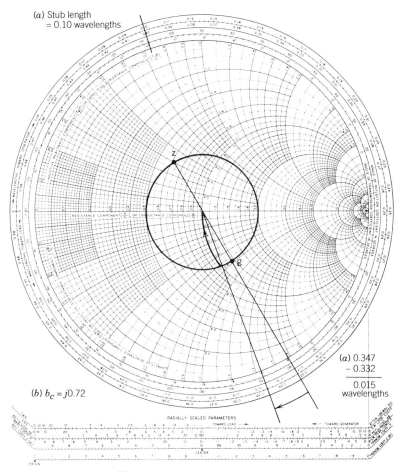

(a) Stub length = 0.10 wavelengths

(b) $b_c = j0.72$

(a) 0.347
− 0.332
_____
0.015
wavelengths

Figure A.4 Solution to exercise 5.4

5.5. Match out the mismatch using a quarter-wave transformer. (See Figure A5.)

   a. How many wavelengths toward the generator from the mismatch must the starting edge of the transformer be placed?

   0.42

   b. What must the normalized impedance of the transformer be?

   $z_T = .707$

   c. Sketch the matching transformer.

   (See sketch on Figure A5)

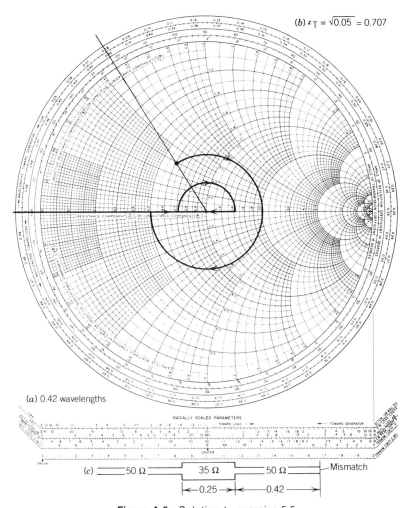

$(b)\; z_T = \sqrt{0.05} = 0.707$

(a) 0.42 wavelengths

Figure A.5   Solution to exercise 5.5

5.6. Match out the mismatch using a lumped series capacitor and a lumped parallel inductor. (See Figure A6.)

   a. What is the normalized reactance of the capacitor?

   $x_C = 0.9$

   b. What is the normalized susceptance of the inductor?

   $b_L = 0.8$

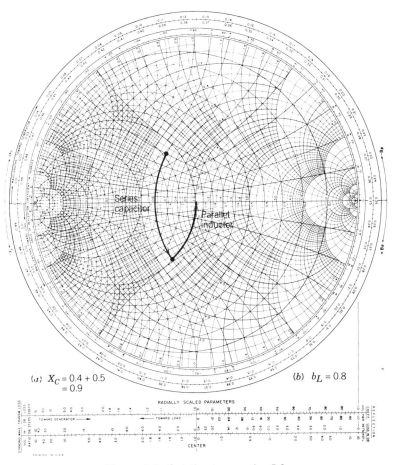

(a) $X_C = 0.4 + 0.5$
   $= 0.9$

(b) $b_L = 0.8$

RADIALLY SCALED PARAMETERS

Figure A.6 Solution to exercise 5.6

CHAPTER 6

6.1. Complete the following table comparing transmission lines at 4 GHz. (Numbers refer to Figure 6.5.)

Transmission Line	CW Power Handling	Attenuation (dB/100 ft)
1. 0.141 in. coax	50 W	25
3. $\frac{7}{8}$ in. heliax	700 W	3
6. WR 187 Waveguide	3 MW	1.5
13. Microstrip	50 W	300

6.2. Complete the following table comparing transmission lines at 10 GHz. (Numbers refer to Figure 6.5.)

Transmission Line	CW Power Handling	Attenuation (dB/100 ft)
1. 0.141 in. coax	50 W	45
7. WR 90 waveguide	730 kW	5
13. Ridged waveguide	100 kW	10
15. Microstrip	50 W	500

6.3. Complete the following table comparing transmission lines at 30 GHz. (Numbers refer to Figure 6.5.)

Transmission Line	CW Power Handling	Attenuation (dB/100 ft)
1. 0.141 in. coax	50 W	80
9. WR 280 waveguide	95 kW	20

6.4. Arrange the following transmission lines in order of decreasing wavelength at 10 GHz.

1. Rectangular waveguide
2. Air-filled coaxial line
3. Teflon-filled coaxial line
4. Microstrip on alumina

(Number 1 is the largest wavelength and Number 4 the smallest.)

6.5. Complete the following table.

Cable Type	RG 233	RG 214	Semirigid 0.141 in. dia	Semirigid 0.085 in. dia
Cable OD (in.)	0.216	0.425	0.141	0.085
Flexible or Semirigid	Flex	Flex	Semirigid	Semirigid
Impedance (ohms)	50	50	50	50
Attenuation at 10 GHz (dB/100 ft)	85	47	44	73
Power-handling at 10 GHz (W)	10	37	160	48

(From Table 6.5.)

6.6. What waveguide would you use for microwave equipment operating at each of the following frequencies?

Frequency (GHz)	EIA Waveguide Designation
16	WR 62
1	WR 770
3	WR 284
10	WR 90
30	WR 28

(From Table 6.6.)

6.7. Compare the following table.

<table>
<tr><th colspan="6">EIA Waveguide Designation</th></tr>
<tr><th></th><th>WR 1500</th><th>WR 284</th><th>WR 137</th><th>WR 28</th><th>WR 10</th></tr>
<tr><td>Frequency range (GHz)</td><td>0.49–0.74</td><td>2.65–3.95</td><td>5.85–8.20</td><td>26.5–40</td><td>75–110</td></tr>
<tr><td>Outer dimensions (in. × in.)</td><td>15.25 × 7.75</td><td>3 × 1.5</td><td>1.5 × 0.75</td><td>0.36 × 0.22</td><td>0.18 × 0.13</td></tr>
<tr><td>Attenuation at highest frequency (dB/100 ft)</td><td>0.05</td><td>0.75</td><td>2.3</td><td>15.0</td><td>70</td></tr>
<tr><td>Power rating (MW)</td><td>93</td><td>3.2</td><td>0.71</td><td>0.03</td><td>0.004</td></tr>
</table>

(From Table 6.6.)

6.8. What is the difference between stripline and microstrip?
Stripline has dielectric on both sides of conductor strip. Microstrip has dielectric on only one side.

6.9. Compare the dielectric constants of the following stripline materials (from Table 6.3).

Material	Dielectric Constant
Teflon fiberblass	2.55
Ceramic-filled Teflon fiberglass (epsilam 10)	10.2
Alumina	9.6
Beryllia	6.8

6.10. The free-space wavelength at 5 GHz is 6.0 cm. What is the wavelength in stripline (with dielectric on both sides of the conductor) with the following materials?

$$\lambda = \frac{\lambda}{\sqrt{\varepsilon}}$$

Material	Wavelength at 5 GHz
Teflon fiberglass	$6.0/\sqrt{2.55} = 3.76$ cm
Alumina	$6.0/\sqrt{9.6} = 1.94$ cm

6.11. What is the width of a 50-Ω stripline using 0.030-in.-thick teflon fiber-glass?
From Figure 6.8,
$$\sqrt{\varepsilon}\, Z_0 = \sqrt{2.55} \times 50\ \Omega = 80\ \Omega$$

$$\frac{w}{b} = 0.75 \qquad b = 2 \times 0.030 \quad \text{So} \quad w = 0.045\ \text{in.}$$

6.12. What is the width of a 50-Ω microstrip line using 0.030-in.-thick Teflon fiberglass?
From Figure 6.9,
3×0.030 in. $= 0.090$ in.

6.13. What is the guide wavelength for the microstrip line of Exercise 6.12 at 5 GHz?
From Figure 6.9,
$$\lambda = \frac{\lambda_0}{1.5} = \frac{6.0\ \text{cm}}{1.5} = 4.0\ \text{cm}$$

6.14. What is the width of a 50-Ω microstrip line using 0.025-in.-thick alumina?
From Figure 6.9,
0.025 in.

6.15. What is the guide wavelength for the microstrip line of Exercise 6.14 at 5 GHz?
From Figure 6.9,
$$\lambda = \frac{\lambda_0}{2.5} = \frac{6.0\ \text{cm}}{2.5} = 2.4\ \text{cm}$$

CHAPTER 7

7.1. Describe what each of the signal control components does.
  1. Termination (load)
   A perfect termination for a transmission line.
  2. Directional coupler
   A component that samples the microwave signal traveling in one direction down a transmission line.
  3. Combiner
   A component that combines microwave signals from separate transmission lines into one common transmission line and allows no coupling between the separate lines.
  4. Isolator
   A component containing ferrite material that allows microwave signals to pass in one direction through the component, but absorbs microwave signals passing in the other direction.

5. Circulator

A three-port component containing ferrite material that allows microwave signals to pass from one port to the second, but routes signals coming into the second port into a third, and routes signals coming into the third port back into the first port.

6. Detector

A component that converts microwave power to a dc voltage. The detector's dc output is proportional to the microwave power.

7. Filter

A component that allows a certain range of frequencies to be transmitted and reflects or absorbs all other frequencies.

8. Multiplexer

A component that removes a desired band of frequencies from the signal in a transmission line and returns the other frequencies back into the transmission line.

9. Fixed attenuator

A component that reduces the level of a microwave signal passing through it by a constant amount.

10. Variable attenuator

A component that reduces the level of the microwave signal passing through it by an adjustable and controlled amount.

11. Switch

A component that turns microwave power on and off or switches the power from one transmission line to another.

12. Phase shifter

A component that controls the phase of a microwave signal.

7.2. Complete the following table of directional coupler characteristics.

Input Power (dBm)	Coupled Power (dBm)	Output Power (dBm)	Coupling (dB)	Insertion Loss (dB)
0	−20	−1	20	1
30	10	29	20	1
50	10	48	40	2
13	−7	12	20	1
10	7	6	3	4
0	−10	−1	10	1

7.3. A power of 23 dBm is incident into a 10-dB directional coupler in the correct direction. What is the power at the coupled port?
13 dBm

7.4. An input signal is divided in a hybrid into two equal outputs. What will be the phase difference between the output signals if the following hybrids are used?

3-dB quadrature hybrid	$\dfrac{90°}{}$
3-dB 180° hybrid	$\dfrac{180°}{}$
Wilkinson combiner	$\dfrac{0°}{}$

7.5. A signal with an amplitude P_1 is applied to one of the input arms of a 3-dB quadrature hybrid. A signal P_2, which is 90° out of phase with P_1, is applied to the other input arm. What is the signal in each of the output arms of the 3-dB quadrature hybrid?

$P_1 + P_2, \qquad P_1 - P_2$

7.6. Complete the following table of isolator characteristics.

Input Power (dBm)	Transmitted Power in Forward Direction (dBm)	Transmitted Power in Reverse Direction (dBm)	Forward Insertion Loss) (dB)	Isolation (dB)
10	9	−13	1	23
25	24	5	1	20
0	−1	−20	1	20
10	9	−15	1	25
−10	−11	−30	1	20

7.7. Complete the following table of circulator characteristics.

Input Power in Arm 1 (dBm)	Output Power in Arm 2 (dBm)	Output Power in Arm 3 (dBm)	Insertion Loss) (dB)	Directivity (dB)
0	−1	−20	1	20
10	9	−13	1	23
−5	−6	−25	1	20
15	14	−10	1	25
14	13	−6	1	20

7.8. An isolator has an insertion loss of 1 dB and an isolation of 17 dB. If a 23-dBm signal is put through the isolator in the wrong direction, how much power is transmitted out of the isolator?
6 dBm

7.9. The isolator in Exercise 7.8 is placed in front of a short. Assuming the isolator has a perfect match, what is the return loss of the isolator and short combination?
18 dB

7.10. A circulator has an insertion loss of 1 dB and a directivity of 23 dB. If a 5-dBm signal is put in arm 1 of the circulator, what is the output power from arm 2 (the correct direction of transmission) and arm 3?
arm 2   4 dBm   arm 3 -18 dBm

7.11. A circulator has an insertion loss of 1 dB and a directivity of 21 dB. If a 0-dBm signal is put into the input port and port 2 is shorted, how much power leaks back into the input port?
-22 dBm

7.12. Two circulators, which have directivities of 19 dB and 23 dB, respectively, are combined to form an isolator. What is the isolation of the isolator?
42 dB

7.13. The following list refers to the filters in Figure 7.32.

1. What type of passband does filter 1 have?	bandpass
2. What type of passband does filter 2 have?	low-pass
3. What is the insertion loss of filter 1 at 4 GHz?	40 dB
4. What is the insertion loss of filter 1 at 6 GHz?	0 dB
5. What is the insertion loss of filter 1 at 10 GHz?	40 dB
6. What is the insertion loss of filter 2 at 4 GHz?	0 dB
7. What is the insertion loss of filter 2 at 6 GHz?	0 dB
8. What is the insertion loss of filter 2 at 10 GHz?	45 dB

7.14. What type of filter can be electronically tuned?
YIG

7.15. Three fixed attenuators with an insertion loss of 3, 6, and 10 dB are connected. What is the total insertion loss of the combination?
19 dB

7.16. What is the power level of the smallest signal that can be detected above the noise by a Schottky diode?
-60 dBm

7.17. What is the power level of the largest signal that will still be in the square-law range of a Schottky diode?
-20 dBm

CHAPTER 8

8.1. Figure 8.24 shows the microwave output power as a function of micro-wave input power for a bipolar transistor amplifier at 2 GHz. The dc input power is 100 mA at 10 V. Determine the following:

a. The output power at an input power of -30 dBm	0 dB
b. Gain at an input power of -30 dB	30 dB
c. Efficiency at an input power of -30 dBm	0.1%
d. Output power at saturation	20 dBm
e. Gain at saturation	20 dB
f. Efficiency at saturation	10%
g. Output power at 1-dB compression	14 dBm
h. Gain at 1-dB compression	29 dB
i. Efficiency at 1-dB compression	2.5%

8.2. Figure 8.25 shows the swept output power of a FET amplifier at several input power levels. Using this data, plot transfer curves for this amplifier at 10 GHz and 12 GHz.
See Figure A7.

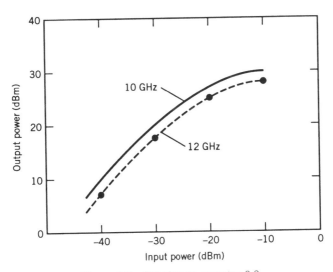

Figure A.7 Solution to exercise 8.2

8.3. Figure 8.26 shows the phase of the output power from a bipolar transistor amplifier (relative to the input power) as a function of input power level.

a. What is the phase change when the input power level is changed from −10 to 0 dBm?

20°

b. What is the AM-PM coefficient at −5 dBm?

2°/dB

8.4. Figure 8.27 shows the output of a bipolar transistor amplifier with an input signal at 3 GHz, as measured on a spectrum analyzer. What are the power levels and the frequencies of the following components in the output.

Output Component	Frequency (GHz)	Power (dBm)
Fundamental	3	0
Second harmonic	6	−15
Third harmonic	9	−30
Spurious	7	−40

8.5. Figure 8.28 shows the fundamental output power and the second harmonic output power of a FET amplifier as a function of the fundamental input power at 8 GHz. The dc input power is 100 mA at 10 V. Determine the following:

a. The harmonic frequency	16 GHz
b. Fundamental power at saturation	20 dBm
c. Second harmonic power at fundamental saturation	13 dBm
d. Separation between fundamental and second harmonics at fundamental saturation	7 dB
e. Second harmonic power at second harmonic saturation	16 dBm
f. Fundamental power at 1-dB compression	16dBm
g. Second harmonic at 1-dB compression of the fundamental	−10 dBm
h. Fundamental to second harmonic separation at 1-dB compression	26 dB
i. Fundamental efficiency at 1-dB compression	4%
j. Linear gain of the fundamental	45 dB

8.6. Figure 8.29 shows the intermodulation products of a FET amplifier operated with input signals at 3.0 and 3.2 GHz as a function of the total power of the two input signals:

a. What is the fundamental power at saturation?

28 dBm

b. What is the separation between the fundamental and the third-order intermod at fundamental saturation?
11 dB

c. What is the fundamental power at 1-dB compression?
24 dBm

d. What is the separation between the fundamental and the third-order intermod at 1-dB compression of the fundamental?
27 dB

8.7. For the intermodulation data of Figure 8.29:

a. What are the frequencies of the two third-order intermodulation products?
2.8 GHz and 3.4 GHz

b. What are the frequencies of the two fifth-order intermodulation products?
2.6 GHz and 3.6 GHz

8.8. What would be the best microwave semiconductor device for each of the following amplifiers?

a. A 10-W amplifier at 1 GHz	Bipolar transistor
b. A $\frac{1}{2}$ W amplifier at 20 GHz	Field-effect transistor
c. A $\frac{1}{2}$ W amplifier at 100 GHz	IMPATT

8.9. What S-parameter approximately represents the following:

a. Input reflection coefficient	S_{11}
b. Transistor gain	S_{21}
c. Isolation	S_{12}
d. Output reflection coefficient	S_{22}

8.10. S-parameters for a bipolar transistor amplifier, taken from a manufacturers data sheet, were given. Plot S_{11} and S_{22} on a Smith chart at the following frequencies.

a. 500 MHz   **b.** 1 GHz   **c.** 2 GHz   **d.** 4 GHz
See Figure A8.

8.11. Calculate the following gains at 1 GHz using S-parameters.

a. Unmatched transistor gain	16.9 dB
b. Unilateral gain with matched transistor	20.0 dB
c. Maximum stable gain	22.5 dB

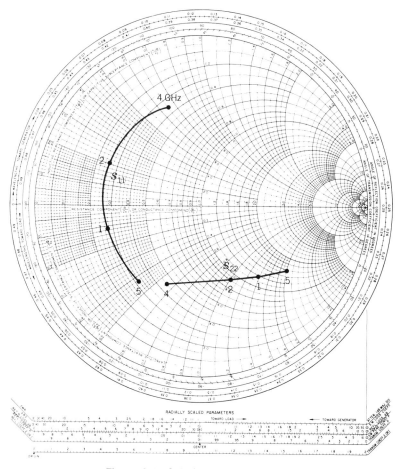

Figure A.8   Solution to exercise 8.10

8.12. Calculate the following gains at 4 GHz using *S*-parameters.

a. Unmatched transistor gain   5.4 dB
b. Unilateral gain with matched transistor   8.6 dB
c. Maximum stable gain   12.9 dB

CHAPTER 9

9.1. What is the purpose of an oscillator?
The purpose of an oscillator is to generate a microwave signal.

9.2. Name the two parts of an oscillator and the purpose of each part.
A resonator to control the frequency.
An active device to generate the microwave power.

9.3. What are six types of resonators used for microwave oscillators?
Lumped L/C; rectangular metal cavity; coaxial metal cavity; microstrip line; dielectric resonator; surface acoustic wave (SAW) resonator

9.4. What are the four frequency stability specifications of microwave oscillators?
Frequency pushing; frequency pulling; frequency change with temperature; frequency change with shock and vibration

9.5. What is frequency pushing?
Frequency pushing is the variation in oscillator frequency caused by power supply voltage or current changes.

9.6. What is frequency pulling?
Frequency pulling is the variation in oscillator frequency with changes in load SWR.

9.7. What are the four specifications for electronic tuning of a microwave oscillator?
Tuning sensitivity; linearity; settling time; posttuning drift

9.8. What is settling time?
Settling time is the time it takes an electronically tuned oscillator to come within a specified percentage of its final frequency after the electronic control voltage has been changed.

9.9. What is posttuning drift?
Posttuning drift is the frequency change of an electronically tuned oscillator at a specified time after it has reached its desired frequency.

9.10. From Figure 9.17 what is the phase noise of each of the following oscillators at 10 kHz from the carrier?

Oscillator	Phase Noise (dBc in 1-Hz band)
FET in cavity	110
Bipolar with DRO	−100
Bipolar in cavity	−120
VCO phased-locked to harmonic of 100-MHz crystal oscillator	−135

9.11. From Figure 9.17, what is the phase noise of a bipolar transistor in a metal cavity in a 100-Hz band at 10 kHz from the carrier?
−100 dBc

9.12. What is the advantage of a varactor-tuned oscillator compared with a YIG-tuned oscillator?
Fast tuning in microseconds

9.13. What are the two advantages of a YIG-tuned oscillator compared with a varactor-tuned oscillator?
Wide tuning range; linear tuning with frequency control signal

9.14. Name two uses of harmonic multipliers?

  1. To generate a high frequency microwave signal from a lower frequency microwave signal.
  2. To generate a stable microwave reference frequency from a quartz crystal-controlled oscillator.

9.15. Drawing a block diagram of a phased-locked loop.
See Figure 9.19.

CHAPTER 10

10.1. Complete the following table. (Use Figure 10.7)

Noise Power Density (dBm/MHz)	Noise Temperature (K)	Noise Figure (dB)
−130	7.25	0.11
−106.9	1500	7.9
−112.2	438	4
−118.6	100	1.29
−117	145	1.76

10.2. Arrange the following low-noise receivers in order of decreasing noise: paramp, FET, bipolar, transistor, HCMT, mixer.
Mixer
Biopolar transistor
FET
HEMT
Paramp

10.3. Complete the following table for a low-noise amplifier at room temperature.

B (MHz)	S_{in} (dBm)	N_{in} (dBm)	$(S/N)_{in}$ (dB)	G (dB)	NF (dB)	S_{out} (dBm)	N_{out} (dBm)	$(S/N)_{out}$ (dB)
3	−80	−109	29	30	4	−50	−75	25
20	−84	−101	17	10	2	−74	−89	15
10	−90	−104	14	25	3	−65	−76	11
1	−90	−114	24	60	4	−30	−50	20
4	−80	−108	28	40	3	−40	−65	25

10.4. a. A source at room temperature (T = 290°K) delivers a signal of −90 dBm with a bandwidth of 2 MHz. What is the SNR of the system at this point?
21 dB

b. An amplifier is inserted to boost the signal to −20 dBm. What is the gain of the amplifier?
70 dB

c. The amplifier has a noise figure of 6 dB. What is the SNR at the amplifier output?
15 dB

d. What is the noise power in dBm at the amplifier output?
−35 dBm

10.5. The SNR at a receiver input is 30 dB. If the receiver has a 5-dB noise figure, what is the SNR at the output?
25 dB

10.6. How much noise power (in dBm) is generated by a noise source at 75°K over a 20-MHz bandwidth?
−114 dBm −6 dB + 13 dB = −107 dBm

10.7. Name the three parts of a mixer.
Local oscillator; combiner; Schottky diodes

10.8. A mixer has a signal frequency of 10 GHz and an IF frequency of 10 MHz. What is the local oscillator frequency?
10.01 GHz or 9.99 GHz

10.9. With the mixer characteristics of the previous problem, what is the image frequency?
10.02 GHz or 9.98 GHz

10.10. What mixer type has the lowest spurious signals?
Double-balanced

10.11. Name the four parts of a parametric amplifier.
Circulator; combiner; varactor diodes; pump oscillator

CHAPTER 11

11.1. What is the difference between a *hybrid* and a *monolithic* microwave integrated circuit?
In a hybrid microwave integrated circuit the microstrip circuitry and other parts, such as diodes and transistors, are fabricated separately and then assembled together.

In a monolithic microwave integrated circuit the microstrip, the diodes and transistors, and all interconnections are fabricated together.

11.2. What are five advantages of *hybrid* MICs compared to MMICs?

1. All components can be made with the best materials, since different materials can be used for each component.
2. High power capability can be obtained.
3. Trimming adustments are possible.
4. Standard diodes and transistors can be used.
5. They are economical for small quantities.

11.3. What are four advantages of MMICs compared to hybrid MICs?

1. Minimum mismatches and minimum signal delays.
2. Many devices can be fabricated into one MMIC.
3. No wire bond reliability problems.
4. Economical for large quantities.

11.4. What is the major advantage of each of the following substrate materials?
Teflon-fiberglass is easily machined.
Alumina's high temperature capability allows devices to be attached by soldering.
Ferrite circulators can be fabricated.
Active devices such as diodes and transistors can be fabricated in Gallium Arsenide.

11.5. What seven components can be fabricated as part of the microstrip line?
Matching stubs and transformers; directional couplers; combiners; microstrip resonators; filters; inductors and capacitors; thin film resistors

11.6. What six components are fabricated separately and added to the microstrip circuitry to form a hybrid MIC?
Bond wires; chip resistors; chip capacitors; dielectric resonators; circulators; diodes and transistors

11.7. What is the advantage of a chip capacitor compared with a microstrip capacitor?
A chip capacitor provides much higher capacitance values.

11.8. What is the advantage of a dielectric resonator compared with a microstrip resonator?
A dielectric resonator provides a much higher Q.

11.9. What is a bond wire?
A bond wire is a very small (0.0007 in. or 20 μ) wire used for connecting components in a hybrid microwave integrated circuit.

11.10. How are tuning and tweaking accomplished in a hybrid MIC?
By making and breaking bond wire connections.

11.11. In what two ways are components mounted in a hybrid MIC?
Thermosetting epoxy adhesive; eutectic soldering

11.12. In what two ways are components connected in a hybrid MIC?
Thermocompression wire bonding; conductive epoxy adhesive

11.13. Why is GaAs used in MMICs instead of alumina? Silicon?
For alumina, diodes and transistors cannot be fabricated in alumina. For silicon, it has too high a dielectric loss and cannot be used for microstrips.

11.14. What is the purpose of each of the following components in an MMIC?
Air bridge connects components that have another component between them.
Overlay capacitor provides greater capacitance than can be obtained with microstrip capacitors.
Via hole connects from the top of the substrate to the ground plane on the bottom.

CHAPTER 12

12.1. What advantage do microwave tubes have over microwave semiconductor devices?
Tubes can provide much greater power (over 10,000 times more
power) than microwave semiconductor devices.

12.2. What is the major disadvantage of microwave tubes?
Their limited lifetime.

12.3. What is the advantage of using a low power semiconductor oscillator
driving a high power tube amplifier, instead of a high power tube
oscillator?
Frequency is much easier to control in the low power semiconductor
oscillator.

12.4. Name the five types of microwave amplifier tubes.
Gridded tube; Klystron; Helix traveling wave tube; coupled-cavity
traveling wave tube; crossed-field amplifier

12.5. What average power can be obtained from the following microwave
tubes at 1 GHz? (*Hint:* Use Figure 12.2.)

Gridded tube	20 kW
Klystron	2 MW
Helix TWT	2 kW
Coupled cavity TWT	200 kW
CFA	100 kW

12.6. What average power can be obtained from the following microwave
tubes at 10 GHz?

Klystron	200 kW
Helix TWT	300 W
Coupled-cavity TWT	50 kW
CFA	3 kW

12.7. What is the major advantage of a gridded tube?
Low cost

12.8. What is the major advantage of a klystron?
High power

12.9. What is the major advantage of a helix TWT?
Wide bandwidth

12.10. What is the major advantage of a coupled-cavity TWT?
Good bandwidth and high power achieved together

12.11. What is the major advantage of a crossed-field amplifier?
High efficiency

12.12. What is the major disadvantage of a gridded tube?
Gridded tubes work only at low frequency end of the microwave band.

12.13. What is the major disadvantage of a klystron?
Narrow bandwidth

12.14. What is the major disadvantage of TWTs?
Low efficiency

12.15. What are two disadvantages of a CFA?
Low gain and high noise

12.16. What limits the frequency capability of gridded tubes?
Transit time of the electrons

12.17. How do klystrons, TWTs, and CFAs avoid the transit time problem?
They use velocity modulation.

12.18. What is the function of each of the following klystron parts?
Electron Gun
  Forms the electron beam.
Focusing magnet
  Focuses the electron beam through the kylstron cavities.
Beam collector
  Collects the electron beam after microwave power has been extracted from it.
Input cavity
  Modulates the microwave signal onto the electron beam.
Intermediate cavity
  Enhances the velocity modulation.
Output cavity
  Extracts microwave power from the bunched electron beam.

12.19. Name the four types of focusing magnets used with a klystron or a TWT.
Electromagnet yoke; solenoid; permanent magnet; periodic permanent magnet.

12.20. What is the function of each of the following TWT parts?
Electron gun
  Forms the electron beam.
Focusing magnet
  Focuses the electron beam through the interaction structure.

Beam collector
Collects the electron beam after microwave power has been extracted from it.
Microwave interaction structure
Provides interaction between the microwave signal and the electron beam in order to generate microwave power.

12.21. What is the purpose of the windows in microwave tubes?
To allow the microwave power to get into and out of the tube, but to prevent air from getting in to destroy the vacuum.

12.22. What type of magnetic focusing is most commonly used in a TWT?
Periodic permanent magnet (PPM) focusing

12.23. What is the advantage of a helix interaction structure over a coupled-cavity interaction structure?
Wide bandwidth

12.24. What is the advantage of a coupled-cavity interaction structure over a helix interaction structure?
High power

12.25. What is the purpose of a depressed collector in a TWT?
To increase efficiency

12.26. What is the function of each of the following CFA parts?
Electron gun
Forms and accelerates the electron beam.
Interaction structure
Provides interaction between the microwave signal and the electron beam.
Sole electrode
Provides the crossed electric field for the interaction process.
Magnet
Provides the crossed magnetic field for the interaction process.

12.27. What is the difference in function of the magnet field in a TWT and in a CFA?
The magnetic field in a TWT is used only to focus the electrons through the tube.
The magnetic field in a CFA is part of the crossed-field interaction process that allows high efficiency to be obtained.

12.28. What is an emitting-sole CFA?
A separate electron gun is not used in an emitting-sole CFA, but the sole electrode, which provides the crossed electric field, also provides the electron beam.

12.29. What is a magnetron?
A magnetron is a high power oscillator tube.

12.30. Name three advantages of the magnetron?
High efficiency; high peak power; low cost

CHAPTER 13

13.1. Define the following antenna requirements:
1. Gain
The amount by which an antenna concentrates its radiation in a given direction relative to what would have been obtained if the antenna had not been used.
2. Receiving area
The effective area of the antenna for receiving microwave power.
3. Beamwidth
The angular width of the antenna beam between the points on either side of the axis where the transmitted power has been reduced by one half.
4. Polarization
The direction of the electric field in the electromagnetic wave radiated from an antenna. May be horizontal, vertical, or circular.
5. Bandwidth
The frequency range over which an antenna has the required gain, area, or other characteristics.
6. Sidelobes
Radiation from an antenna at other angles than the desired direction.
7. Antenna pattern
A measure of the microwave power radiated from an antenna as a function of angular direction from the antenna axis.

13.2. What is an isotropic antenna?
A hypothetical antenna radiating or receiving equally well in all directions.

13.3. What is the gain of an isotropic antenna?
0 dB

13.4. Complete the table.

Frequency (GHz)	Antenna Area (m²)	Gain (dB)
1	1	21.4
10	1	41.4
1	0.71	20
10	0.070	30

13.5. Name the five types of antenna elements.
Dipole; slot; horn; spiral; helix

13.6. How long is a half-wave dipole at the following frequencies?

Frequency (GHz)	Length of Half-Wave Dipole (m)
2	0.075
5	0.030
12	0.012

13.7. How long is a half-wave slot at the following frequencies?

Frequency (GHz)	Length of Half-Wave Slot (m)
6	0.0250
10	0.0150
20	0.0075
100	0.0015

13.8. An antenna is formed of an array of 20 dipoles. Each dipole has a gain of 1 dB. What is the total gain of the array?
14 dB

13.9. An antenna is formed of four helix antennas, each of which has a gain of 14 dB. What is the total gain of this antenna array?
20 dB

13.10. Complete the following table.

Antenna Diameter (m)	Frequency (GHz)	Gain (dB)	Effective Area (m^2)	Beamwidth (deg)
1	4	30.3	0.47	4.5
10	4	50.3	47	0.45
30	4	59.8	424	0.15
1	10	38.2	0.47	1.8
3	10	47.8	47	0.6
10	10	58.2	424	0.18

Antenna Diameter (m)	Frequency (GHz)	Gain (dB)	Effective Area (m²)	Beamwidth (deg)
.5	30	41.8	0.12	1.2
1	30	47.8	0.47	0.6
3	30	57.3	47	0.2
3	4	39.8	47	1.5

13.11. What is the gain at 6 GHz of a parabolic antenna with a diameter of 2 m?
39.8 dB

13.12. A parabolic antenna has a gain of 20 dB at 4 GHz. What is the gain of this antenna at 10 GHz?
28 dB

13.13. A parabolic antenna has a gain of 30 dB. How much must its diameter be increased to raise its gain to 40 dB?
3.16 times

13.14. What is the purpose of a phased array radar?
To move the antenna beam without moving the antenna

13.15. How far apart must the antenna elements of a phased array radar be placed?
$\frac{1}{2}$ wavelength

13.16. A square phased array radar is to have a gain of 30 dB.
  a. What must be the width dimensions in wavelengths?
  13
  b. If the antenna elements are spaced $\frac{1}{2}$ wavelength apart, how many elements are required?
  676

Define the following types of phased array antennas in Exercises 13.17 through 13.19.

13.17. Reflective array
The total transmitted power is radiated from a feed horn onto the array surface. The phase is shifted at each element of the array and reflected to form the beam.

13.18. Lens array
The total transmitted power is radiated from a feed horn onto the array surface. The phase is shifted at each element of the array and transmitted out the other side.

13.19. Active array

Each element of the antenna array contains an amplifier and a phase shifter. A small input signal is distributed to each array element, amplified and radiated. The total radiated power is the sum of the powers from all the elements.

CHAPTER 14

14.1. Define the communication systems terms in the following list:
1. Baseband

The electrical signal to be transmitted. This electrical signal may represent audio, video, or digital data.
2. Carrier

The electrical signal used to carry the baseband information from one location to another.
3. Modulation

Variation of the carrier amplitude, frequency, or phase by the baseband signal, so that the baseband signal can be transmitted via the carrier.
4. Demodulation

Removal of the baseband information from the carrier at the receiving end, so that the baseband signal can be used.
5. Frequency modulation (FM)

Modulation of the baseband onto the carrier by varying the frequency.
6. Amplitude modulation (AM)

Modulation of the baseband signal onto the carrier by varying the amplitude.
7. Sidebands

Frequencies present in the amplitude modulated or frequency modulated carrier due to the modulation process.
8. Bandwidth

The frequency range over which the modulation sidebands exist.
9. Single sideband (SSB)

An amplitude modulation process where the carrier and one set of modulation sidebands are removed to reduce bandwidth requirements.
10. Multiplexing

A method for combining several signals, all of which occupy the same frequency range, into a baseband so that they do not interfere with each other.
11. Frequency-division multiplexing (FDM)

A multiplexing technique where each signal to be multiplexed is shifted from its original frequency and then combined.

12. Pulse code modulation (PCM)
The baseband signal is sampled in time, and each sample is represented by a digital code.

13. Time-division multiplexing (TDM)
A multiplexing method where several signals are combined by forming PCM codes of each and then interleaving the PCM pulses.

14. Carrier-to-noise ratio (C/N)
The ratio of the received microwave carrier to the microwave noise in the receiver.

15. Signal-to-noise ratio (S/N)
The ratio of the received signal to noise after the signal has been demodulated to get out the baseband information.

14.2. The frequency range of a single telephone signal is from 300 Hz to 3400 Hz.

14.3. A single telephone signal is fitted into a frequency band from 0 Hz to 4000 Hz.

14.4. How many telephone signals are frequency-division-multiplexed together to form an FDM basic group?
12

14.5. What is the frequency range of an FDM basic group?
60–108kHz

14.6. The frequency range of a single TV signal is from 0 Hz to 4 MHz.

14.7. How many digital bits are required for a single letter in an asynchronous ASCII code with parity?
10

14.8. What signal-to-noise ratio is required for satisfactory telephone service?
50 dB

14.9. What signal-to-noise ratio is required for satisfactory television reception?
40 dB

14.10. Coding of an analog signal into a series of digital pulses is called pulse code modulation (PCM).

14.11. The combining of many telephone signals together by first coding each into a series of digital pulses and then interleaving the pulses series of each signal is called time-division multiplexing (TDM).

14.12. What must be the PCM sampling rate on a 4-kHz analog signal?
8 kHz

14.13. What is the quantization signal-to-noise ratio if an eight-digit PCM code is used?
50 dB

14.14. What is the digital bit rate for a standard PCM telephone channel?
64 kb/s

14.15. A special TDM system combines six telephone channels using a six-digit PCM code. What is the bit rate for this TDM?
288 kb/s

14.16. A high-speed TDM system combines four broadcast TV channels using a six-digit PCM code. What is the bit rate of this TDM?
192 Mb/s

14.17. What is the bit error rate (BER) if the received signal-to-noise ratio is 10 dB? (Use Figure 14.14.)
3×10^{-6}

14.18. In the TDM hierarchy:
   a. How many telephone channels are combined into the smallest TDM Group?
   24
   b. What is the bit rate for this TDM group?
   1.544 Mb/s
   c. What is this signal formating?
   T1

14.19. In the TDM hierarchy:
   a. How many telephone channels are transmitted in a T2 line?
   96
   b. What is the T2 bit rate?
   6.312 Mb/s

14.20. Name four advantages of PCM compared with analog.
Noise immunity; less expensive multiplexing; less expensive switching; compatibility with datacommunications

14.21. What is the disadvantage of PCM compared with analog?
Greater bandwidth required

14.22. The changing of digital signals from a computer into frequency tones for transmission down a conventional voice grade telephone circuit is done in a modem.

14.23. What is the advantage of cable communication systems compared with a broadcast system?
Higher capacity

14.24. Name two advantages of broadcast systems compared with cable systems?
Cable does not have to be laid; multipoint distribution

14.25. What is the advantage of FM compared with AM?
Signal-to-noise improvement

14.26. What is the advantage of AM compared with FM?
Less transmission bandwidth required

14.27. Why is the requirement for a wide bandwidth a disadvantage in communication system?
Less information can be transmitted

14.28. Draw the spectrum of a 4-GHz carrier amplitude modulated by a 4-MHz baseband signal.
See Figure A9.

14.29. Draw the spectrum of a 4-GHz carrier frequency modulated by a 4-MHz baseband signal if the carrier deviation is 4 MHz.
See Figure A9.

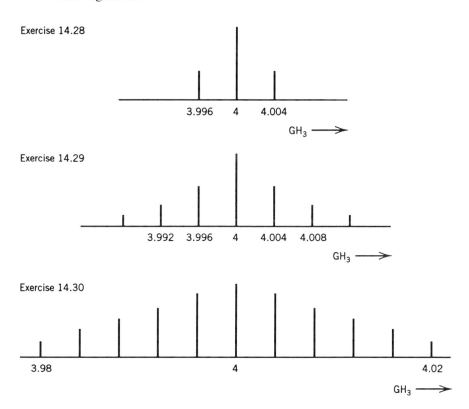

Figure A.9 Solution to exercises 14.28 through 14.30

14.30. Draw the spectrum of a 4-GHz carrier frequency modulated by a 4-MHz baseband signal if the carrier deviation is 20 MHz.
See Figure A9.

14.31. What bandwidth is required for the AM signal of Exercise 4.28?
8 MHz

14.32. What are the modulation index and bandwidth for the FM signal of Exercise 4.29?
Modulation index: 1
Bandwidth: 16 MHz

14.33. What are the modulation index and bandwidth for the FM signal of Exercise 4.30?
Modulation index: 5
Bandwidth: 48 MHz

14.34. What is the approximate S/N improvement factor for an FM signal with a modulation index of 5?
20 dB

14.35. Name four methods for modulating digital signals onto a microwave carrier.
Amplitude level; frequency shift keying; biphase; quadraphase

CHAPTER 15

15.1. Define the microwave relay terms in the following list:
   1. Baseband microwave radio
   A microwave relay system that starts with the baseband signal in the transmitter and modulates this signal onto the microwave carrier, receives the modulated microwave carrier, and, after amplification, demodulates the carrier to obtain the baseband signal at the receiver.
   2. RF multiplexing
   A technique for combining several transmitters, each operating in a different frequency channel, into a common transmission line so that all the transmitters can be connected to the same antenna. RF multiplexing can be used in the same way to connect several receivers to the same receiving antenna.
   3. Baseband repeater
   A microwave relay repeater that receives the modulated microwave carrier and obtains the baseband signal from it, and then remodulates the baseband signal onto another carrier and retransmits the new carrier with the baseband modulated onto it.

4. Heterodyne repeater

A microwave relay repeater that receives the modulated microwave carrier, amplifies it at the IF frequency without removing the baseband information, and then up-converts the IF frequency to a new microwave frequency for retransmission.

5. RF repeater

A microwave relay repeater that amplifiers the received microwave signal at the microwave frequency and retransmits it without converting the received signal to IF or baseband frequencies.

6. Regenerative repeater

A repeater for digital signals that receives the noisy digital bit stream and reconstructs it so that a perfectly timed series of perfectly shaped digital pulses is retransmitted.

7. Free-space path loss

The ratio of the area of a one square wavelength antenna to the area over which the transmitted power has been spread (which is a sphere with a radius equal to the transmitter receiver separation).

8. Multipath fading

The reduction of the received signal in a microwave relay due to the signal from a multipath being out-of-phase and canceling the direct transmitted signal.

9. Frequency diversity

A technique for reducing multipath fading by transmitting the same information in two frequency channels. This technique is based on the fact that if a fade occurs at one frequency it will not occur at the same instant of time at a different frequency.

10. Space diversity

A technique for reducing multipath fading in which two receivers, spaced approximately 100 wavelengths apart, are used. This technique is based on the fact that if a signal fades at one antenna it will not fade at the same instant of time at the other antenna, since the path lengths are different.

11. Diffraction microwave relay system

A microwave relay system that achieves transmission distances beyond the curvature of the earth by diffracting the microwave signal off of hills and other obstacles in the microwave path.

12. Troposcatter microwave relay system

A microwave relay system that achieves transmission distances beyond the curvature of the earth, up to several hundred miles, by receiving the microwave signal scattered from air masses and clouds in the troposphere.

13. Local area network

A transmission system used to send telephone, video, and data between buildings located a few hundred feet to a few miles apart.

14. Common carrier frequency band
Those microwave relay frequencies allocated to long-distance telephone companies for the transmission of telephone, video, and data for any customer.

15. Studio transmitter link
A microwave relay that sends radio and television program material from a distant location to the studio, and from the studio to the transmitter.

16. Channelization of bandwidth
The dividing of the common carrier, or other microwave relay bands into channels. Separate transmitters and receivers are used for each channel. The channel, typically 20 MHz wide, can handle 1000 telephone calls or one television program.

17. Dual polarization
A technique for transmitting different information at the same frequency and in the same direction from a microwave relay transmitter to a receiver by using two polarizations of the transmitted microwave fields.

15.2. Why is a microwave relay often called a terrestrial microwave relay?
Because it is located on the surface of the earth.

15.3. Why is a microwave relay often called a line-of-sight relay?
Because the microwave signal travels in approximately a straight line from the transmitter to the receiver.

15.4. What are the frequency ranges of the three most commonly used common carrier frequency bands?
3.7–4.2 GHz
5.925–6.425 GHz
10.7–11.7 GHz

15.5. What is the frequency range of the most common industrial microwave relay band?
6.575–6.875 GHz

15.6. What two frequency bands are most often used for local area microwave networks?
18.58–19.16 GHz
21.2–23.6 GHz

15.7. What are typical transmitter-receiver spacings for microwave relay?
10–50 mi

15.8. What limits the transmitter-receiver spacing for a microwave relay?
The curvature of the earth.

15.9. What are two advantages of a digital microwave relay system?
1. The digital signal is immune to noise.

2. The digital signal is regenerated at each repeater so the noise is not additive.

15.10. What is a radome?
An antenna covering that the transmitted or received microwave power can pass through, used to protect the antenna and the antenna feed from weather.

15.11. What is a horn-reflector antenna?
A microwave relay antenna consisting of a horn that radiates onto a reflector. The horn-reflector antenna is made as a single unit and has lower sidelobes than a parabolic dish antenna.

15.12. What signal-to-noise ratio is required of the baseband signal when it has been demodulated from the received microwave carrier?
50 dB for telephone systems and 40 dB for television

15.13. What carrier-to-noise ratio is required for microwave relay systems?
10–20 dB

15.14. What are two methods for improving the received carrier-to-noise ratio up to the required baseband signal-to-noise ratio?
1. FM improvement factor with frequency modulation.
2. Digital coding of the analog signal before transmission.

Exercises 15.15. through 15.19 refer to a microwave relay system with the following characteristics.

Frequency	11 GHz
Transmitter-receiver spacing	20 mi
Transmitter power	10 W
Transmitter antenna gain	30 dB
Receiver antenna gain	30 dB
Path attenuation	30 dB
Receiver noise figure	3 dB
Bandwidth	20 MHz

15.15. What is the free-space path loss?
144 dB

15.16. What is the received carrier power?
−74 dBm

15.17. What is the noise power in the receiver?
−98 dBm

15.18. What is the carrier-to-noise ratio?
24 dB

15.19. For how many minutes per day will multipath fading be greater than 18 dB?

14 min

CHAPTER 16

16.1. Define the communication satellite terms in the following list.

1. International satellite

A satellite communication system designed for communications from one continent to another.

2. Domestic satellite

A satellite communication system designed for communicating within a given continent or geographical area, such as the United States.

3. MARISAT

A satellite communication system designed for ship-to-shore communication.

4. Direct broadcast satellite

A satellite communication system designed to broadcast television directly from the satellite to a home TV receiver.

5. Synchronous orbit

The orbit of satellites around the earth's equator in which the satellite takes 24 h to completely orbit the earth, so that the satellite appears in a stationary location relative to the rotating earth.

6. ERP

The ERP is the effective radiated power of the satellite transmitter. It is equal to the product of the transmitter power times the transmitter antenna gain, and this single quantity completely characterizes the satellite transmitter.

7. G/T

G/T completely characterizes the earth station receiver of a satellite communication system, and is equal to the ratio of the receiver antenna gain to the total noise temperature of the receiver (including the antenna noise temperature plus the receiver amplifier noise temperature.

8. Spin-stabilized satellite

A satellite that maintains its orientation in space relative to the earth by spinning about its axis at a rate of about 50 revolutions per minute.

9. Three-axis stabilized satellite

A satellite that maintains a fixed orientation in space relative to the earth by continually correcting its orientation by firing small jets.

10. Transponder
A single channel of a satellite communication system. Each transponder in a satellite has its own separate transmitter. A transponder channel is typically 40 MHz wide.

11. Antenna footprint
Received power contours from the satellite on the earth surface. The satellite antenna is designed, often with multiple feed horns, to direct its power onto particular geographic areas, and this pattern of received power on the earth's surface is the antenna footprint.

12. Global beam
An antenna beam from the satellite that covers the entire one third of the earth's surface that can be seen from the satellite.

13. Spot beam
The antenna beam from the satellite that covers a limited geographic region, such as a part of a country, an entire country, a continent, or a hemisphere. Some satellites have a variety of spot beams, each covering a larger or smaller geographic area.

16.2. What is the synchronous orbit for a communication satellite in miles?
22,300 mi

16.3. Why is the downlink the most difficult path in a communication satellite system?
Because the transmitter power in the satellite is limited by the power that can be obtained from the solar cells.

Exercises 16.4. through 16.10. refer to a domestic communication satellite downlink with the following characteristics.

Frequency:	11 GHz
Satellite transmitter power:	1 W
Satellite transmitter antenna gain:	34 dB
Earth station antenna gain:	63 dB
Noise temperature of receiving system:	300°K
Transponder bandwidth:	40 MHz

16.4. What is the free-space path loss?
205 dB

16.5. What is the ERP?
64 dBm

16.6. What is the received carrier power?
−78 dBm

16.7. What is the total receiver noise power?
−98 dBm

16.8. What is C/N?
20 dB

16.9. What is G/T?
38 dB

16.10. What is the sum of ERP and G/T?
102 dB

CHAPTER 17

17.1. Define the radar system terms in the following list:

    1. Range-measuring radar

        A radar that measures the distance between the radar and a target by determining the time required for a microwave signal to travel from the radar to the target and back.

    2. Velocity-measuring radar

        A radar that measures the velocity of a target by determining the change in the frequency of the reflected microwave signal due to the target's velocity. This change in frequency is called the Doppler shift, and so this type of radar is often called a Doppler radar.

    3. Pulse Doppler radar

        A radar that measures both the distance to a target and its velocity by measuring both the change in frequency of the reflected signal and time required for the microwave signal to travel from the transmitter to the target and back to the radar.

    4. Moving-target-indicator (MTI) Radar

        A radar that measures the range to the target and distinguishes between moving and stationary targets.

    5. Pulse-compression radar

        A radar that obtains high range resolution by varying the frequency of the transmitted microwave signal and then compresses the pulse in the receiver.

    6. Search radar

        A radar that scans its beam over a given volume of space to determine if targets are present.

    7. Tracking radar

        A radar that directs its beam into a given area of space to determine accurately the angular location of a target.

    8. Monopulse radar

        A tracking radar that determines the angular location of a target with a single pulse.

    9. Conical-scan radar

        A tracking radar that rotates its beam in a conical pattern to determine the angular location of the target. Several pulses are required to obtain the angular information.

10. Phased array radar
    A radar that moves its beam through space by changing the phase of a multiplicity of transmitting elements that make up the antenna, rather than mechanically moving the entire antenna.
11. Synthetic aperture radar
    A radar using a single antenna element, which obtains extremely high angular resolution by moving this single element over a long distance, storing the information obtained from the antenna at each position, and then recombining the received information.

17.2. What three things can a radar measure?
Range; velocity; angle (Azimuth and elevation)

17.3. Which of the above can radar measure with the best accuracy?
Velocity

17.4. Which of the above are the most difficult for a radar to measure accurately?
Angle

17.5. Complete the table.

Range (ft)	Time for Pulse to Go out and Return
25,000	50 μs
10,000	20 μs
5	10 ns

17.6. Complete the table.

Microwave Frequency (GHz)	Doppler Shift (Hz)	Velocity (mph)
10.1	1500	38.6
10.1	1970	50
24	1500	16.3
14	50,000 (50 kHz)	929

17.7. What range resolution can be obtained with a 2-us pulse?
1000 ft

17.8. What range resolution can be obtained with the following pulse-compression radar?

Pulse length	100 μs
Frequency sweep during pulse	50 MHz
20 ft	

17.9. What is the beamwidth of a 5-m antenna at 3 GHz?
1.2°

17.10. What lateral resolution in feet can be obtained with the antenna of Exercise 17.9 at a range of 25 mi?
0.5 mi

17.11. What is the beamwidth of a 10-cm-diameter missile-seeking antenna at 10 GHz?
18°

17.12. What lateral resolution in feet can be achieved with the antenna of exercise 17.11 at a range of $\frac{1}{2}$ mi?
750 ft

17.13. How long must the flight path be for a 5-GHz synthetic aperture radar to achieve a 10-ft resolution at 50 mi?
0.98 mi

17.14. How many "looks" at the target must the synthetic aperture radar of Problem 17.13 make?
52,800

CHAPTER 18

18.1. Define the electronic warfare terms in the following list.

   1. Electronic warfare (EW)

   Electronic warfare is the techniques and equipment used to render electronic target location and weapon control ineffective. Electronic warfare includes tactics such as flight plans to avoid detection, electronic surveillance, spying techniques to determine how enemy weapon systems work and their deployment, and the jamming of infrared, optical, and radar target location and control systems.

   2. Electronic countermeasures (ECM)

   Electronic countermeasures is sometimes used interchangeably with the term electronic warfare. In the strict sense, electronic countermeasures applies only to techniques and equipment used to render electronic target location and weapon control systems ineffective.

3. Electronic counter-countermeasures (ECCM)

Electronic counter-countermeasures are the techniques and equipment used by the radar to reduce the effectiveness of the enemy's jamming against it.

4. Electronic support measures (ESM)

Electronic support measures are the equipment used to determine the electronic characteristics of the enemy target location and weapon control systems, and to determine what should be jammed by the electronic countermeasures.

5. Stealth

Stealth is an electronic countermeasure technique in which the aircraft is completely redesigned to greatly reduce its radar cross section.

6. Antiradiation missile (ARM)

An antiradiation missile uses the radar transmissions as a beacon and homes on the radar to destroy it.

7. Chaff

Chaff is metal foil launched from a target to create false radar targets.

8. Decoy

A decoy is a small unarmed missile launched from aircraft, or a buoy launched from ships, to create false radar targets.

9. Barrage jamming

Barrage jamming is the transmission of noise-modulated high power microwave signals to raise the noise level in the enemy radar receiver so that range and velocity information cannot be obtained.

10. Stand-off jamming

Stand-off jamming is another name for barrage jamming, and stresses that the barrage jamming source must be out of the range of enemy missiles so that they cannot home on the jamming.

11. Deceptive jamming

Deceptive jamming is the repeating back of an amplified version of the radar signal with its characteristics changed to give false information to the radar.

12. Range-gate pull-off (RGPO)

A deceptive jamming technique where the received radar signal is amplified and sent back to the radar delayed in time from the actual reflected radar signal so that the radar thinks the target is further away that it really is.

13. Range-gate pull-in (RGPI)

A deceptive jamming technique where the jamming pulse is sent to the radar before the actual echo from the target so that the radar thinks that the target is closer to the radar than it really is.

14. Velocity-gate pull-off (VGPO)
    A deceptive jamming technique where the jamming signal is given a false Doppler shift so that the radar thinks the target velocity is different than it really is.

15. Inverse conical-scan jamming
    A deceptive jamming technique where the radar signal is transmitted with an amplitude variation that is out of phase with the conical scan information so that the radar thinks that the target is at a different angular location.

16. Home-on jammer
    A ECCM technique where the radar launches a missile to home on the barrage jamming or deceptive jamming signal. The jammer then serves as a beacon to guide the attacking missile to it.

17. Channelized receiver
    An ESM receiver consisting of many mixer–local oscillator–IF amplifiers in parallel to instanteously determine the frequency of the radar signals.

18. Bragg cell
    An ESM receiver using optoacoustical effects to instantly determine the time of arrival and the frequency of radar signals.

18.2. List four stealth techniques that reduce the radar reflections from an aircraft.
    No sharp corners; cover exposed jet turbines; single-frequency radome; absorptive coating on aircraft skin

18.3. What is the advantage of decoys compared with chaff for creating false targets?
    Decoys can create a Doppler shift in the reflected microwave signal. Chaff remains stationary after it is launched and so can be distinguished from the target if the radar uses MTI or pulse Doppler techniques.

18.4. How does noise jamming prevent the radar from determining target range and velocity?
    Noise jamming raises the noise level in the radar receiver so that the returned echo pulse or the returned echo Doppler shifted frequency cannot be detected in the noise.

18.5. What are four types of ESM receivers?
    Crystal video; superheterodyne; channelized; Bragg cell

18.6. Why is the crystal video receiver ineffective in a modern electronic warfare scenario?
    The crystal video receiver cannot determine the frequencies of the received radar signals.

18.7. What is the disadvantage of a superheterodyne receiver in a modern electronic warfare scenario?
The superheterodyne receiver has a low probability of detection and may miss radar signals at one frequency while it is analyzing signals at another frequency.

18.8. How does range-gate pull-off work?
The deceptive jammer receives the radar signal and transmits it back later than the echo pulse to make the radar think the target is further away than it really is.

18.9. How does range-gate pull-in work?
The radar estimates when the radar pulse will be received at the target and what its frequency will be, and then transmits a deceptive pulse ahead of the reflected signal from the target.

18.10. How does velocity-gate pull-off work?
In this deceptive jamming technique, the CW radar signal is received and retransmitted with a false Doppler shift to make the radar think the target is moving at a different velocity than what it really is.

18.11. How does deceptive angle jamming of a conical-scan radar work?
The jammer transmits an amplified version of the echo signal with amplitude modulation that is out of phase with the actual conical-scan modulation to make the radar think the target is at a different angular location.

18.12. What is the basic advantage that a jammer has over a radar?
Power, since the jammer power needs to travel in only one direction from the jammer to the radar. The radar signal must travel out from the radar to the target and back from the target to the radar.

18.13. What is the basic advantage that a radar has over a jammer?
The radar knows the characteristics of the signal that it is transmitting. The jammer must measure the radar signal and determine these characteristics; thus, if the radar can rapidly change its characteristics, the jammer does not have time to respond.

CHAPTER 19

19.1. What position accuracy can be obtained with the GPS?
Position accuracy is determined to within 2 m.

19.2. How does the GPS receiver determine its location?
The GPS satellites transmit a pulse-coded signal that gives the satellite

locations and the time of transmission. The receiver measures the time of reception and calculates its range from the satellite. With range information from three satellites and the time reference from the fourth satellite, the receiver calculates its position.

19.3. What is the microwave transmission frequency of GPS satellites?
1.575 GHz

19.4. What is the frequency range of cellular telephones?
806–949 MHz.

19.5. In what two ways can the channel capacity of cellular mobile radio be increased compared with VHF mobile radio?
1. Use of microwave frequency range.
2. Reuse of frequency by dividing service area into cells and reusing each frequency in nonadjacent cells.

19.6. What is the frequency of microwave ovens?
2.450 GHz

19.7. How are microwaves used to generate cancer treatment X rays?
The microwaves are used to accelerate electrons in a linear accelerator, so that the high-energy electrons can generate X rays as they hit the X-ray target.

19.8. Name three scientific applications of microwaves.
Radio astronomy; nuclear research; triggering of thermonuclear power generator

INDEX

539